Occupational Health Promotion

Occupational Health Promotion

Health Behavior in the Workplace

George S. Everly, Jr., Ph.D.
Robert H. L. Feldman, Ph.D.
and Associates

With a Foreword by

Stephen M. Weiss, Ph.D.
Chief, Behavioral Medicine Branch,
The National Heart, Lung, and Blood Institute
National Institutes of Health

John Wiley & Sons
New York *Chichester* *Brisbane* *Toronto* *Singapore*

Library of Congress Cataloging in Publication Data:
Main entry under title:

Occupational health promotion.

Includes bibliographies and indexes.
1. Industrial hygiene. 2. Psychology, Industrial.
I. Everly, George S., 1950– . II. Feldman, Robert
H. L. [DNLM: 1. Preventive Health Services—United States.
2. Occupational Health Services—United States. 3. Health
Promotion—United States. WA 412 E93o]
RC967.028 1985 613.6′2 84-20863
ISBN 0-471-89533-4

Printed in the United States of America

10 9 8 7 6 5 4 3 2 1

To all of those in business who have learned
the message of Hygeia, a desideratum sine qua non;
And, to those pioneers who shared their wisdom and
changed my life in the process. (GSE)

To Suzanne, Alicia and Daniel,
Sylvia, and in memory of Abraham. (RHLF)

Even the most cost-conscious society should recognize that money spent on human capital is the single most important investment it can make.

NEWSWEEK, October 18, 1982.

FOREWORD: The Case for Worksite Health Promotion

Stephen M. Weiss, Ph.D.

Since publication of the Canadian national health plan (the Lalonde Report) in 1974, disease prevention and health promotion have taken center stage in health planning throughout the industrialized world. The virtual conquest of the acute infectious disorders has left the chronic degenerative diseases (heart disease, cancer and stroke) as the major causes of morbidity and mortality in the United States. Inasmuch as we all will eventually die from *something,* this fact in itself would not be disheartening were it not for the distressingly high levels of *premature* morbidity and mortality associated with these figures. Combined with accidents, the chronic diseases account for 75% of the premature mortality experienced in the United States (Beary, 1981).

John Knowles sums up the situation in his book, *Doing Better and Feeling Worse: Health in the United States* (1977), ". . . over 99% of us are born healthy and suffer premature death and disability only as a result of personal misbehavior and environmental conditions." The reknowned health economist Victor Fuchs agrees: ". . . the greatest potential for improving the health of the American people is to be found in what they do—or don't do—for themselves. Individual decisions about diet, exercise, and smoking are of critical importance" (Fuchs, 1975).

To illustrate the magnitude of the problem, using coronary heart disease as an example, let me share with you a little story from Jack Farquhar's book, "The American Way of Life Need Not Be Hazardous to Your Health" (1978),

> Let us suppose that on January 1, two jumbo jets had a midair collision, and both crashed with a loss of *500* lives. On that same day four more jumbo jets malfunctioned and crashed, with *over 1,500* persons aboard, with all sustaining serious injuries. A unique feature compounding this tragedy was that all of these people were between 30 and 65, the prime of life.
>
> Now suppose, on January 2nd the same disaster struck!—and the day after that it happened again—and so on throughout the year. A total of 183,000 *premature* deaths and 570,000 injuries would result, the same number of heart disease and stroke morbidity and mortality victims claimed by these disorders at an annual cost exceeding $40 billion.

When we speak of prevention, we really mean *delay*. We have not learned to *prevent* disease or death, but we are extremely concerned about the premature (i.e., under 65 years) departures from this earth. Since 1900, we have seen an increase of 20 years in the lifespan of the average American. However, a close inspection of these figures reveals that the major changes have taken place during the childhood years, the result of immunization, penicillin, and other pharmacologic discoveries to combat the acute infectious diseases. For the 45-year-old male, however, the increase in life span since 1900 has amounted to only about three years. Thus, the progress toward ameliorating the effects of the chronic degenerative disease has left much to be desired.

On the basis of such statistics, one might infer that the solution of the problem of premature mortality must indeed be complex. Yet, if one considers the data collected on Alameda County residents over a seven-year period (Belloc and Breslow, 1972), ascribing to seven relatively simple personal habits on a daily basis (eg., 7–8 hours of sleep, regular breakfast, no smoking, moderate alcohol usage, regular exercise, etc.) can increase average lifespan for a 45-year-old male by 11.5 years (7 years for 45-year-old females). Belloc and Breslow concluded that ". . . the association between mortality and health practices was stronger than between mortality and physical health status or income level."

Whether one accepts this data at face value or not, it clearly demonstrates that health behaviors are related to premature mortality from chronic disease and that public health prevention efforts must seriously address the issue of health behavior change.

The prevention of chronic disease may be the only way we can ethically stabilize the present unacceptable acceleration of health care expenditures. These costs have escalated from $10 billion in 1950 to over $354 billion in 1983, for over 11% of the Gross National Product (doubling the percentage within 15 years). U.S. industry currently pays 25% of these costs, and the percentage is likely to increase in future years, if present trends continue.

The cost issue clearly must be brought under control. Yet how can we deny expenditures that have life-saving implications? The implementation of emergency medical services, coronary-care units, and the recently developed sophisticated scanning and imaging techniques have undoubtedly made major contributions to disease detection and treatment. Coronary bypass surgery, as an example of a technology unheard of 20 years ago, was performed over 170,000 times in the United States in 1983 at a total cost of over $3 billion. Although its life-prolonging capabilities are still in dispute, the enhanced quality of life through pain reduction and increased physical activity will undoubtedly increase the utilization of this reparative technique (and its refinements) in accordance with the demand for such services.

Obviously, these innovations are expensive—and their successors will undoubtedly be even more so. Is that sufficient reason to deny them to those in need? Although in some subsistence cultures human life is valued somewhat below that of the family cow, we have made our choices on this issue very clear (e.g., kidney dialysis).

So, what is to be done? This reminds me of the story of the little village by a river in which one day were heard the cries of a drowning man floating down the river. Through heroic effort, the villagers managed to save him. The next day the villagers spotted two more people floating along in similar straits. They too were rescused. Gradually more and more people were discovered floating down the river. The villagers began to devise increasingly innovative means of rescuing them. Specially fitted boats, trained observers, and safety nets were organized—the villagers became increasingly adept at rescuing potential drownees.

The numbers continued to increase, however, threatening to overwhelm the resources of the village. Although very proud of their rescue capabilities, the villagers realized they could not continue to cope with the problem with their present systems. Then, and only then, did someone propose, "Why don't we walk upriver to find out who or what is throwing all these people into the river in the first place?"

This little story illustrates both the importance of good "rescue" facilities as well as the potential for reducing demand on those facilities (through prevention) as the most legitimate means of controlling costs without sacrificing quality of care.

When we speak of prevention efforts, we use terms such as "risk factor" and "risk factor reduction" to describe both the problem and its potential resolution. Epidemiological studies have identified "markers" and related behavior patterns that have demonstrated consistent associations with increased probability of disease (Keys et al., 1971; Belloc and Breslow, 1972; Rosenman et al, 1975). The term *probability* is important to keep in mind as there is no guarantee that a given individual will remain healthy (or become ill) by following certain regimens or engaging in certain behaviors. Risk factors for disease are essentially "probability statements" that should be considered by the individual as information relevant to personal decision making regarding health enhancing (or inhibiting) behavior patterns.

From a public health and national health planning standpoint, however, developing risk factor reduction strategies should be a *key* element in designing a comprehensive approach to improving the nation's health while containing and ultimately reducing national health expenditures. Taking a closer look at some of the more obvious risk factors, we find elevated blood pressure, smoking, high serum cholesterol, and the Type A behavior pattern to be four of the most widely accepted risk factors for coronary heart disease and stroke. Smoking, diet, and environmental pollutants are some of the risk factors associated with cancer, while substance abuse (particularly alcohol), excessive speed, seat belts, and utilization of available safety equipment on the job have obvious implications for accident prevention.

Using just one risk factor as an example (smoking), the statistics are quite impressive. It is estimated that cigarette smoking is responsible for approximately 325,000 premature deaths each year from cancer and other diseases of the lungs, heart, and circulatory system (Surgeon General, 1979). A smoker is at twice the risk of developing coronary heart disease as compared to a nonsmoker; approximately 85% of the 111,000 deaths associated with lung cancer alone are directly linked to smoking (Surgeon General, 1982). Chronic obstructive pulmonary diseases such as emphysema have also been associated with smoking. Over the past 20 years, an impressive literature on smoking prevention, cessation, and maintenance of cessation has developed. The efforts to date on smoking prevention in schools have been encouraging (Evans, 1983). The majority of smokers who have attempted to stop smoking have been able to do so. Unfortunately, the recidivism rate has been quite discouraging, with approximately 30% being able to maintain cessation over a one-year period (Orleans, 1980; Hunt et al., 1971). Thus, program planning and evaluation must address improving the *maintenance* of cessation through such strategies as provision of appropriate social support, extended contingency incentive programs, "buddy" systems, etc.

The development of such programs must also focus on "health promotion" as a logical extension of the disease prevention perspective. Although considerable overlap exists in defining the terms, health promotion also includes assisting well people to feel even better, perhaps better than they've ever felt in their lives. Therefore, one can speak of health enhancement without reference to disease. This is consistent with the World Health Organization's definition of health as being ". . . a state of complete physical, mental and social well-being, and not merely the absence of disease or infirmity." Thus, in addition to the more quantifiable measures of utilization of the health care system, one should also address "quality of life" measures that have implications on a broad variety of issues associated with life and job satisfaction.

Health promotion programs have been developed within three major domains:

1. schools
2. worksites
3. communities

Although all three have significant public health implications, the unique constellation of factors pertinent to the worksite merit special consideration.

First, the 100,000,000 employed persons in our society are those at greatest risk for premature morbidity and mortality (almost all under 65) and are the backbone of our society's productivity. The employers of these 100 million workers have a major stake in the overall welfare of their employees, as health status (and employee health costs) contribute in a major way to the productivity and profitability (or lack thereof) of their businesses. Absenteeism, for example, is a major source of lost productivity. Smokers are absent five days per year more than nonsmokers, accounting for 80 million workdays lost per year. The prospect of both reducing employee health costs *and* increasing the health status of the workforce is a powerful incentive to management.

As most employed persons have families, the inclusion (directly or indirectly) of those family members into the health promotion milieu increases both the coverage and potential benefit to perhaps another 100,000,000 persons. Thus in addition to the workforce, worksite programs have the potential to extend to nearly the entire U.S. population.

Finally, the design, implementation, and evaluation of health promotion programs may involve considerable financial investment over several years before one can hope to see evidence of benefit (Knobel, 1983). The private sector has a long tradition of financing research and development to assure their long-term standing in the marketplace, so such concepts are not foreign to their traditional approaches to new ideas. They do need to have some evidence, however, of the potential efficacy and benefit of these programs to justify commitment of stockholder assets to such enterprises. Thus we must look to the experience of a few pioneering companies over the past five to ten years for such evidence.

As noted in Figures F.1 and F.2, the content and process of an "ideal" worksite health promotion program provides some sense of the potential scope and nature of such efforts. Actual programs may incorporate only two or three of the components noted in Figure F.1; the intensity and scientific validity of the evaluation efforts depicted in Figure F.2 vary greatly. Nonetheless, sufficient data does exist across programs to provide a basis for rational decision making by corporate managers and union leaders who have a large stake in employee welfare issues.

As many of the succeeding chapters of this volume will address these issues in the context of their topical discussions, a detailed elaboration of the cost-benefit issue is not warranted here. By way of example, a brief look at the data from one type of program (smoking cessation) should suffice for our purposes.

Although occupational health directors have long suspected that smoking employees made greater use of health services than nonsmokers (and therefore engendered higher health-related costs to the company), the most convincing and consistent data on this issue has come from recent evaluations of worksite health promotion programs. On the basis of several such studies it was determined that

a. Smokers cost their companies between $600–650 more than nonsmokers in health care costs per year (insurance, absenteeism, use of occupational health facilities, air filter clogging, repainting and cleaning of offices, etc.). One study estimated this cost to be as high as $4600 per year when considering reduced life span, passive and sidestream smoke effects on family and coworkers, costs of "segregating" smokers, etc. (Goldbeck and Kiefhaber, 1981)

b. Smokers have twice the death rate and ten times the risk of developing lung cancer as nonsmokers (Fielding, 1982)

c. Hospital stays are 114% longer for smokers than for nonsmokers (Naditch, 1984).

On the other hand, the risk for heart disease decreases to that of a nonsmoker within five to ten years following cessation (Surgeon General, 1979); the risk of lung cancer decreases in a similar pattern ten to fifteen years following cessation (Salonen, 1980). The average cost of a worksite smoking cessation program comes to about $200–$500 per quitter (assuming a quit rate of 40–50% after one year—about average for the better worksite programs) (Fielding, 1982).

As is readily apparent, successful programs make good health economic and social-consciousness sense by any yardstick. One caveat: the quality of the program itself, the wholehearted support of top management (particularly the CEO) and the

Typical Programs May Include

- Smoking cessation
- Hypertension detection and control
- Nutrition
- Weight control
- Exercise/fitness
- Stress management
- Employee assistance programs
    - Alcohol
    - Drugs
    - Personal/emotional problems
    - Accident prevention
        - On the job
        - Off the job

Figure F.1. Health Promotion at the Worksite—Content

A typical sequence of events might include

1. Screening (eg., Health Hazard Appraisal)
2. Health counseling session(s)
3. Choice of one or more modules, as appropriate

smoking cessation	weight control	stress management	exercise/ fitness	etc.

4. Followup counseling/maintenance (repeat Health Hazard Appraisal)

Program Evaluation

Figure F.2. Health Promotion at the Worksite—Process

willingness to follow through by assuring a supportive social and physical environment for the maintenance of cessation are all essential ingredients to developing and continuing a successful program (Goldbeck and Kiefhaber, 1981).

In conclusion, our best hope for improving the health status of our citizens while containing spiraling health care expenditures rests with our capacity to develop effective disease prevention/health promotion programs. There are several paths to achieve this worthy goal; the worksite approach to be discussed in succeeding chapters appears reasonable, possible, affordable, and do-able. It is encouraging to realize that an increasing number of major corporations (recently estimated at over 1500) that now have significant health promotion programs for their employees appear to be in agreement with this statement. Given the recent data suggesting a \$3–\$5 return/savings for every dollar spent on worksite health promotion, it is not difficult to appreciate their enthusiasm.

References

Beary, C. A. *Good Health for Employees and Reduced Health Care Costs for Industry*. Washington, D.C. Health Insurance Institute, 1981.

Belloc, N. and Breslow, L. "Relationship of physical health status and health practices." *Preventive Medicine,* **1,** 409–421, 1972.

Evans, R. I. "Social Development Approaches to Deterrence of Risk Factors in Cardiovascular Disease." In *Behavior and Arteriosclerosis*. Herd, J. A. and Weiss, S. M., eds. New York, Plenum Press, 1983.

Farquhar, J. W. *The American Way of Life Need Not Be Hazardous to Your Health*. New York, Norton, 1978.

Fielding, J. E. "Effectiveness of Employee Health Improvement Programs." *J. Occup. Med.,* **24,** 11, 1982.

Fuchs, V. *Who Shall Live? Health, Economics and Social Choice*. New York, Basic Books, 1974.

Goldbeck, W. B. and Kiefhaber, A. K. "Wellness: The new employee health benefit." *Group Practice Journal,* 20–26, 1981.

Hunt, W. A. et al. "Relapse rates in addiction programs." *J. Clin Psychol.* 27, 455–456, 1971.

Keys, A. et al. "Mortality and coronary heart disease among men studied for 23 years." *Arch. Int. Med.* **128,** 201, 1971.

Kiefhaber, A. and Goldbeck, W. "Smoking: A Challenge to Worksite Health Management." In *Proceedings, The National Conference on Smoking or Health: Developing a Blueprint for Action*. New York, American Cancer Society, 1981.

Knowles, John, ed. *Doing Better and Feeling Worse: Health in the United States*. New York, Norton, 1977.

Lalonde, M. *A New Perspective on the Health of Canadians*. Ottawa, Ministry of National Health and Welfare, 1974.

Naditch, M. "The Staywell Program." In *Behavioral Health: A Handbook of Health Enhancement and Disease Prevention,* Matarazzo, J. D., Weiss, S. M., Herd, J. A., Miller, N. E. and Weiss, S. M., eds. New York, Wiley, 1984.

Orleans, C. S. "Quitting smoking: promising approaches and critical issues." *Institute of Medicine, Health and Behavior: A Research Agenda Interim Report No. 1: Smoking and Behavior,* 1980.

Rosenman, R. H. et al. "Coronary heart disease in the Western Collaborative Group Study: final follow-up experience of 8½ years." *JAMA,* **233,** 872–877, 1975.

Salonen, J. T., "Stopping smoking and long-term mortality after acute myocardial infarction." *British Heart Journal,* **43,** 463–469, 1980.

"Smoking and Health." *A Report of the Surgeon General*. Office on Smoking and Health, USDHEW (PHS) 79-50066 Washington, D.C. GPO, 1979.

"The Health Consequences of Smoking: Cancer." *A Report of the Surgeon General*. Office on Smoking and Health. USDHHS (PHS) 82-50179, GPO, 1982.

Preface

We stand poised at the crossroads of Jonas Salk's Epoch A and Epoch B, and the course we follow will surely affect not just our generation, but several generations to come. According to Salk, to turn towards Epoch A means to be satisfied viewing health as merely the absence of disease and to passively watch as the quality of our lives declines under the pervasive influences of environmental pollution and a general neglect of our mental and physical well-being. To turn towards Epoch B, on the other hand, means to actively pursue health, to seek optimum health and well-being by taking responsibility for our environment and ourselves, as well as for future generations. The pursuit of health in Epoch B merely begins at the absence of disease. Its boundaries may be limitless.

This book is designed to help its reader turn towards Epoch B. Furthermore, it is designed to help the reader assist others in pursuing Epoch B, should they choose such a course. It is an underlying assumption of this book that the pursuit and promotion of health may logically emerge from, or be fostered by, the occupational environment—where the majority of the adult population spends one third of its life. The pursuit of Epoch B is not just the pursuit of good health, it is the pursuit of good business. Indeed, the pursuit of good health is good business as the reader will learn as he or she reads this volume.

In this text, we will provide examples of how health can be promoted on the basis of behavioral interventions within the occupational environment; at the same time this text will serve as a practical guide for the actual development of occupational health promotion programs. Our book takes a comprehensive approach to the promotion of health in the workplace. Numerous perspectives and fields are presented and a truly interdisciplinary approach is offered. The chapters within this text address traditional health promotion topics such as stress management, weight control, and physical fitness, as well as unique, nontraditional topics such as the prevention of back injuries and how to manage dual-career relationships.

Our text is divided into three sections, and an appendix has been included. Part One examines the basic foundations of occupational health promotion. The reader is introduced to the concepts of health, and more specifically the concept of occupational health promotion. This section also provides a unique ideological and historical perspective on occupational health promotion. In addition, the critical issue of how to assess and enhance health compliance in the workplace is examined.

Part Two of the book presents general guidelines for program development as it relates to some of the more commonly utilized aspects of occupational health promotion. A wide variety of health promotion programs are presented and model programs are included. Part Three considers the planning and evaluation of occupational health promotion programs. Included in this section is an examination of cost-effectiveness and cost-benefit analysis in occupation health promotion. In addition, a discussion of future directions in occupational health promotion is presented. In the appendix we present three examples of on-going multi-dimensional health promotions—Control Data's STAYWELL, IBM's Health Education Program, and Johnson & Johnson's LIVE FOR LIFE Program. In our opinion, these programs represent differing, yet highly desirable examples of how health promotion efforts may be realized in the workplace.

Occupational Health Promotion is an interdisciplinary book. The fields of health psychology, health education, medicine, nursing, social work, business, physical education, and medical sociology are represented in the chapters. Therefore, it is written for a variety of audiences. The book may be used as a text for advanced undergraduates (juniors and seniors) as well as graduate students. Courses in occupational health, program planning and evaluation, and health planning could benefit from this book. In addition, practitioners and administrators working in corporate, business, and labor settings can use this book in planning, development, implementation, and evaluation of occupational health promotion programs.

In the fiscal analysis, however, no matter who the reader, it is our hope that it will assist them in the pursuit of Epoch B.

May 31, 1984

GEORGE S. EVERLY, JR., PH.D.
Baltimore, Maryland

ROBERT H. L. FELDMAN, PH.D.
College Park, Maryland

Acknowledgments

The authors would like to thank the following individuals for their contributions to this project:

Dr. Robert Karsch, for his early interest and suggestions;

Dr. Lawrence Green, for his constructive comments as the project was just beginning;

Raymond T. O'Connell, for his initial interest and encouragement as an editor;

All of our contributors who gave of their time and expertise;
Those too numerous to mention who assisted in the technical preparation of the manuscript;

The author (RHLF) wishes to thank Dr. John J. Burt for his encouragement and support, Mrs. Janet Siarnicki for her patience and fortitude in typing the manuscript, Dr. S. Stephen Kegeles who introduced me to the field of health behavior and the application of psychology to health, and Dr. Suzanne Laidlaw Feldman for her assistance, assurance, endurance, and kindness.

Finally, the author (GSE) would like to offer his deepest gratitude to those who taught him that which made this book possible: Dr. Daniel A. Girdano, Dr. Dorothy E. Dusek, Dr. Stephen M. Weiss, Dr. Calvin F. Fuhrmann, and Dr. Sharlene M. Weiss.

G. S. E.
R. H. L. F.

Contents

PART 1 THE FOUNDATIONS OF OCCUPATIONAL HEALTH PROMOTION 1

1. An Introduction to Health and Occupational Health Promotion—*George S. Everly, Jr.* 3

Introduction 3
What Is Health? 6
The Determinants of Health and Disease 7
Health Promotion as a Mechanism of Mediation 10
Is the Workplace Appropriate for Promoting Health? 12
Summary 14
References 15

2. Ideological and Historical Rationales for Occupational Health Promotion: From Asclepius to Hygeia—*Arthur L. Anderson* 17

Introduction 17
Religion 19
Nationalism 22
Capitalism 26
Summary 30
References 31

3. The Assessment and Enhancement of Health Compliance in the Workplace—*Robert H. L. Feldman* 33

Introduction 33
Measuring Compliance 34
Factors that Affect Worksite Compliance 35
Methods to Enhance Compliance 37
Summary 43
References 43

PART 2 GUIDELINES FOR PROGRAM DEVELOPMENT IN OCCUPATIONAL HEALTH PROMOTION 47

4. The Development of Occupational Stress Management Programs—*George S. Everly, Jr.* 49

Introduction 49
Scope of the Problem 49
The Nature of Human Stress 54
Stress Management 57
The Goals of a Stress Management Program 57
Review of Research on Stress Management 59
Guidelines for Program Development 62
Summary 68
Resource Guide for Stress Management 69
References 69

5. Behavioral Approaches to Weight Reduction and the Treatment of Obesity—*George S. Everly, Jr.* 74

Introduction 74
Scope of the Problem 74
Factors that Contribute to and Maintain the Obese Condition 76
Goals of a Weight Reduction Program 78
Review of Research on Weight Control 80
Guidelines for Program Development 81
Summary 87
Resource Guide for Weight Control 88
References 88

6. The Development of a Smoking Cessation Program—*Sharlene M. Weiss, Calvin F. Fuhrmann, and George S. Everly, Jr.* 92

Introduction 92
Scope of the Problem 92
Physiological Mechanisms of Action 94
Why People Smoke 95
Goals of a Smoking Cessation Program 97
Review of Research on Smoking Cessation 98
Guidelines for Program Development 102
Summary 110
Resource Guide for Smoking Cessation 110
References 111

7. Health Behavior Strategies for the Control of High Blood Pressure at the Worksite—*Donald E. Morisky* 113

Introduction 113
Scope of the Problem 113
Problems of Awareness, Treatment, and Control 115
Behavioral Strategies for the Patient Setting 115
Behavioral Strategies for the Provider of Health Care 117
Rationale for Hypertension Control Programs at the Worksite 118
Guidelines for Program Development 120
Summary 123
Resource Guide for Hypertension Control Programs 124
References 124

8. Occupational Health Promotion through Physical Fitness Programming—*Donald Weller and George S. Everly, Jr.* 127

Introduction 127
Scope of the Problem 128
Review of Research on Physical Activity as a Health Promotion Strategy 129
Goals of an Exercise Program 131
Guidelines for Program Development 132
Summary 141
Resource Guide for Physical Fitness Programs 142
References 143

9. Employee Assistance Programs—Alcohol Abuse and Related Concerns—*Melvin Sandler and Steven A. Sobelman* 147

Introduction 147
Scope of the Problem 148
Occupational Alcoholism Programs (OAP): Predecessor to the Employee Assistance Program 150
Employee Assistance Program (EAP) 153
Guidelines for Program Development 154
Summary 163
Appendix: United Airlines EAP Structure 164
Resource Guide for EAPs 166
References 166

10. Health Promotion for the Working Couple: A Health Promotion Program for Dual-Career Couples—*Eileen C. Newman* 169

Introduction 169
Scope of the Problem 171
Goals of a Dual-Career Marriage Assistance Program 176
Guidelines for Program Development 176
Summary 184
Resource Guide for Dual-Career Relationships 184
References 186

11. Promoting Occupational Safety and Health—*Robert H. L. Feldman* 188

Introduction 188
Scope of the Problem 189
Goals for an Occupational Safety and Health Program 190
Solving the Problems of Workplace Injuries and Illnesses 191
Guidelines for Program Development 196
Summary 201
Resource Guide for Occupational Safety and Health 203
References 205

12. Back Injury Prevention Programs—*Carole Lewis and Kay Schaefer* 208

Introduction 208
Scope of the Problem 209

Research Review on Back Injury Prevention 210
Guidelines for Program Development 213
Summary 218
Resource Guide for Prevention of Back Injury 219
References 220

13. Time Management Training: Overcoming a Major Barrier to Occupational Health Promotion—*George S. Everly, Jr.* 221

Introduction 221
Scope of the Problem 222
Time Management, Self-Responsibility, and Health Promotion 223
Guidelines for Program Development 224
Summary 228
References 228

PART 3 THE PLANNING AND EVALUATION OF OCCUPATIONAL HEALTH PROMOTION PROGRAMS 231

14. The Planning and Evaluation of Occupational Health Promotion Programs: An Introduction—*Robert H. L. Feldman* 233

Introduction 233
Planning Occupational Health Promotion 233
Evaluating Occupational Health Promotion 237
Summary 248
Resource Guide for Planning and Evaluation 248
References 249

15. Planning Health Promotion Intervention and Evaluation: A Behavioral Perspective—*Sam Berkowitz* 251

Introduction 251
Steps in Implementing a Behavioral Intervention and Evaluation 252
Specific Behavioral Procedures 257
Summary 263
References 264

16. Psychological Assessment in Occupational Health Promotion—*Catherine J. Green* 265

Introduction 265
General Criteria Used to Evaluate Assessment Instruments 266
Selected Instruments 267
Summary 282
References 282

17. Cost-Effectiveness and Cost-Benefit Analyses of Occupational Health Promotion—*Roberta B. Hollander, Joseph J. Lengermann and Nancy M. DeMuth* 287

Introduction 287
Cost-Effectiveness Analyses of Worksite Health Promotion Programs 289
Cost-Benefit Analyses of Worksite Health Promotion Programs 295
Summary 296
References 298

18. Future Directions in Occupational Health Promotion—*Robert H. L. Feldman and George S. Everly, Jr.* 301

Future Directions 301
The Healthy Work Environment and the Culture of Health 303
Concluding Remarks 306
References 306

Appendix Three Examples of Multidimensional Occupational Health Promotion Programs: 309

Control Data's STAYWELL—*Murray P. Naditch 310*
IBM Health Enhancement Activities—Kent W. Peterson 317
The Johnson & Johnson's LIVE FOR LIFE PROGRAM—Curtis S. Wilbur 320

Authors 329

Contributors 330

Index 333

Occupational Health Promotion

PART One

THE FOUNDATIONS OF OCCUPATIONAL HEALTH PROMOTION

In Part One of this text, the first of three parts, we will examine the basic foundations of occupational health promotion. Chapter 1 introduces the reader to the concepts of health, health promotion, and more specifically, the concept of occupational health promotion. Chapter 2 provides a unique ideological and historical perspective on occupational health promotion. Such perspectives may serve as rationales for the development of occupational health promotion programs, but more importantly serve to yield insight into the current status of occupational health promotion in this country. Finally, Chapter 3 addresses a critical consideration in the development of health promotion programming, that is, compliance. In that chapter, we discuss the nature of health compliance and how to increase it.

CHAPTER 1

An Introduction to Health and Occupational Health Promotion

George S. Everly, Jr.

It is 8 o'clock in the morning. Richard has just finished the one-mile jog around the indoor track that he completes three times a week as part of an ongoing weight reduction program. At the same time, Bob is having his blood pressure taken as part of the hypertension screening program he participates in. JoAnn is in the midst of leaving her son at the child care center. Barbara is attending a course designed to help her stop smoking. Finally, Michael is just now leaving the counselor's office who is assisting him with the drinking problem he has. What do all of these individuals have in common? They are all at work. And each one of these individuals has chosen to take advantage of the resources made available to him or her through the occupational health promotion program provided by their respective employers. The experiences of Richard, Bob, JoAnn, Barbara, and Michael are but a small sampling of the opportunities that can be made available to individuals who are interested in improving their personal and occupational health through organized and systematic efforts now referred to as "occupational health promotion."

INTRODUCTION

Recent publications in the fields of psychology, medicine, and education have heralded the forthcoming age of Health Promotion—an age where prevention is considered as important as treatment and where life-style and health-related behavior are viewed as new and promising points of intervention for affecting human health status. Within the genre of health promotion, occupational health promotion efforts may be the most promising means of promoting health in this country today. This concept of integrating health behavior-change interventions with traditional occupational health and human resource development is being practiced in a growing number of occupational settings. The purpose of this text is to examine health promotion within the occupational environment. More specifically, we will examine how behavioral programs can foster the development and practice of health-promoting behaviors in the workplace.

To begin, scientists investigating human health and disease are now repostulating the tenets upon which disease theory has traditionally been built. For generations the delivery of health care services has been built upon the *one germ-one disease* theory of illness originally postulated by Louis Pasteur and others.

This theory of disease

> was the most powerful single idea in the history of medicine, being responsible for (1) a massive and effective assault on acute disease through immunization and treatment, (2) a refocusing of attention onto the disease process with an attendant loss of interest on the part of the medical profession in the nonbiological aspects of illness, (3) a period of optimism in which it seemed possible to eradicate all illness from human populations, and (4) a fundamental reorganization of medical training in the modern medical school. (Twaddle and Hessler, 1977, p. 11)

It is difficult to exaggerate the gains and benefits to human health achieved as a result of the germ theory of disease. It represented one of those truly watershed ideas and scientific models that resulted in a cascade of research, knowledge, therapies, and improved health. However, the germ theory of disease could also represent an intellectual trap. Why? Because it could lead us to ignore the fact that what was disabling and killing Americans in the year 1900 is no longer the primary cause of death in the United States today. A recent examination of mortality statistics for the United States will reveal that chronic, degenerative diseases have replaced the germ-based infectious diseases as the leading causes of death in America (Public Health Service, 1979a). Furthermore, there is a growing body of evidence that life-style and health-related behavior are *the* major determinants of these chronic diseases (American Psychological Association, 1976; Pelletier, 1979; Public Health Service, 1979a; Green, 1981). Thus the germ theory of illness is being reconsidered in deference to a multifactorial or multideterminant explanation of disease. According to the 1976 American Psychological Association Task Force on Health Research (1976, p. 264):

> New knowledge . . . has increased the recognition that the etiology of poor health is multifactorial. The virulence of infection interacts with the particular susceptibility of the host. Predisposition of the individual to succumb selectively to various assaults on physical integrity not only is related to particular early life experiences but also is associated with economic and social status, especially as reflected in living and working conditions.

Thus we see growing evidence that factors such as life-style, personality, working conditions, and other related behavioral variables play the major role in the determination of human health and disease. If this conclusion is correct, as current evidence suggests, then there appears to be a strong rationale for the development and implementation of behavior change interventions designed to foster health-promoting behavior on the part of the individuals of our society.

The workplace, where most individuals spend at least one third of their life, is considered to be an ideal place to promote health behavior. Occupational health promotion offers the potential of affecting not only the health of the individual worker, but the health of the organization and community as a whole. The purpose of this text is to provide examples of how health can be promoted on the basis of behavioral interventions within the occupational milieu.

While the occupational environment may seem to be an unusual place to promote health and health behavior, consider the following: It is generally accepted that this country's health care costs exceeded $300 billion in 1984, and the cost is increasing at a 12–15% rate each year. American corporations will pay over $80 billion of those costs, which is in excess of 25%. The impact is such that, although estimates vary, on the average, corporations are thought to spend at least 10% of their operating budgets on employee health-related expenses (American Institute of Certified Public Accountants, 1981).

Employee health-related expenses are influenced by a number of subtle, as well as obvious, health factors and health behaviors. For example:

1. Smoking in the amount of one pack of cigarettes a day has been estimated to cost $600 in health-related expenses per person (Goldbeck and Kiefaber, 1980). The annual health-related costs of smoking have been estimated to be in excess of $11 billion per year (Kristein, 1980). Smoking two packs a day doubles the absenteeism rate among such smokers (National Chamber Foundations, 1978).
2. Alcoholism increases employee absenteeism by 2.5 times and increases the accident rate by 3–5 times (National Chamber Foundation, 1978; O'Connor, 1984).
3. Over 12,000 occupationally related accidental deaths and over 2 million disabling injuries have occurred in the workplace within a single year (National Safety Council, 1982).
4. Approximately one third of the American work force is obese. Obesity increases the likelihood of diabetes mellitus, hypertension, hernias, and gallbladder disease (National Chamber Foundation, 1978).
5. The cost of recruiting replacements for executives lost due to heart disease is approximately $700 million per year (Goldberg, 1978).
6. Health insurance rates for many corporations are "experience-related," therefore, increased employee health impairment and resultant benefit utilization will increase subsequent health insurance costs to the organization (Hamilton and Martin, 1977).
7. Finally, with employment health benefits becoming more important in labor negotiations, it may be anticipated that corporations will be asked to

absorb increasing percentages of employee health care costs. Such a condition must surely impact the entire corporation. As Hamilton and Martin (1977, p. 19) note:

> This relentless upward spiral of health benefit costs is taking its toll throughout industry. Each year corporations are forced to allocate a larger share of their operating expenses just to provide employee health benefits, resulting in higher consumer prices or lower profits, or both.

Thus, we see that the health of the American worker is a factor in corporate existence that can no longer be taken for granted. Furthermore, its apparent influence on corporate well-being must certainly be translated into an effect on the community within which the corporation resides. Therefore, occupational health is a factor that warrants attention not only from the highest levels of management and union organizations, but from the highest levels of civic organization as well.

The subsequent discussions within this text are based on the proposition that occupational health influences not only the worker, but the corporation and the community as well. Therefore, we would extend this proposition to include the notion that the advocacy and active promotion of occupational health and health behavior is a legitimate concern of individual workers, unions, management, the community, and nation as a whole.

Within this text we shall discuss the development and implementation of occupational health promotion programs. We shall make general recommendations on establishing the components of such programs. Initially, however, it seems prudent to devote some time to defining certain key terms and concepts upon which occupational health promotion activities are ultimately based.

WHAT IS HEALTH?

The most generally accepted definition of health is that recommended by the World Health Organization: "Health consists of a state of complete physical, mental, and social well-being, and not merely the absence of disease or infirmity." It is important to note that health may be thought of as more than just the absence of disease or illness. Indeed, health may be thought of in the context of increasing levels of healthfulness, or "wellness" to use a term currently enjoying popularity (Dunn, 1961; Ardell, 1977). This higher-level health, or "wellness," represents a state in which an individual enjoys increased health and more effective levels of functioning beyond that rather neutral state of merely being illness free (see Figure 1.1).

Higher-level health may be operationalized as increased levels of fitness, reduced risk of disease, increased coping ability, increased emotional stability, reduced chronic stress levels, and increased energy, to mention only a few examples.

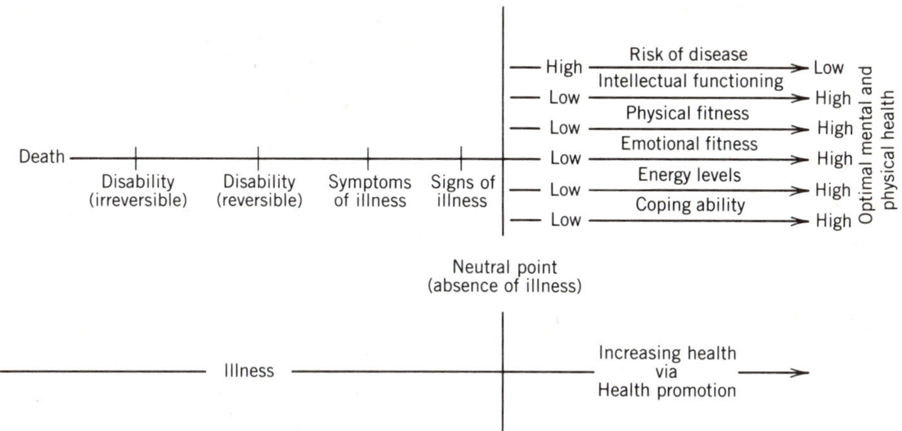

Figure 1.1. From illness to health.

These concepts of health are by no means new. In 1934 Jesse Williams (1934) offered the viewpoint that defining health as simply freedom from disease was in effect establishing a standard of mediocrity for the pursuit of health. Ivan Illich (1976) has even argued that health is not truly at its optimum until the individual demonstrates responsible, autonomous, personal coping behavior.

THE DETERMINANTS OF HEALTH AND DISEASE

What determines the health of any given individual? The answer to this question has been debated for centuries and probably will continue to be debated far into the future. While complete agreement on what determines health may be impossible to reach, we have made some progress toward identifying the major determinants of human health and disease.

The Canadian Minister of Health and Welfare, Marc Lelonde (1974) introduced the concept of the ''Health Field'' to attempt to describe the major determinants of health and disease. From his perspective, human health is said to be mostly a function of four factors:

1. Human biology.
2. Environment.
3. Life-style.
4. Health care organization.

One contribution that this model makes is that it views the role of human biology (usually considered the greatest influence of human health) within a more evenly balanced framework from which to view human health and disease.

Subsequent work by the U.S. government was apparently greatly influenced by the Canadian efforts. In attempting to identify the major influences on human health and disease processes, the U.S. government typically includes four major factors:

1. Life-style.
2. Biology.
3. Environment.
4. Health care services.

These factors are the same as those identified by the Canadian efforts. Table 1.1 below indicates the estimated percent contribution that each of these four influences have on the 10 leading causes of death in the United States.

Table 1.1 clearly indicates the important role that life-style plays in the eight leading causes of death in the United States.

Just how important of an influence is one's behavior/life-style on health status? Looking back at Table 1.1 we see that life-style is estimated to have the greatest single effect on the 8 leading causes of death in the United States. Furthermore, combining all 10 causes of death together we see that a person's life-style is estimated to account for 51% of all 10 major causes of death in the United States. Although the data in Table 1.1 are rough estimates, Lawrence Green, noted public

Table 1.1 Estimated Contribution of Four Factors to the 10 Leading Causes of Death before Age 75 (Expressed in Percentages)

	Factors			
Causes of Death	*Life-style*	*Environment*	*Biology*	*Health Care Services*
Heart disease	54	9	25	12
Cancer	37	24	29	10
Motor vehicle accidents	69	18	1	12
Other accidents	51	31	4	14
Stroke	50	22	21	7
Homicide	63	35	2	0
Suicide	60	35	2	3
Cirrhosis	70	9	18	3
Influenza/pneumonia	23	20	39	18
Diabetes	34	0	60	6
All 10 causes together	51.5	20.1	19.8	10

Source: U.S. Center for Disease Control. "Ten Leading Causes of Death in the United States." Atlanta: CDC, July 1980; personal communication, CDC, June 1984.

health scholar, reaches a similar conclusion. Green (1981) concludes that of the major determinants of human health, one's behavior is the most dominant force, followed by the environment, and finally medical technology. Furthermore, Green argues that one's behavior greatly influences the latter two factors:

> Behavior influences health directly through preventive health behavior and life-style in general. The environment is behaviorally influenced through social action and through peoples' exposure of themselves to the environmental risks. The effect of medical care on health depends on illness behavior and sick-role behavior, the appropriate utilization of services, and the effective following of medical regimens. (Green, 1981, p. 20)

In a paper addressing the role of the behavioral sciences in health, Richard Dwore and Joseph Matarazzo (1981, p. 4) state: "Scientific evidence indicates that individual behavior and life-style indirectly and directly play a major role in health outcomes." Such seems to be the case in other industrialized countries as well. In his analysis of health factors in England and Wales, Thomas McKeown (1978) concludes that the two most important factors resulting in improvements in health status in those areas were modifications of individuals' health-related behavior and changes in the environment. The pioneering work of biologist Rene Dubos (1968) supports the intercultural importance of life-style/behavior as a determinant of health status. Dubos argues that life-style/behavior factors serve as a major determinant of health status not only in industrialized countries but in nonwestern cultures as well. Finally, according to the U.S. Surgeon General's Report on Health Promotion and Disease Prevention (Public Health Service, 1979a) life-style alterations enable individuals to reduce risk factors that may predispose to illness and at the same time promote emotional and physical health. Three epidemiological investigations appear to bear this conclusion out. Longitudinal research by Erdman Palmore (1971) examined the variable of human longevity. Palmore concluded that four factors were recurrent in functional human longevity:

1. The maintenance of a satisfying social role.
2. A positive outlook on life.
3. Physical fitness.
4. Avoidance of smoking.

In a 5½-year longitudinal study of over 7000 individuals, Belloc and Breslow (1972) uncovered seven factors that were significantly correlated with life expectancy and health:

1. Moderate exercise two or three times a week.
2. Eating three regular meals a day.
3. Eating breakfast every day.
4. Not smoking.

5. Moderate or no alcohol consumption.
6. Maintaining moderate weight.
7. Seven or eight hours of sleep each night.

Finally, in a 9-year study of mortality in 4725 adults Berkman and Syme (1979) found that individuals who were married, maintained close contact with relatives and friends, attended religious services, and participated in formal and informal group activities had less than one-half the age-adjusted risk of death compared to those who did not possess these traits. It is important to note that the factors predisposing to health that emerged from the previously cited research are all behavioral/life-style factors.

To return to the question asked earlier in this section, how important is life-style in the pursuit of health? Knowles (1977, p. 58) provides one answer to this question by noting:

> . . . over 99 percent of us are born healthy and made sick as a result of personal misbehavior and environmental conditions.

Health economist Victor Fuchs (1974) concludes that the greatest current potential for improving health lies in the behavior of the individual. Everly and Newman's review (1982) on behavioral precursors of disease concluded in agreement with Fuchs' while emphasizing the critical role that "individual self-responsibility" plays in the determination of human health and disease.

While it is clear that human health and disease are a function of a multifactorial array of interacting variables, an individual's behavior/life-style may be the single most important determinant in the majority of cases. If this is so, as current evidence suggests, then we are armed with a powerful rationale for helping people (1) change their health-related behavior patterns and (2) promote a life-style which is generally conducive to health. Such can be the goals of health promotion.

HEALTH PROMOTION AS A MECHANISM OF MEDIATION

Having offered a definition of health and its major determinants, let us now turn to a discussion of health promotion.

"Health care professionals from a wide variety of fields are beginning to confront the issue of changing peoples' day-to-day habits and behavior patterns. Those without a social science background sometimes make the naïve assumption that once a person understands how to enhance his health status . . . a change in behavior will follow" (Henderson, Hall, and Lipton, 1979, p. 141). There is growing belief among health professionals, however, that simply providing health-related information to individuals is *insufficient* to longitudinally alter their health-related behavior (Stone, Cohen, and Adler, 1979). For example, even the newest, most advanced exercise facilities in the country will rust from lack of use unless people view such activities as personally relevant and practical. Therefore, there clearly exists a need to develop and implement a repertoire of behavior change programs that can affect a significant and lasting increase in health-promoting behavior in our society. Such interventions would act by showing people that health-promoting behavior is personally relevant by making health-promoting behavior practical and reinforcing at the same time, whether such behavior includes preventive physical exercise, dietary alterations, stress management, or improving compliance in medication or rehabilitation regimens.

In its most generic form, then, health promotion may be viewed as any intervention designed to facilitate and/or support personal health maintenance/improvement behavior, especially when one keeps in mind that health is best represented as a continuum and is more than just the absence of disease.

Lawrence Green (1981, p. 22) has offered health promotion as "the combination of health education and related organizational, economic, and environmental supports for behavior conducive to health."

The Proceedings of the National Conference on Health Promotion Programs in Occupational Settings (Public Health Service, 1979b, p. 32) described health promotion as a "system of components (e.g., exercise, nutrition, stress management, etc.) that attempts to integrate the concepts of disease prevention and life-style modification with the more traditional practice of treating diseases after they occur." The emphasis of such efforts as described in those proceedings would be upon utilizing a variety of educational/behavioral strategies to facilitate life-style modification and health maintenance. In addition, emphasis is placed upon the individual assuming *reasonable* responsibility for his or her own life-style and health.

In an attempt to integrate these perspectives, we believe that health promotion may best be understood as a "mechanism of mediation." From a process perspective, a mechanism of mediation may be thought of as a medium to bring about an outcome or result. Therefore, health promotion may be viewed as the mediating mechanism consisting of any and all interventions (strategies and tactics) that can be used to link a body of health information (as it is currently held by health professionals) to the desired outcome of initiating and propagating personal health maintenance/improvement behaviors on the part of the public, with the ultimate goal of improving the general level of health.

Three core assumptions underlying health promotion efforts would be

1. Prevention is preferable to treatment or rehabilitation.
2. Training people to be healthy is less costly in time and resources than treating people once they are ill.
3. Healthful life-styles may ultimately reduce risk, improve health, and improve the quality of life (adapted from the National Chamber Foundation, 1978).

Having examined health promotion in general, now let us turn to the concept of "occupational health promotion." Occupational health promotion simply entails the utilization of health promotion interventions as described above but as implemented in the occupational milieu or under the auspices of one's employing organization.

Occupational health promotion may also be thought of as striving to improve *personal health* from the individual's (and union's) perspective, while striving to improve the *human resource* from the organization's perspective. Thus we see a new perspective added in considering this last statement—that of the organization's vested interest in the health of its employees, that is, its human resource.

IS THE WORKPLACE APPROPRIATE FOR PROMOTING HEALTH?

Why promote health on the job? "The occupational or worksite is a promising setting in which to provide health information, risk appraisal, education, and promotion efforts to adults" (Public Health Service, 1979c). A report issued by the National Chamber Foundation (1978, p. ix) agrees with this conclusion:

> The workplace may be a setting particularly well-suited to effective and widespread health promotion programs. Through the programs, business can help employees and their families achieve and maintain better health. The programs have the potential to benefit not only the individual participants, but also the sponsoring company. . . .

Let us examine issues in the development of occupational health promotion programs from three individual perspectives: (1) that of the organization in general, (2) that of management, and (3) that of the employees.

At the Newsweek Conference on Cost-Effective Strategies for Corporate Healthcare, Richard Egdahl (1980, p. 1) concluded that ". . . companies will have to recognize that the health and well-being of employees influence the company's overall productivity."

Clearly the organization has a vested interest in the health and welfare of its employees. As described in an earlier section, business and industry pays approximately 25% of the health-related costs in this country and chances are that that percentage will increase. Similarly, most corporations' health insurance premiums are indeed "experience-related." Therefore, the greater the corporation's utilization of health insurance, the greater the subsequent cost of that same coverage. Finally, it has been suggested that the overall health of an employee greatly affects that employee's job performance, overall productivity and job satisfaction (Public Health Service, 1979b). While compellingly logical, such latter relationships have yet to be clearly demonstrated. In summarizing these views from an overall corporate perspective the following statement appeared in the *Proceedings of the National Conference on Health Promotion in Occupational Settings* (Public Health Service, 1979b, p. 35).

> A sizable portion of the costs of health care and of disability and death are paid through employers; hence, a sizable portion of the benefits from effective disease control and health promotion will tend to return to employers. The potential economic return provides a . . . rationale for providing health promotion programs.

Despite the aforementioned points, many corporate managers may be hesitant to initiate occupational health promotion programs. The two major objections seem to be (1) decision makers feel health promotion activities could be perceived as "interfering in the personal lives of the employees" and/or (2) decision makers may fail to perceive an administrative or budgetary precedent for the development of occupational health promotion programs within their existing organizational structures. Many decision makers are hesitant to propose anything that cannot in some way fit into existing administrative and budgetary procedures.

In reviewing these objections several points must be made clear. First, occupational health promotion programs should be *voluntary*. It will be recalled from an earlier section that emphasis is placed upon the individual employee assuming responsibility for his or her actions and life-style. Occupational health promotion simply allows individuals to make *informed* personal health decisions and further assists them in improving their personal health *if* they choose to do so. Second, depending upon the organization's existing structure, health promotion programming could readily be developed under the structure of medical divisions or even training departments. Another consideration, however, would be to consider health promotion as nothing less than Human Resource Development (HRD). In doing so health promotion would be integrated within any larger personnel or HRD department—a department most organizations are already highly committed to. Of course, any health screening procedures should come under the direct supervision of some health professional.

Having examined health promotion program development from organizational and management perspectives, let us consider the employee's point of view. As already noted, unions as well as management are increasingly interested in health promotion. Numerous union contracts are requesting, even demanding, preventive medical care and health promotion programs. Historically, workers' demands have been shifting from increased compensation to safe working conditions, adequate health insurance, and most recently, to health promotion programs in the workplace. Generally speaking, employees want to stop smoking, lose weight, stay fit, and reduce stress in order to live longer and enjoy their longer lives. Indeed, employees appear ready for occupational health promotion efforts. The National Conference on Health Promotion in Occupational Settings (Public Health Service, 1979b, pp. 34–35) enumerated three employee-related considerations that contribute to the utilization and overall effectiveness of occupationally based health promotion:

Convenience Generally, an industrial health program will be located near to or at the job site, and utilizing it will require a minimum of scheduling and travel effort.

No Cost Worksite programs benefit from a psychological attitude widely prevalent among workers, namely,that what is received free at work is really part of one's pay and is an entitlement that one does not wish to lose. Consequently there is a strong desire to utilize free services at the worksite.

Presumed Quality Many workers share an attitude that if the company provides something, then the quality must be good or the company would not have "bought" it.

Summary

In this initial chapter we have examined the human condition called "health" and its major determinants. We have concluded that life-style plays a major role in human health and disease. Indeed, life-style may be the single most influential factor. If current evidence is correct, and life-style is a major determinant of health, then one can easily see the need for programs that promote health through an inclusion of, if not emphasis upon, behavior change interventions. Our current notion of what health promotion is meets this need.

We have also addressed the more specific issue of occupational health promotion. In doing so we must conclude that the workplace seems to be an environment ideally suited for health promotion activities. We strongly believe that occupational health promotion may be thought of as striving to improve *personal health* vis-à-vis individual employees, while at the same time striving to improve the *human resource* vis-à-vis the organization.

> The escalating cost of health benefits has been a major incentive for industry to become involved in preventive health care. It is becoming more apparent, however, that containing the costs of health benefits is only one motive for exploring the feasibility of a health promotion program at the worksite. Additional benefits of such a program may include reduced absenteeism, increased productivity, increased ability to cope with problems on and off the job, and enhanced functional efficiency of the corporate organism as a whole. (Public Health Service, 1979b, p. 2)

This then is the challenge: to promote the health of the individual while at the same time promoting the health of the organization and the community at large.

The application of behavioral principles and programs appears to be a viable means of meeting this challenge. For example, according to the American Psychological Association (1976, p. 272), "There is probably no specialty field within psychology that cannot contribute to the discovery of behavioral variables crucial to a full understanding of susceptibility to physical illness, adaptation to such illness, and prophylactically motivated behaviors.'' Reviews by Dwore and Matarazzo (1981) and Stainbrook and Green (1982) conclude that the behavioral sciences in general have combined a significant scientific base with professional experience for successfully changing human behavior. The time has clearly come to integrate the behavioral programs into health care and health promotion.

The remainder of this text is devoted to a discussion of how to design, implement, and evaluate effective occupational health promotion interventions. In considering the design of health promotion programs, however, it is important to keep in mind that such programs may potentially consist of numerous and diverse component elements allowing the potential for the development of very simple to very complex health promotion programs. In this text we shall present a wide range of potential program component options. It is beyond the scope of this text to make specific recommendations as to the final constituents of any occupational health promotion program. To be most effective, the scope of any occupational health promotion program should be tailored to meet the specific *needs* and *resources* of the specific organization under consideration.

REFERENCES

American Institute of Certified Public Accountants. ''Dear employees: Don't get sick.'' *CPA Client Bulletin,* November 1981, 2.

American Psychological Association. ''Contributions of psychology to health research: patterns, problems and potentials.'' *American Psychologist,* 1976, **31,** 263–273.

Ardell, D. *High Level Wellness.* Emmaus, Pa.: Rodale, 1977.

Belloc, N., and Breslow, L. ''Relationship of physical health status and health practices.'' *Preventive Medicine,* 1972, **1,** 409–421.

Berkman, L. and Syme, S. ''Social networks, host resistance, and mortality: a nine-year follow-up of Alameda County residents.'' *American Journal of Epidemiology,* 1979, **109,** 186–204.

Dubos, R. *Man, Medicine and Environment.* London: Pall Mall, 1968.

Dunn, H. L. *High Level Wellness.* Arlington, Va.: Beatty, 1961.

Dwore, R. and Matarazzo, J. ''The behavioral sciences and health education.'' *Health Education,* **12,** 1981, 4–7.

Egdahl, R. ''Corporate and Organizational Strategies for Health Care Cost Containment.'' Paper presented at the Newsweek Conference on Cost-Effective Strategies for Corporate Healthcare, Chicago, Ill., November 10–11, 1980.

Everly, G. and Newman, E. ''Individual self-responsibility and the determinants of health and disease.'' Paper presented at the Health Education Risk Reduction Conference. Division of Public Health, State of Delaware, Newark, Delaware, February 12, 1982.

Fuchs, V. *Who Shall Live?* New York: Basic, 1974.

Goldbeck, W. and Kiefhaber, A. ''Wellness . . . The new employee benefit.'' *Voluntary Effort Quarterly.* 1980, **2,** 1–3.

Goldberg, P. *Executive Health.* New York: McGraw-Hill, 1978.

Green, L. ''Emerging federal perspectives on health promotion.'' *Health Promotion Monographs,* July 1981, Whole No. 1.

Green, L. et al. *Health Education Planning: A Diagnostic Approach.* Palo Alto, Ca.: Mayfield, 1980.

Hamilton, W. and Martin, S. "Benefits fever." *The Wharton Magazine,* 1977, **1,** 19–25.

Henderson, J., Hall, S., and Lipton, H. Changing self-destructive behaviors. In *Health Psychology,* G. Stone et al. San Francisco: Jossey-Bass, 1979, 141–160.

Illich, I. *Medical Nemesis—The Expropriation of Health.* New York: Pantheon, 1976.

Knowles, J. J. *Doing Better and Feeling Worse: Health in the United States.* New York: Norton, 1977.

Kristein, M. "How much business can expect to earn from smoking cessation." Paper presented to the National Interagency Council on Smoking and Health, January 9, 1980 (Chicago).

Lelonde, M. *A New Perspective on the Health of Canadians.* Ottawa, Canada: Minister of National Health and Welfare Information, 1974.

Matarazzo, J. "Behavioral health and behavioral medicine: Frontiers for a new health psychology." *American Psychologist,* 1980, **3,** 807–817.

McKeown, T. "Determinants of health." *Human Nature,* April 1978, 60–67.

National Chamber Foundation. *How Business Can Promote Good Health for Employees and Their Families.* Washington, D.C.: National Chamber Foundation, 1978.

National Safety Council. *Accident Facts.* Chicago, NSC, 1982.

Ng, L., David, D., Manderscheid, R., and Elkes, J. "Toward a conceptual formulation of health and well-being." In *Strategies for Public Health,* L. Ng and D. Davis, eds. New York: Van Nostrand Reinhold, 1981.

O'Connor, R. O. "Sobering up the corporation." *Corporate Fitness and Recreation,* 1984, **3,** 16–21.

Palmore, E. "Longevity predictors—implications for practice." *Postgraduate Medical Journal,* July 1971, **50,** 160–164.

Pelletier, K. *Holistic Medicine.* New York: Delta, 1979.

Public Health Service. *Healthy People: The Surgeon General's Report on Health Promotion and Disease Prevention.* Washington, D.C.: U.S. Government Printing Office, 1979a.

Public Health Service. *Proceedings of the National Conference on Health Promotion Programs in Occupational Settings.* Washington, D.C.: U.S. Government Printing Office, 1979b.

Public Health Service. *Promoting Health: Issues and Strategies.* Washington, D.C.: U.S. Government Printing Office, 1979c.

Stainbrook, G. and Green, L. "Behavior and behaviorism in health education." *Health Education,* 1982, **13,** 14–19.

Stone, G., Cohen, R., and Adler, N. *Health Psychology.* San Francisco: Jossey-Bass, 1979.

Twaddle, A. and Hessler, R. *A Sociology of Health.* Saint Louis: Mosby, 1977.

Williams, J. *Personal Hygiene Applied.* Philadelphia: Saunders, 1934.

CHAPTER 2

Ideological and Historical Rationales for Occupational Health Promotion: From Asclepius to Hygeia

Arthur L. Anderson

In Chapter 1 the nature of health and health promotion was discussed with an emphasis on understanding the concept of occupational health promotion. In this chapter we will examine the development of occupational health promotion as we know it today.

INTRODUCTION

Growing children are forever asking parents and other adults why we do things in a certain way. When children ask why and we provide answers, we are inevitably involved in ideology. Ideology refers to the ideas and beliefs people use to explain and legitimatize the way they behave (Mannheim, 1936). A key word here is legitimate, since simply explaining our behavior is not satisfactory. Local manufacturers could explain that they knowingly exposed their employees to excessive radiation and thereby improved their "bottom line," but most people today—including manufacturers—would regard that as an illegitimate way to turn a profit. Legitimacy has to do with the rightness of things (Berger, 1967), and ideology consists of those ideas and beliefs that make it seem right in our minds.

In the fields of health and health promotion we find an enormous variety of ideologies in different historical periods. While we would expect differences in a comparison of different societies, sometimes we note great differences within even a single society. Rene Dubos in *Mirage of Health* sugests that ancient Greek society embodied two contradictory ideologies of health symbolized in the goddess Hygeia and the god Asclepius.

> Hygeia was probably an emanation, a personification of Athena, the goddess of reason. Although identified with health, she was not involved in the treatment of the sick. Rather, she was the guardian of health and symbolized the belief that men could remain well if they lived according to reason. . . . Throughout the classical world, Hygeia continued to symbolize the virtues of a sane life in a pleasant environment. . . . From the fifth century B.C. on, her cult progressively gave way to that of the healing god, Asclepius. (Dubos, 1959, pp. 129–130)

As Dubos notes, we are all quite familiar with Asclepius for he is the god holding a staff coiled with a snake which symbolizes an ideology of health that is disease- and healing-oriented. Indeed, health is not defined as a positive ideal but as the absence of disease. Unlike the Hygiean ideal of the golden mean of moderation in all things, Asclepius implies a life-style in which one pushes life to its limits and then turns to a physician for repair of damaged body and psyche. It is no accident that Asclepius became the symbol for modern western medicine since he reflects both education in our medical schools, and American culture as well. Specifically, almost no American medical school offers required courses in preventive health, health promotion, nutrition, physical fitness, or—in short—how to live. The core of medical education is consistent with the Asclepian ideal of understanding disease and how to cure it, and that is what we expect from medicine.

In his discussion of Hygeia and Asclepius, Dubos takes some scholarly liberties with ancient mythology and Greek history in setting up this contrast in ideologies, but it is a very useful set of ideal types for the purposes of this chapter.[1] Like ancient Greece, American society embodies the twin—complementary as well as opposing—ideologies of Hygeia and Asclepius. Based on how we spend our health dollars there is no question, however, that the Asclepian ideology is the dominant one. For example, in 1982 close to $290 billion, or about 10% of the nation's gross national product, was spent on health in America, but some estimates suggest that less than 1% of that was spent on health promotion and education.

There is reason to believe, however, that Hygeia and the health promotion she symbolizes are on the ascendency in American society. There are multiple ideological supports for that trend, and the purpose of this chapter is to analyze what appear to be three important ideological rationales for the practice of health promotion. The three areas we will examine are religion, nationalism, and capitalism.

While these three ideological rationales provide legitimation and support for everything from health food stores to corporate fitness centers to the modern hospital to what makes physicians behave as they do, they also undergird current efforts in occupational health promotion. Moreover, as the following sections will make clear, any one ideological rationale is rarely found in pure form, just as motivation

[1]Except for eating somewhat better—not by design but because of what was available—the ancient Greeks pursued life-styles remarkably similar to ours today, and Asclepius was therefore their most popular god of health.

for any human behavior is rarely based on a single feeling or factor. Instead, the ideological rationales for health promotion intertwine and intermix—often in curious ways—in areas such as corporate occupational health promotion efforts, and in Congressional legislative efforts to promote occupational health.

If the reader has any question as to why a chapter on ideological rationales is included in a textbook on occupational health promotion, there is a straightforward answer to this question. Such rationales provide answers to why we are engaged in such activity to begin with. Moreover, occupational health promotion professionals need all of the ideological rationales they can muster to support their efforts. As we shall see, in the governmental or public sector, nationalism and political support are legitimations enough for health promotion. In the voluntary and professional sector, good will and altruism are legitimations enough. In the private-sector workplace, however, business managers demand more than political support or altruism. Business managers demand facts that will prove in economic terms that occupational health promotion makes good, capitalist economic sense. As this chapter will make clear, this is not an easy legitimation to earn. So much for introductions. Let us now examine the three major ideological rationales for occupational health promotion in present-day American society.

RELIGION

The Judeo-Christian tradition has powerfully shaped most of the values Americans live by. Even that sizable minority of Americans (40%) who identify themselves as Jews and Christians but have no religious affiliation or particular religious practices would find on careful self-examination that often they are living their daily lives by values that in Max Weber's terms (Weber, 1958, p. 182) represent, at the very least, "the ghost of dead religious beliefs."

It would be frivolous to think one could summarize in this brief section the profound ways the Judeo-Christian tradition has shaped how we think about life, death, and health. The following two passages from the sacred writings (The Holy Bible, 1952) of this tradition are, however, illustrative:

> So God created man in his own image, in the image of God he created him; male and female he created them. . . . then the Lord God . . . breathed into his nostrils the breath of life; and man became a living being. Genesis 1:27–2:7

> Do you not know that your body is a temple of the Holy Spirit within you, which you have from God? You are not your own; you were bought with a price. So glorify God in your body. I Corinthians 6:19–20

The biblical view of time and history is a lineal not a cyclical one. In this framework, the biblical view of life is that we only pass this way once. We have but one life to live and it is sacred. More than that,

> in the Judeo-Christian tradition—and especially in the Christian phase—life, for the individual, is defined in the first instance as a *gift*, directly or indirectly, from God (Parsons et al., 1973, p. 5).

This idea of life as a sacred gift was combined in the Christian tradition with the idea of stewardship. Christian stewardship meant that God has placed us in what is metaphorically often called the vineyard of life, and it is our religious duty to practice fiduciary responsibility toward everything that God has entrusted to our care during the brief span of our lives.

At the individual level this meant practicing personal stewardship, but it went beyond that. Every believer also has a responsibility to practice stewardship on a social level toward his or her community, and this came to be known as civic duty. Social welfare, philanthropy, involvement in community affairs and politics, establishing and serving voluntary associations—all such efforts represented Christian stewardship extended to the social level in the form of civic duty.

For Americans imbued with this religious ideology it required no leap in logic to see the relevance of stewardship to health and health promotion. The stewardship of life involved "the protection of the gifts of life [and this] involves practical efforts to control the causes of unnecessary and premature death" (Parsons et al., 1973, p. 21). At the social level this meant establishing institutions that express and practice the religious impulse in the form of civic duty.

So much for the religious rationale. What were some of the practical applications of this ideology? The historical evolution and application of this religious ideology would require multiple volumes to adequately document, but some illustrations are certainly appropriate to our discussion here.

Benjamin Franklin (1706–1790) provides an interesting case study from eighteenth-century America. He was a child of the Enlightenment, acutely interested in science, and the epitome of a pragmatic, capitalist mentality. Franklin had deeply internalized a secularized version of values central to the religious version of the Protestant ethic: thrift, the work ethic, and noteworthy for our discussion—stewardship and civic duty.

Benjamin Franklin's secularized sense of the religious values of stewardship and civic duty motivated him to help found everything from the United States Postal Service to the University of Pennsylvania to a local fire department. Among his many accomplishments, Franklin in 1751 played a key role in establishing the first successful general hospital in American history, because, as he wrote (Rosen, 1963, p. 23), "Many must suffer greatly, and some probably perish that might other wise have been restored to health and comfort, and become useful to themselves, their families, and the publick for many years after." Franklin further writes of the need for health promotion in the community in general. Franklin's motivation for doing this—stewardship and civic duty; his goal—preserving useful and productive citizens in American society.

Moving forward from the eighteenth to the nineteenth century, among the nascent public health movement—as well as among many physicians—we find that "a certain number, especially among this pioneer generation, found a central motivation for their driving activism in a pervading spirit of millennial piety" (Rosenberg and Rosenberg, 1968, p. 34). Evangelical Protestantism infused the religiously oriented humanitarianism that motivated John H. Griscom and Robert M. Hartley to become giants in the nineteenth-century public health movement in places such as New York City. As John Adronaux (Rosenberg and Rosenberg, 1968, p. 27), another articulate spokesman for the movement, put it,

> there is religion in cleanliness, in ventilation, and in good food. . . . Disease, like sin . . . is permitted to exist; but conscience and revelation on the one hand, and reason and science on the other, are the kindred means with which God has armed us against them.

The religious idea of the sacred gift of life, plus the stewardship of promoting health was very much a factor in the early public-health movement of the nineteenth century.

It is, however, in the lives of the brothers, Dr. John Harvey and William Keith Kellogg, at the end of the nineteenth century that we see one of the better examples of a concrete expression of the Judeo-Christian ideology for health promotion. Dr. John Harvey Kellogg, an avid Seventh-Day Adventist, was superintendent of the world-famous Battle Creek Sanitarium in Michigan, where his brother William served as business manager. Vegetarianism ruled the dietary offerings for patients at the sanitarium. In attempts to enhance such a diet, John and William serendipitously in the sanitarium kitchen came on the discovery of wheat and later corn flakes by mashing, rolling, and baking wheat and corn.

It should be noted that Seventh-Day Adventists "accept at face value Paul's statement that man's body is the 'temple of the Holy Spirit.' Believing this, they refrain from all harmful indulgences that might weaken their efficiency and the sincerity of their witness, as workers of God" (Maxwell in Rosten, 1975, p. 251).

While Dr. John Harvey Kellogg remained at the Battle Creek Sanitarium, entrepreneurial William moved their cereal discovery across town in 1906 and founded what became the largest ready-to-eat cereal-manufacturing concern in the world—the W. K. Kellogg Company. Apart from being an enterprising capitalist, there is strong evidence that William Keith Kellogg was quite literally packaging evangelically inspired health promotion in boxes of "nutritious cereal" for the sake of improving the breakfast eating habits of millions of Americans. In 1929 William devoted $50 million to found the W. K. Kellogg Foundation (present assets $750 million), the purpose of which was "the promotion of the health, education and welfare of mankind" (Moskowitz, 1980, p. 52).

More recently in the twentieth century there have been attempts by noted theologians to systematically articulate theological definitions of health. ''Seminal thinkers who have tackled this concern have been Paul Tillich, Seward Hiltner, James M. Lapsley, and Ivan Illich'' (Calian, 1978, p. 45). In the writings of such thinkers the sacredness of the gift of life has been reaffirmed. More than that, health promotion in all areas—including the occupational area—is frequently stated as participating in and advancing God's process of salvation that begins in this life and culminates in the next.

In a culture as secular as America appears to be today, it would be both presumptuous and inaccurate to assert that the average physician, nurse, public health professional, or occupational health promotion specialist carries on their work in such a theological context. However, it is probably equally presumptuous and inaccurate to assert that the preservation and promotion of health by health professionals and laypeople alike has no connection with this religious tradition. For example, the Hippocratic tradition does not explain the modern physician's dedication to preserving life at all costs. Hippocrates advised his students to walk away from borderline and terminal patients since they usually die and that is not good for a physician's reputation. We do not walk away. We regard life as a sacred gift and spare little in trying to preserve it. That is our religious tradition. Much that passes for secular, humanistic altruism has as its wellspring the Judeo-Christian religious tradition.

In sum, in various forms the Judeo-Christian religious tradition continues to be one of the most powerful ideological rationales for health promotion in general, and for occupational health promotion specifically. The past two decades of American history have manifested a renewal of the behavioral connection between religion and health. The revival of evangelical Protestantism—some estimates indicate there are now over 40 million self-declared ''born again Christians'' in America—the Catholic charismatic movement, and one should also add, the small but growing interest in Oriental religions—all have combined to strengthen the religious rationale for health promotion in America.

NATIONALISM

> Nationalism is a state of mind in which the supreme loyalty of the individual is felt to be due the nation-state. . . . Although objective factors are of great importance for the formation of nationalities, the most essential element is a living and active corporate will. It is this will which we call nationalism, a state of mind inspiring the large majority of a people and claiming to inspire all its members. It asserts that the nation-state is the ideal and the only legitimate form of political organization and that the nationality is the source of all cultural creative energy and of economic well-being. (Kohn, 1955, pp. 1–2)

> If we add that man must learn to face himself as he faces all others, we imply that so far in history he has made every effort *not* to see that mankind is one species . . . [we call

> this] "pseudo-speciation". . . . The term denotes the fact that while man is obviously one species, he appears and continues on the scene split up into groups (from tribes to nations, from castes to classes, from religions to ideologies) which provide their members with a firm sense of distinct and superior identity—and immortality. (Erikson, 1969, p. 431)

The psychology that feeds nationalism is as old as humankind and is grounded in what anthropologists call ethnocentrism, or psychologists such as Erik Erikson call pseudospeciation. This psychology did not express itself in nationalism until the rise of the nation-state, and for that reason nationalism is a modern phenomenon.

Nationalism as an ideological rationale for promoting health had emerged already in early seventeenth-century Europe. The connecting of nationalism and health rests on the assumption that a strong and superior nation requires healthy citizens, and that it is therefore the responsibility of the state to play an active role in maintaining and promoting the health of its people. Such early nationalistic health promotion

> culminated in the monumental work on medical police of Johann Peter Frank [in Germany]. . . . In 1766 he conceived a plan for the measures to be taken by government for the protection of individual and group health—that is, for medical police. . . . In practice, it meant a program of social action for health and welfare intended primarily to augment the power of the state (Rosen, 1972, pp. 31–33).

In America there were, however, countervailing values that strongly retarded the expression of nationalism in health promotion. After all, "the concept of medical police as developed by Frank and other German writers was authoritarian and paternalistic" (Rosen, 1972, p. 32). American beliefs in individualism, the right to privacy, personal liberty, and democracy caused Americans to be wary, suspicious, and at times hostile, toward any institution that manifested authoritarianism or paternalism. The very idea of Frank's "medical police" would be repugnant to most Americans. We sometimes have ambivalence toward regular police, much less state-sponsored hygienists monitoring our personal health.

In short, the values of Americans militated against nationalistically inspired intervention in health promotion and care. This resulted in health promotion and care remaining essentially an activity for individual and voluntary associations throughout most of the nineteenth century. The voluntary hospital—which as an institutional form still represents over half of America's hospitals—is a good example of this.

While it is true that even in the area of the nineteenth-century public health movement often religiously inspired voluntarism was the major form health promotion would take, other powerful forces were afoot—specifically nationalism—that would enable Americans to reconcile the authority of the state with health promotion and care.

In the early decades of the twentieth century this new synthesis of nationalism and health promotion is strikingly illustrated in the work of the Hygiene Reference Board of the Life Extension Institute. One accomplishment of the Institute was sponsoring the publication of the book, *How to Live,* which between 1915 and 1933 went through nineteen editions—no small achievement for any book at that point in American publishing history. It is no exaggeration to say that *How to Live* was probably the most important and exhaustive handbook on health promotion in American history up to that time. The Foreword to the first edition was written by no less a figure than the president of the United States, William Howard Taft (Fisher and Fisk, 1915, p. x), whose comments are most revealing to the theme of this section.

> Many years ago, Disraeli, keenly alive to influences affecting national prosperity, stated: ''Public health is the foundation on which reposes the happiness of the people and the power of a country. The care of the public health is the first duty of a statesman.'' It may well be claimed that the care of individual and family health is the first and most patriotic duty of a citizen.

(While President Taft made explicit connections between health promotion and nationalism as it affected the economic well-being of the nation, later American presidents would relate health promotion to national defense—an executive sentiment that would ultimately result in the President's Council on Physical Fitness. In large part this was motivated by the significant numbers of rejections for military induction in World War I, but especially during World War II, when more rigorous standards of fitness were applied to armed forces recruits.)

What evolved in the twentieth century from Taft's time forward was the increasing role that local, state, and especially the federal government would play in health care and promotion. Nationalism—along with the religious factor we have already mentioned—provided the rationale to legitimate this activity. Later in the century this trend would result in enacting such legislation as Public Law 94-317, ''The National Consumer Health Information and Health Promotion Act of 1976,'' which in a detailed and explicit manner provides for the federal government to engage in research, education, and programs dealing with health promotion (*Preventive Medicine U.S.A., 1976*).

Even more germane to the present text is the government's direct involvement in the area of occupational health. The federal government got involved in occupational safety and health for the first time in 1890 when federal legislation governing safety in coal mines was passed. Occupational health promotion in this early instance took the form of promoting health through preventive measures aimed at reducing accidents, injuries, and death. The next major thrust was the adoption by the states of worker's compensation laws. The effect of these laws on occupational health promotion was indirect. That is, to avoid the penalties of increased premiums, litigation, and costly settlements, it was in the vested interest of employers to promote health and safety. Relative to today's philosophy, standards, and expectations, this resulted in the simplest and most elementary kinds of health promotion efforts, such as wearing safety helmets in high-risk work areas.

What we see in retrospect in the twentieth century is a complex, gradual, and very uneven kind of development in occupational health.

> Some states had acted to establish and enforce safety and health standards, but only a few had modern laws and adequate resources for administration and enforcement. At least eight states had no identifiable programs in occupational health at all. . . . Health and safety programs merited up to $2.70 per worker per year in some states, while others spent less than one cent. . . . Federal programs were spotty, concentrated on selected industries such as longshoring, railroading, federally contracted construction and service suppliers, atomic energy, and mining. Federal financial and technical support had risen and fallen over the years, but it never had been enough to have any real significance. (Bureau of National Affairs, 1971, p. 14)

While the above characterizes the general picture up to 1950, since that time a ground swell had begun to develop that would eventually result in Public Law 91-596, the Occupational Safety and Health Act of 1970 (OSHA). Support for this sweeping new legislation developed over time from multiple sources: individual congressmen and senators, professionals in the occupational health and promotion areas, labor unions, foundations, and researchers in the field. The OSHA regulations signal a fundamental shift in governmental involvement in the area of occupational health promotion.

> Earlier in this century, governmental response to problems raised by the rapidly increasing occupation injury rate took the form of workmen's compensation legislation at the state level. This legislation was aimed at settling the question of responsibility and payment for injury arising out of the course of employment in a rapidly industrializing nation. In contrast, the federal Coal Mine and Safety Act of 1969 and Occupational Safety and Health Act of 1970 represent a very different use of the law—namely, as a means of exercising social control over technology . . . [A new assumption and policy is involved here, namely,] it is prevention rather than treatment-oriented medicine that is required to reduce occupational disease—and new health-care legislation is, in fact, moving in the direction of preventive strategies. (Ashford, 1976, p. 535)

What was the ideological rationale used to legitimate this gradual—and in the case of OSHA, dramatic—movement of the state into the area of occupational health promotion? The answer is clearly nationalism as we have defined it in this section. Congressional hearings testimony prior to the enactment of OSHA are revealing in this respect. While the costs of occupational injuries and disease are documented ad infinitum, it is most often put in terms of the cost to the nation's economy. Some testimony even shows how problems in occupational health weaken our competitive position as a nation in the international economy. In short, what appear on the surface as cost-benefit analyses of occupational health problems are invariably related to the problems and consequences this poses for the nation.

By way of summary we would note that, unlike Germany, American society historically was not a promising place to fuse the ideology of nationalism with occupational health promotion. Increasingly in the twentieth century we have found ways to merge these two. Presently one would have to conclude that nationalism is one of the strongest ideologies providing a rationale for occupational health promotion. Moreover, with federal and state governments bearing 40% of the cost for health care in the United States, the government holds enormous economic leverage to increase efforts in both general and occupational health promotion, and to legitimate such efforts in the ideological terms of nationalism.

CAPITALISM

In his comprehensive study, *The Economics of Industrial Health,* Joseph F. Follman, Jr., discusses (Follman, 1978, p. 179) the role of employers in occupational health care, prevention, and promotion. He quotes Kenneth H. Klipstein, president of the American Cyanamid Company, as saying, "In assuming responsibility for occupational health, industry only serves its own enlightened self-interest." One could aptly summarize Mr. Klipstein's and other business leaders' comments cited by Follman in terms of the following syllogism:

> A healthier work force is a more productive work force. A more productive work force yields greater profits. Therefore, promoting health is promoting profits.

We have seen in our discussion of nationalism that this syllogism holds merit, since no nation can ignore the health of its workers without imperiling its national economic viability in the long run. However, when we ask whether the syllogism holds true in reality for individual business enterprises, the answer is not so simple.

As any economist knows, one of the cornerstones of capitalism as an economic system and ideology is income maximization, otherwise known as the profit motive (Heilbronner, 1968). Capitalism applied to everyday business activity means a business person aims to get the largest possible return in earnings from the smallest possible investment of capital and work. In this context, it is only under certain specific conditions that it makes any capitalist sense at all to be concerned for the health of workers, or to make any investment in things as esoteric as occupational health promotion. If one is operating a business in an area where there is a limited pool of labor or limited number of workers with the skills one needs, then having a dangerous plant or office that is injuring or killing workers is, in economic terms, unenlightened. Such behavior makes no more sense than failing to maintain plant and equipment.

If, however, there are ready replacements for injured or dead workers, then investments in the health promotion of one's present work force makes no economic sense. Simply hire new workers as others are injured or killed. Investments in health promotion in this instance reduce one's bottom line and are a drain on profitability. In other words, with all due respect to Mr. Klipstein of American Cyanamid, occupational health promotion is only in certain instances an expression of "enlightened self-interest," whereas much of the time such efforts fail to serve at all the self-interest, that is, profitability, of individual businesses in a capitalist economy.

In terms of the religious rationale described earlier, this line of reasoning sounds callous and amoral. Capitalism, however, is an economic system and ideology. It has never pretended to be an ethical or moral system. It deals with the production, distribution, and consumption of the goods and services we need or want to survive. And the kind of calculated economic reasoning described above is precisely what has informed the behavior of most American business managers for most of American history (Navarro, 1975). Indeed, in the first 250 years of American business it is difficult to find references by American business managers to health issues or occupational health promotion. The most that we find is some paternalistic concern on the part of a minority of business managers for the general welfare of their work force.

To be sure, business managers rarely behave in practice according to the pure capitalist model. Business executives, like other Americans, have religious and humanitarian values as well, and these values have ameliorated and modified the harshness of the capitalist model. Based, however, on observations of actual business behavior over time, one would have to conclude that the capitalist model has been the dominant operating model for American business. It would be surprising if things were otherwise, since business is an economic institution and this is a capitalist economy and society.

In view of this it is appropriate to ask, when and why did health become an occupational issue? It became an issue when it was made an issue—first, through the political process at the state and federal levels, and second, when trade unionism—only slowly to be sure—went beyond bread and butter unionism to push for health insurance as a compensation benefit,and later pushed for worker safety and health issues. Rare is the American business firm or business manager who volitionally chose to allocate resources for occupational health promotion or the health care and safety of their workers. Such benefits were gained through the enforcement of government regulation, or through negotiations at the collective bargaining table with labor unions.

These assertions should in no way be construed as a critique of the role American business has played in the area of occupational health. To be critical of business is a failure to understand the fundamental logic of capitalism and the profit motive. Health care and occupational health promotion cost money. For one business firm to move from its economic *raison d'être* into the area of social welfare or humanitarianism is to place itself at a competitive disadvantage with firms who do not choose this policy but decide instead to take their earnings as profit, or invest those earnings in capital investment for the sake of earning further profit.

For this reason, health issues—if they were to be raised at all—had to be raised primarily by institutions external to business itself. If government raised the issues and enacted regulation, or if trade unions raised the issues, the costs of paying for health care and health promotion would be borne equally by all affected companies and industries, and no one company would be at a competitive disadvantage because of having to internalize and absorb such costs.

Workers' compensation is an interesting case in point. Workers' compensation arose because people in American society came to feel generally that Americans were often engaged in dangerous occupations that yielded a profit for business. However, when workers were injured, disabled, or killed in such occupations, society, not business, was expected to pay the costs of such casualties. Workers' compensation laws represented a political attempt by society to force business—which had profited from the labor of workers in high risk occupations—to absorb the costs if workers were injured, disabled, or killed.

It is noteworthy that the adoption of workers' compensation laws gave rise to a new role in American business—the company doctor, who was often later retitled the corporate medical director. For most of the twentieth century the job of corporate medical directors has not been primarily health promotion. Their major job has been to minimize the legal culpability of the company to compensation claims and other litigation. In short, their job is not health promotion, but instead protecting the health of the bottom line of the company. That has been the primary reason for physical examinations, namely, screening out future compensation claimants, or making note of weaknesses in workers. This continues as the primary motivation in recent proposals by business firms to engage in highly sophisticated genetic testing of prospective employees.

In addition to the negative effect of the economic logic of capitalism, health care and promotion by business has faced another obstacle. It is the same obstacle encountered in trying to relate nationalism to health promotion. Health is a private matter for most Americans, and recent attempts to legally protect the privacy and content of company personnel files only underscore the point. For even the most "modern, enlightened, humanistic" managers to actively engage in occupational health promotion is perceived by many employees—including other managers—as paternalistic at the very least, illegitimate, and an invasion of privacy. When, therefore, business firms have engaged in health promotion they have done so gingerly, and most often under the guise of personnel management.

A company may have employed an in-house recreation or fitness director, who might also hold membership in associations such as the National Employees Service and Recreation Association, or the more recently organized American Association of Fitness Directors in Business and Industry. The company may sponsor athletic programs, in-house health spas, or stress management programs—usually for managers and executives. Such programs are, however, most often presented in the guise of a benefit—not direct health promotion—and usually are justified to company financial analysts as good for morale and therefore good for the bottom line. The truth remains that currently the most powerful occupational health promotion measures are those imposed from without through forced compliance with government regulations spelled out in legislation such as OSHA, or union contractual obligations.

At this point in the discussion the reader may well wonder why the subject of capitalism is even introduced here as an ideological rationale for health promotion.

The fact is that the past two decades of American history have witnessed some dramatic changes in the health care system—changes that are of more than passing interest to American business. From 1960 to 1982 expenditures for health care in America have grown over tenfold, from $27 billion in 1960 to $290 billion in 1982. No major sector of the economy has either grown or inflated at those rates, making health care today the third largest industry in the country behind housing and food. Even controlling for inflation, we note that the percentage of the gross national product spent on health care increased modestly from 3.5% in 1930 to 4.4% in 1955, whereas in the next 25 years—1955–1980—health care as a percent of gross national product dramatically increased from 4.4% to 9.6%.

The relevance of these developments to business is very direct. Business is presently paying for 27% of the nation's total health bill. Moreover, at the present time, 35% of employee compensation consists, on an industry-wide average, of benefits—the largest single component of which is health insurance. To illustrate what this means in the life of a single American corporation, from 1974 to 1980 General Motors' health insurance bill went from $630 million to $1.5 billion annually, or from $1922 to $2909 per employee. As of 1982 this represents over $350 added cost for every vehicle General Motors sells. The health insurance factor alone greatly exacerbates the competitive disadvantage American business experiences in international competition generally, and vis-à-vis the Japanese challenge in particular.

The result of these developments is that health and health care have become a center-stage concern for the total economy of the United States. As medical economist Victor Fuchs notes (Fuchs, 1974, p. 17) in *Who Shall Live?*, "No country is as healthy as it could be; no country does as much for the sick as it is technically capable of doing." Nevertheless, as Fuchs notes, all issues in society—including health—get down to basic economic issues of resource allocation.

In bringing this national economic thinking down to the level of individual businesses the implications are clear. In less than a decade the average business has doubled its health expenditures per employee. Not only is there no evidence that employees are twice as healthy as a result, there is little evidence they are in any measurable way healthier at all. There has to be a better, more cost-effective approach. While that cry of alarm may first have been uttered by a company benefits manager, increasingly corporate executives are beginning to think about the health care costs of doing business in America.

This realization is resulting in receptiveness to alternatives to the present system. Over 20 years ago Kaiser pioneered the development of such an alternative with the establishment of the Kaiser-Permanente Health Plan. This early health maintenance organization was successful at cutting corporate expenditures for health and promoting health at the same time through preventive medicine. So successful was the concept and the plan that it is being marketed nationally, and today is the largest health maintenance organization program in the United States.

There has to be a better, more cost-effective approach to provide for the health needs of employees and Americans generally. This calculating, capitalist mode of thinking is one of the greatest potential supports for occupational health promotion in American society today. To exploit and build on the capitalist rationale, occupational health promotion professionals require one basic and crucially important thing. They need quantifiable health status indicators and a thorough knowledge of cost-benefit analysis. Capitalistically oriented business managers will buy occupational health promotion but their fiduciary responsibility to stockholders and the profit motive is such that they tend to buy it on their terms. "What in measurable terms does it do for the bottom line?" Too often health promotion has been sold on faith, its public relations value, or it has been leveraged with guilt. Managers are not paid for faith, public relations value is always difficult to assess, and guilt may motivate token efforts at health promotion, but it will not attract real commitment or money.

In summary, if occupational health professionals can sell health promotion in quantifiable bottom-line terms they will be able to tap into and get support from one of the most powerful ideological rationales in this society—American capitalism. "The business of America is business," said Calvin Coolidge. If occupational health promotion is to have a lasting effect in business, it must make sense in terms business understands.

Summary

In this chapter we have traced ideological and historical roots of occupational health promotion. In doing so we have found that the diverse areas of religion, nationalism, and capitalism all lend support and credibility to the forthcoming "age" of occupational health promotion.

It was said earlier that rarely is any one rationale the sole motivation for any new enterprise and this is the case for occupational health promotion. Instead, these multiple rationales intermix and intertwine in unique ways in different individuals and in different social settings as to form the basic motives for occupational health promotion.

One thing more must be said. In the area of occupational health promotion, the rationales we use to legitimate our efforts seem to be moving us slowly in a new direction—away from Asclepius in the direction of Hygeia—and that is a very exciting and hopeful prospect.

REFERENCES

Alford, R. R. *Health Care Politics*. Chicago: University of Chicago Press, 1975.

Ashford, N. A. *Crisis in the Workplace: Occupational Disease and Injury*. Cambridge, Mass.: The MIT Press, 1976.

Berger, P. L. *The Sacred Canopy*. Garden City: Doubleday, 1967.

Calian, C. S. "Theological and Scientific Understandings of Health." *Hospital Progress*, December 1978.

Dubos, R. *Mirage of Health*. New York: Harper & Row, 1959.

Ehrenreich, B. and Ehrenreich, J. *The American Health Empire: Power, Profits, and Politics*. New York: Vintage Books, 1971.

Erikson, E. H. *Gandhi's Truth* New York: Norton, 1969.

Fisher, I. and Fisk, E. L. *How to Live*. New York: Funk & Wagnalls. 1915.

Fuchs, V. R. *Who Shall Live? Health, Economics, and Social Choice*. New York: Basic, 1974.

Follman, J. F., Jr. *The Economics of Industrial Health*. New York: AMACOM, 1978.

Holy Bible, The. New York: Thomas Nelson & Sons, 1952.

Heilbronner, R. L. *The Making of Economic Society*. Englewood Cliffs, N.J.: Prentice-Hall, 1968.

Job Safety and Health Act of 1970, The. Prepared by the Editorial Staff of the Bureau of National Affairs. Washington, D.C.: Bureau of National Affairs, 1971.

Kohn, Hans. *Nationalism*. New York: Van Nostrand, 1955.

Lubove, R. *The Struggle for Social Security:* 1900–1935. Cambridge, Mass. Harvard University Press, 1968.

MacAvoy, P. W., ed. *OSHA Safety Regulation: Report of the Presidential Task Force*. Washington, D.C.: American Enterprise Institute for Public Policy Research, 1977.

Mannheim, K. *Ideology and Utopia*. New York: Harcourt, Brace & World, 1936.

Maxwell, A. S. "what Is A Seventh-Day Adventist?" In *Religions of America,* Leo Rosten, ed. New York: Simon and Schuster, 1975.

Milio, N. *Promoting Health Through Public Policy*. Philadelphia: Davis, 1981.

Mitnick, B. M. *The Political Economy of Regulation*. New York: Columbia University Press, 1980.

Moskowitz, M., Katz, M. and Levering, R. eds. *Everybody's Business*. San Francisco: Harper & Row, 1980.

Navarro, V. "The Political Economy of Medical Care—An Explanation of the Composition, Nature, and Functions of the Present Health Sector of the United States." In *Health and Medical Care in the U.S.: A Critical Analysis,* Vicente Navarro, ed. Farmingdale, N.Y.: Baywood, 1975.

Newman, S. G. *Health and Social Evolution.* London: Allen & Unwin, 1931.

Parsons, T., Fox, R. C., and Lidz, V. M. "The Gift of Life and Its Reciprocation." In *Death in American Experience,* Arien Mack, ed. New York: Schocken, 1973.

Preventive Medicine U.S.A. New York: Prodist, 1976.

Regulating Business: The Search for an Optimum. San Francisco: Institute for Contemporary Studies, 1978.

Rosen, G. "The Evolution of Social Medicine." In *Handbook of Medical Sociology,* H. E. Freeman, S. Levine, and L. G. Reeder, eds. Englewood Cliffs, N.J.: Prentice-Hall, 1972.

Rosen, G. "The Hospital: Historical Sociology of a Community Institution." In *The Hospital in Modern Society,* E. Freidson, ed. Glencoe: The Free Press, 1963.

Rosen, G. *Preventive Medicine U.S.A.: Preventive Medicine in the United States, 1900–1975, Trends and Interpretations.* New York: Prodist, 1976.

Rosenberg, C. E. and Rosenberg, C. S. "Pietism and the Origins of the American Public Health Movement." *Journal of the History of Medicine,* January, 1968.

Smith, R. S. *The Occupational Safety and Health Act.* Washington, D.C.: American Enterprise Institute for Public Policy Research, 1976.

Weber, M. *The Protestant Ethic and the Spirit of Capitalism.* New York: Scribner, 1958.

Weidenbaum, M. L. *Business, Government, and the Public.* Englewood Cliffs, N.J.: Prentice-Hall, 1981.

White, L. J. *Reforming Regulation.* Englewood Cliffs, N.J.: Prentice-Hall, 1972.

Wilson, J. Q., ed. *The Politics of Regulation.* New York: Basic, 1980.

CHAPTER 3

The Assessment and Enhancement of Health Compliance in the Workplace

Robert H. L. Feldman

Occupational health promotion programs aim to increase healthy behaviors that result in improved worker health. These programs have shown varied degrees of success in establishing changes in workers' health behavior. Since health promotion programs require monetary and time commitments from both management and workers, it is essential to examine ways of improving these programs, that is, increasing health compliance. Therefore, the focus of this chapter is on an examination of the factors that affect health compliance with occupational health promotion programs. Also, the chapter will consider methods and techniques of enhancing health compliance with workplace programs. A comprehensive program will be presented based upon the most effective components of current and past occupational health promotion programs.

INTRODUCTION

The rates of success of occupational health promotion programs exhibit wide variation. Kimberly Clark's rehabilitation program for workers with chemical dependency problems reports a 65% success rate (Cunningham, 1982). A program for the control of hypertension designed for members of the United Storeworkers Union employed by Gimbel's and Bloomingdale's department stores in New York obtained about 80 percent satisfactory control of blood pressure (Alderman, 1976). In contrast, less than 25% of workers from 81 industries in the Chicago area were able to maintain control of their blood pressure in a program sponsored by the Chicago Heart Association (Schoenberger, 1976). Since health compliance shows such variation it is important to examine the components which make up these programs and how health compliance is measured. The types of health behaviors involved in health promotion programs are listed in Table 3.1.

Table 3.1 Health Behaviors in Occupational Health Promotion Programs

Alcohol Abuse Control
Back Exercises and Posture
Breast Self-Examination
Dental Care
Diet and Weight Control
Drug Abuse Control
Exercise and Physical Activity
First Aid and CPR
Health Monitoring
Hypertension Control
Medication Taking
Smoking Cessation
Stress Control
Appointment Making
Appointment Keeping
Information Seeking

MEASURING COMPLIANCE

Health compliance is defined as the extent to which an individual's behavior coincides with health or medical advice (Haynes, 1979). The extent to which a person is able to follow the advice of a smoking clinic and cease smoking is a measure of compliance. The ability of an individual to carry out a diet recommended by a weight control program is another example of compliance. The degree of success of a worker in taking medication to control hypertension advised by a worksite physician is a further example.

Compliance requires congruence between health advice and health behavior. To measure health compliance properly, it is necessary to determine what advice has been recommended to the individual and to what extent the individual has performed the recommended behavior. In terms of medical advice to patients, discrepancies have been found among recorded advice (e.g., a medical record or prescription form), what is reported by the health professional, and what is reported by the patient (Kirscht and Rosenstock, 1980). Therefore, before compliance can be measured, a clear determination of the recommended health advice is needed.

The measurement of health behaviors that are public actions can be accomplished without much difficulty. These public health behaviors include appointment making, appointment keeping, attending meetings, utilizing services, utilizing facilities (e.g., gym, pool, jogging path), utilizing vending machines (food choices), food use in company cafeterias, and so forth. However, the measurement of health behaviors that are private actions pose a number of methodological issues. Private health behaviors include private eating behavior, smoking, alcohol and drug use and abuse, private physical activity, dental behavior, stress reduction techniques, and so forth. Another type of private health behavior is medication taking. This type of health behavior has been assessed by clinical judgments of health care providers, pharmacologic effects of side effects, body fluid analyses (e.g., blood and urine analyses), measurement of remaining medication (e.g., pill counts), and patient interviews (Gordis, 1979; Sackett, 1979). For nonmedication behaviors such as dietary or exercise behavior, physiological assessments have been used, but they are costly and variation in the physical characteristics of individuals lead to unreliable results (Kirscht and Rosenstock, 1980). Therefore, researchers have used interviews and questionnaires, that is, self-report measures of private compliance behavior. Self-reports of health behaviors have sometimes yielded over estimates of desirable practices (Gordis, Markowitz, and Lilienfeld, 1969); however, with nonthreatening questions and particular attention to sensitive areas accurate results occur (Francis, Korsch, and Morris, 1969; Sackett, 1979).

FACTORS THAT AFFECT WORKSITE COMPLIANCE

Worksite health promotion programs have shown greater adherence among its participants and have been less costly than similar community-based programs (Logan, 1981). However, a number of components found in occupational programs can be improved to increase health compliance.

The physical setting where the occupation program is conducted affects health compliance. Workers are less likely to continue in an occupational program that is removed from the workplace than an occupational program that is actually carried out at the worksite (Schoenberger, 1976; Wanzel, 1977). Wanzel's study of a company exercise program for 480 salaried employees found that the major reason for leaving the program was that the exercise facility was too far from the workplace. Problems of transportation, finding a parking place, and other inconveniences reduce participation in off-worksite programs (Haskell and Blair, 1982).

Time and scheduling problems also contribute to compliance problems (Parkinson, 1982; Wanzel, 1977). If an occupation program is offered outside work hours, that is on worker's own time, workers may not be willing to give up their time (e.g., lunch time, after-work hours). This would be especially true for a long program, where commuting is necessary or when the program is conducted during holidays or at vacation time (Parkinson, 1982). Programs held during company time raise scheduling problems and need the cooperation of workers' supervisors. Another aspect of the setting of the occupation program is how long workers have to wait before they can be seen by the health or medical professional. The longer the waiting time the less likely the worker will return again to the program (Geersten, Gray, and Ward, 1973; Raynes and Warren, 1971).

Satisfaction with the occupational program is an important factor determining whether workers will return to a program and carry out the program's recommendations. Workers are more satisfied with programs when health care professionals such as physicians and nurses show care, warmth, and concern in their interactions with workers (Wear, 1980). Increased contact time with health care providers has been shown to increase satisfaction and improvement in blood pressure control (Alderman, Green, and Flynn, 1982). In addition, increasing the number of contacts also positively affects health compliance (Alderman et al., 1982). However, increased contacts should be with the same health care provider, since lack of continuity of care has been shown to decrease satisfaction and compliance (Becker, Drachman, and Kirsch, 1972). Worker satisfaction also increases when workers have the opportunity to actively participate in the program (Levine, Green, and Deeds, 1979).

Psychosocial factors play an important role in health compliance (Feldman, 1982). Social support (the support of significant other people) influences a worker's compliance with health recommendations. In a physical activity program Heinzelmann and Bagley (1970) found that men whose wives had positive attitudes toward the program had greater compliance to the program, i.e., 80% excellent or good adherence, than men whose wives had negative or neutral attitudes, i.e., 40% excellent or good adherence.

Workers' attitudes and beliefs regarding illness, disease, and their relationship to health promotion can influence their participation and involvement in these programs. The major health behavior model concerning attitudes and beliefs that has been used to study health compliance is the Health Belief Model (Kirscht, 1983). According to this model, a person is likely to comply with a health or medical recommendation to prevent a particular illness the more he or she (1) perceives the illness as having serious consequences, (2) perceives him- or herself susceptible to the illness, and (3) perceives benefits gained from complying with the health advice (Becker, Maiman, Kirscht, Haefner, Drachman, and Taylor, 1979). Although the majority of studies support the Health Belief Model, these studies are mostly retrospective studies and therefore are correlational in nature. One predictive study that used the Health Belief Model to examine health compliance in the workplace was conducted by Taylor (1979). In an investigation of Canadian steelworkers he used the Health Belief Model to predict participation in a blood pressure screening and hypertension program. Health behavior compliance was measured by unobtrusive pill counts and by a nonthreatening interview. Taylor found that beliefs regarding the seriousness of hypertension after six months of treatment best explained compliance at six months and best predicted compliance at twelve months.

Table 3.2 Factors That Affect Worksite Compliance

I. Program Setting
   A. Place
   B. Transportation, parking
   C. Time: worker's time vs company time
   D. Scheduling
   E. Waiting time

II. Worker Satisfaction with Health Care
   A. Interactions with health care providers
   B. Contact time
   C. Number of contacts
   D. Continuity of care

III. Psychosocial Factors
   A. Social support
   B. Perceive illness as serious
   C. Perceive self susceptible
   D. Perceive benefits from compliance

IV. Health knowledge

In general, the other beliefs in this study were not found useful in explaining health compliance.

The relationship between knowledge and health behavior compliance shows mixed results (Feldman, 1982). In a review (Podell, 1975) of over 30 studies, no consistent pattern was evident. It appears that knowledge about health and medical matters is necessary for correct compliance, but that information is rarely sufficient to produce adequate cooperation with recommended advice (Becker, 1979). A summary of the factors that affect health compliance in the workplace is found in Table 3.2.

METHODS TO ENHANCE COMPLIANCE

A number of methods and combinations of methods can be used to enhance compliance with worksite health programs. The choice of methods and techniques will depend on (1) the factor or factors that are causing the lack of compliance, (2) the costs, time restraints, and staff available, (3) the motivation and willingness of the program staff to make the recommended changes, and (4) the research findings on health promotion programs (Feldman, 1982). A summary of methods and techniques is found in Table 3.3.

Program Setting

The time, place and scheduling of workplace programs can be arranged so as to maximize health compliance. Arguments have been raised both in favor and against programs on company time versus programs outside of work hours. However, in terms of removing obstacles to worker participation, programs that are held on company time have clear advantages. Some workers may not be able or willing to give up time that is committed to family or other activities. Also, by making programs available on company time, a company is demonstrating that it is committed to worker health (Parkinson, 1982).

The best place to conduct an occupational health program is at the worksite. The less distance that workers have to travel, the less problems that have to be encountered in terms of parking and other arrangements, the more likely they will be able to participate in a program (Haskell and Blair, 1982). If the workplace does not have the facilities, or has a small number of workers, then centrally located establishments such as ''Ys'' and clinics can be used. Thus, the scheduling of health promotion programs should emphasize convenience and attempts should be made to reduce barriers that prevent participation.

Another aspect of scheduling is the setting of appointments to minimize waiting time. Programs that require frequent worker interaction with health care professionals, such as hypertension, alcohol, and stress management programs, should utilize efficient scheduling techniques to increase appointment keeping and worker satisfaction. Additional ways of increasing appointment keeping include postcard reminders and telephone reminders (Cook, Morch, and Noble, 1976; Shepard and Moseley, 1976).

Worker Satisfaction with Health Care

Many of the problems associated with worker dissatisfaction with health care stems from ineffective communication by health care providers. Workers are more satisfied when health care providers express warmth and concern in their interactions. Dr. Roland F. Wear, Jr. (1980), corporate medical director at Campbell Soup Company, reports that the concern of an interested nurse was the most important factor in a successful hypertension control program.

A number of other factors that influence worker satisfaction with health care include continuity of care, increased number of contacts, and increased contact time with health care providers. Studies on the control of high blood pressure have found that increasing the continuity of contact between the hypertensive patient and the health care provider through increased number of contacts and increased contact time leads to an increase in blood pressure control (McGill, 1978). Worksite programs with periodic contact with health care providers offer the opportunity for the repetition and reinforcement of behavioral changes conducive to improved health (Chwalow et al., 1978; Alderman, Green, and Flynn, 1982).

An additional concern to workers participating in a health promotion program is confidentiality. Health and medical records, reports of participation, information on workers' families, and any additional information collected and related to health promotion programs must be held strictly confidential. Workers must be made aware of the confidentiality of their records, and their right to refuse to participate in any program or aspects of any program (Feldman, 1983b).

Psychosocial Factors

Social support components have been incorporated in many health promotion programs. The involvement of spouses, family members, friends, co-workers and significant others in health programs reduces the social isolation of the participant, makes the participant more accountable for his or her actions, and assists the worker in remembering and carrying out the recommended health advice. Studies by Alderman (1977; Alderman and Schoebaum, 1975) found that the support provided by other union members, family, and friends accounted for a large portion of the success of worksite programs to control blood pressure.

Workers' perceptions of health and illness play an important role in whether they participate in health promotion programs. If they have misperceptions regarding their susceptibility to various illnesses or the seriousness of these illnesses, they are less likely to participate in health promotion programs. Carefully designed and pretested messages that focus on the seriousness and consequences of specific illnesses are useful in correcting misperceptions about health and illness (Feldman, 1982).

Health Knowledge

A number of methods have been developed to improve health knowledge among participants in health promotion programs. One method to increase knowledge retention is repetition. Health care instructions that are given both in written and oral form will, in general, increase knowledge retention better than instructions given by only one method (Sharpe and Mikeal, 1974). Demonstrating and rehearsing procedures, which allows the participant to actually practice the techniques, improves the performance of health behaviors, for example, stress reduction techniques, dental care, safety behaviors.

Educational methods have also been developed to improve medication taking for example, hypertension medication. Clearer and more comprehensive labeling, package inserts, unit doses, pill counters, and pill calendars have found to improve medication taking (Eshelman and Fitzloff, 1976).

Psychological Techniques

A number of techniques developed by psychologists have been found useful in improving compliance with various health and life-style recommendations (Feldman, 1983a). Self-monitoring, a process by which individuals observe and record their own behavior, has been shown to be effective for people who lack initial understanding of what they are not doing properly. For example, workers who desire to take part in a weight control program may not be aware of the quantity and quality of the foods that they eat. Self-monitoring is a method of increasing self-awareness (Rimanczyk, 1974).

Contracting, a process by which an individual specifies a set of rules regarding some behavior and formalizes a commitment to adhere to them, has been found to be an effective technique to increase compliance to health recommendations (Dunbar, Marshall, and Howell, 1979). Contracting requires individuals to plan their own programs, therefore individualizing the process. Contracts are written statements that clearly identify specific goals, therefore reducing the possibility that the individuals will forget what to do. Also, contracts provide incentives by stipulating specific rewards for specific behaviors.

Self-Contracts involve the individual administering his or her own rewards. Contracting with a significant other (spouse, family member, friend, or co-worker) has been found to be more effective than self-contracts (Mahoney and Thoresen, 1974). By contracting with another person, the individual makes a formal public commitment to carry out a specific health behavior. Contracting has been successful with blood pressure control programs (Brucker, 1977) and weight control programs (Foreyt, Scott, and Gotto, 1982).

Tailoring, a method by which a program is designed to meet the specific characteristics of the individual in the program, has also been found to assist compliance with health recommendations. Kimberly-Clark Corporation's Health Management Program to control weight utilizes tailoring in that it individualizes health programs based upon each worker's health risk evaluation (Foreyt et al., 1982). Tailoring has also been used in smoking cessation programs (Best, 1975).

Another behavioral technique that has been successfully used in health promotion programs is graduated regimen implementation. Graduated regimen implementation is the process of introducing components of a program sequentially as the individual learns and performs the prior steps in the sequence (Dunbar et al., 1979). The steps are ordered for the person in terms of difficulty. For example, if a person wanted to reduce his or her daily diet by 1000 calories a day, the person could begin by reducing the diet by 250 calories the first day. Once the individual has successfully reduced the diet by 250 calories, the person would reduce the diet further by another 250 calories, and so forth.

Graduated regimen implementation has been often used in exercise and fitness programs. A group of 1200 insurance company employees of the Canada Life

Assurance Company and the North American Life Assurance Company in Toronto, Canada participated in a fitness program where fitness performance was implemented in gradual stages. Due to the general low-fitness level of the participants, as well as the inclusion of older employees, overweight individuals, and employees who were in an organized program for the first time, the graduated implementation of the fitness program proved to be the most effective means of maintaining participation of the employees (Peepre, 1981).

Reinforcement

An important component in the preceding methods of improving health behavior compliance is reinforcement that refers to "any consequence that increases the probability of the behavior being repeated" (Dunbar et al., 1979, p. 185). Reinforcement may be *material reinforcement* such as show tickets, or money, or *social reinforcement* such as acknowledgment, praise, or recognition.

Reinforcement strategies have been widely used in occupational health promotion programs. The Bonnie Bell Cosmetics firm offered monetary rewards to its employees for physical exercise behavior. Employees received one dollar for every mile run, 50 cents for each mile walked, and 25 cents for each mile on bicycle. Ironically, the program was "too successful" and the company ran out of money for the program and had to discontinue this reinforcement strategy (Fixx, 1980).

Weight control programs have used nonfood material rewards: a certain amount of money for each pound lost, and praise for maintaining a diet (Foreyt et al., 1982; Parkinson, 1982). Blood pressure control programs have used material rewards such as trading stamps, books, and tickets to sporting events (Brucker, 1977; Steckel and Swain, 1977). Dow Chemical Company has used lotteries and financial awards in a smoking cessation program. Each month of the program the novice ex-smoker had the opportunity to win a boat and motor valued at $2400. In addition, weekly $1 bonuses were given for abstinence and workers had the chance to win a $50 bonus quarterly. Other programs have used released time, compensation and awards such as T-shirts, plaques, and publicity in their health promotion programs (Parkinson, 1982). Further discussion of psychological and behavior techniques are found in Chapter 15.

Administrative/Organizational Approaches

Support and sponsorship of an occupational health promotion program by a company has certain advantages. Though workers may be critical of some policies of the company for which they work, if the company's reputation in other health and medical matters is positive, that is, the company is rated highly in terms of competency, confidentiality, and concern, then there is less likely to be suspicion of a new health promotion program (Collings, 1982).

Table 3.3 Methods, Techniques, and Approaches to Enhance Compliance with Worksite Health Promotion Programs

Issue	*Recommendation*
Program Setting	
1. Place	At the worksite
2. Time	On company time
3. Scheduling	Convenient to workers
4. Waiting time	Minimize waiting time
5. Appointment keeping	Postcard and telephone reminders
Worker Satisfaction with Health Care	
1. Perceptions of caring	Greater expressions of caring and concern by providers
2. Continuity of care	Increase continuity of care
3. Medical contacts	Increase medical and health contacts
4. Contact time	Increase contact time
5. Confidentiality	Strict confidentiality of all records
Psychosocial Factors	
1. Social support	Involvement of spouses, family, friends, and co-workers
2. Misperceptions about illness	Patient education
Knowledge	
1. Medication taking	Labeling, package inserts, unit doses, pill counters, and calendars
2. Retention	Repetition, written and oral instructions, demonstration of procedures

Other Methods
Psychological
1. Self-monitoring
2. Self-contacting
3. Contacting with a significant other
4. Tailoring
5. Graduated regimen implementation
6. Material reinforcement
7. Social reinforcement
Administrative
1. Company sponsorship
2. Union support
3. Community involvement

Programs sponsored and supported by unions, such as the United Storeworkers Union's hypertension screening and treatment program, have also been successful. Use of an offsite union health center clinic with a well-staffed health team, removal of financial barriers, and participation of workers in the health care process has resulted in a number of effective programs (Alderman et al., 1982). In several work settings collaboration between management and labor unions has also occurred (Foot and Erfurt, 1976). In work settings where unions represent workers, unions should be included in the establishment of company-supported health promotion programs.

The involvement of community resources, especially the worker's own private physician, in worksite-based programs has improved further health behavior compliance. A hypertension program developed by Burlington Industries concerned detection, referral and follow-ups. Workers were first screened at the worksite. Next, the workers' own physicians received a letter with a record of the worker's blood pressure readings. An offer was then made to provide each worker with blood pressure checks, health education, and other services at the workplace in cooperation with the worker's physician (Murphy, 1976).

Summary

The assessment and enhancement of health compliance in the workplace is a complex process. Compliance is affected by the physical setting of the program, worker satisfaction with the program, psychosocial factors, and workers' knowledge of health care. To enhance compliance, a number of methods, techniques, and approaches have been described. Changes in the program setting can be made to reduce obstacles to participation. More effective communication patterns between workers and health care providers can be developed to improve worker satisfaction with health care. Social support groups and well-designed tailored messages can affect the psychosocial factors of compliance. In addition, health knowledge can be increased by a variety of health education techniques and health behavior can be changed by use of psychological and behavioral methods. Finally, administrative and organizational approaches involving management, unions and community-based health resources show promise as a means of increasing health compliance. Thus, by utilizing a variety of approaches, health compliance in the workplace can be enhanced.

REFERENCES

Alderman, M. H. ''Detection and treatment of high blood pressure at the workplace.'' Presented at the Conference on High Blood Pressure Control in the Work Setting, Washington, D.C., October 1976.

Alderman, M. H. ''Three years of work-site based hypertensive treatment.'' Paper presented at the National Conference on High Blood Pressure Control. Washington, D.C., 1977.

Alderman, M., Green, L. W., and Flynn, B. S. ''Hypertension control programs in occupational settings.'' In *Managing Health Promotion in the Workplace: Guidelines for Implementation and Evaluation,* R. S. Parkinson and Associates. Palo Alto, California: Mayfield, 1982.

Alderman, M. and Schoebaum, E. ''Detection and treatment of hypertension at the work site.'' *New England Journal of Medicine,* 1975, **293,** 65–68.

Becker, M. H. ''Understanding patient compliance: The contributions of attitudes and other psychosocial factors.'' In *New Directions in Patient Compliance,* S. J. Cohen, ed. Lexington, Mass: Lexington Books, 1979.

Becker, M. H., Drachman, R. H., and Kirsch, J. P. ''Motivations as predictive of health behavior.'' *Health Service Report,* 1972, **87,** 852–861.

Becker, M. H., Maiman, L. A., Kirscht, J. P., Haefner, D. P., Drachman, R. H., and Taylor, D. W. ''Patient perceptions and compliance: Recent studies of the Health Belief Model.'' In *Compliance in Health Care,* R. B. Haynes, D. W. Taylor, and D. L. Sackett, eds. Baltimore, Johns Hopkins University Press, 1979.

Best, J. A. ''Tailoring smoking withdrawal procedures to personality and motivational difference.'' *Journal of Consulting and Clinical Psychology,* 1975, **43,** 1–8.

Brucker, C. ''Assuring patient compliance by health care contrasts.'' Final summary report on NIH grant No. HL17230. National High Blood Pressure Education Program, 1977.

Chwalow, A. J. et al. ''Effects of the multiplicity of interventions on the compliance of hypertensive patients with medical regimens in an inner-city population.'' *Prevention Medicine,* 1978, **7,** 51.

Collings, G. H., Jr. ''Perspectives of industry regarding health promotion.'' In *Managing health promotion in the workplace: Guidelines for implementation and evaluation,* R. S. Parkinson and Associates. Palo Alto: California: Mayfield, 1982.

Cook, D., Morch, J., and Noble, E. ''Improving attendance at follow-up clinics.'' *Dimension in Health Services,* 1976, **53,** 46–49.

Cunningham, R. M. *Wellness at Work: A report on health and fitness programs for employees of business and industry.* Chicago: Inquiry (Blue Cross Blue Shield Association), 1982.

DiMatteo, M. R. ''A social-psychological analysis of physician-patient rapport: Toward a science of the art of medicine.'' *Journal of Social Issues,* 1979, **35** (1), 12–33.

Dunbar, J. M., Marshall, G. D., and Howell, M. F. ''Behavior strategies for improving compliance.'' In *Compliance in Health Care,* R. B. Haynes, D. W. Taylor, and D. L. Sackett, eds. Baltimore: Johns Hopkins University Press, 1979.

Eshelman, F. N. and Fitzloff, J. ''Effect of packaging on patient compliance with an antihypertensive medication.''*Current Therapeutic Research,* 1976, **20,** 215–219.

Feldman, R. H. L. ''A guide for enhancing health care compliance in ambulatory care settings.'' *Journal of Ambulatory Care Management,* 1982, **5** (4), 1–15.

Feldman, R. H. L. ''Changing stressful behaviors.'' In *Comprehensive Stress Management,* J. S. Greenberg. Dubuque, Iowa: Brown, 1983a.

Feldman, R. H. L. ''Strategies for improving compliance with health promotion programs in industry.'' *Health Education,* 1983b, **14** (4), 21–25.

Fixx, J. *Jim Fixx's Second Book of Running.* New York: Random House, 1980.

Foot, A. and Erfurt, J. ''A model system for high blood pressure control in the work setting.'' In *High Blood Pressure Control in the Work Setting: Issues, Models, Resources.* Proceedings of the National Conference Washington, D.C., 1976.

Foreyt, J. P., Scott, L. W., and Gotto, A. M. "Weight control and nutrition education programs in occupational settings." In *Managing Health Promotion in the Workplace: Guidelines For Implementation and Evaluation,* R. S. Parkinson and Associates. Palo Alto, California: Mayfield, 1982.

Francis, V., Korsch, B. M., and Morris, M. J. "Gaps in doctor-patient communications: Patients' response to medical advice." *New England Journal of Medicine,* 1969, **280.**

Geersten, H. R., Gray, R. M. and Ward, J. R. "Patient noncompliance within the context of seeking medical care for arthritis." *Journal of Chronic Disease,* 1973, **26,** 689–698.

Gordis, L. "Conceptual and methodological problems in measuring patient compliance." In *Compliance in Health Care,* R. B. Haynes, D. W. Taylor, and D. L. Sackett, eds. Baltimore: Johns Hopkins University Press, 1979.

Gordis, L., Markowitz, M., and Lilienfeld, A. "The accuracy of using interviews to estimate patient reliability in taking medications at home." *Medical Care,* 1969, **1,** 49–54.

Haskell, W. L. and Blair, S. N. "The physical activity of health promotion in occupational settings" *Managing health promotion in the workplace; Guidelines for implementation and evaluation,* R. S. Parkinson et al. Palo Alto, California: Mayfield, 1982.

Haynes, R. B. "Introduction." In *Compliance in health care,* R. B. Haynes, D. W. Taylor, and D. L. Sackett, eds. Baltimore: Johns Hopkins University Press, 1979.

Heinzelmann, F. and Bagley, R. W. "Response to physical activity programs and their efforts on health behavior." *Public Health Reports,* 1970, **85,** 905–911.

Kirscht, J. P. "Preventive health behavior: A review of research and issues." *Health Psychology,* 1983, **2** (3), 277–301.

Kirscht, J. P. and Rosenstock, I. M. "Patients' problems in following recommendations of health experts." In *Health Psychology A Handbook,* G. C. Stone, F. Cohen, and N. E. Adler. San Francisco: Jossey-Bass, 1980.

Levine, D. M., Green, L. W., and Deeds, S. G. "Health education for hypertensive patients." *Journal of the American Medical Association,* 1979, **241,** 1700–1703.

Logan, A. G. et al. "Cost-effectiveness of a worksite hypertension treatment program." *Hypertension,* 1981, **3,** 211–218.

Mahoney, M. J. and Thoresen, C. E. *Self-control: power to the person.* Monterey: California: Brooks-Cole, 1974.

McGill, A. M. "A National High Blood Pressure Education Research Program: Abstracts of papers presented at the First International Congress on Patient Counseling." *Patient Counseling and Health Education,* 1978, **1** (35).

Murphy, A. F. "The Burlington Industries industrial hypertension program." In *High Blood Pressure Control in the Work Setting: Issues, Models, Resources.* Proceedings of the National Conference, Washington, D.C., 1976.

Parkinson, R. S. and Associates. *Managing health promotion in the workplace: Guidelines for implementation and evaluation.* Palo Alto, California: Mayfield, 1982.

Peepre, M. "The Canadian Employee Fitness Life-style Project." In *Health Fitness: The Corporate View,* Special Report on Employee Fitness in the 1980s. *Athletic Purchasing and Facilities.* 1981.

Podell, R. N. *Physicians's guide to compliance with hypertension.* West Point, Pa: Merck, 1975.

Raynes, A. E. and Warren, G. "Some characteristics of 'drop-outs' as first contact with a psychiatric clinic." *Community Mental Health Journal,* 1971, **7,** 144–150.

Rimanczyk, R. G. "Self-monitoring in the treatment of obesity: Parameters of reactivity." *Behavior Therapy,* 1974, **5,** 531–540.

Sackett, D. L. "A compliance practicum for the busy practitioner." In *Compliance in health care,* R. B. Haynes, D. W. Taylor, and D. L. Sackett, eds. Baltimore: Johns Hopkins University Press, 1979.

Schoenberger, J. A. "Heart disease in industry: The Chicago Heart Association Project." Presented at the Conference on High Blood Pressure Control in the Work Setting, Washington, D.C., October 1976.

Sharpe, T. R. and Mikeal, R. L. "Patient compliance with antibiotic regimens." *American Journal of Hospital Pharmacy,* 1974, **31,** 479–484.

Shepard, D. S. and Moseley, T. A. "Mailed versus telephone appointment reminders to reduce broken appointments in a hospital outpatient department." *Medical Care,* 1976, **14,** 268–273.

Steckel, S. B. and Swain, M. A. "Contracting with patients to improve compliance." *Hospitals,* 1977, **51,** 81–84.

Taylor, D. W. "A test of the health belief model in hypertension." In *Compliance in health care,* R. B. Haynes, D. W. Taylor, and D. L. Sackett, eds. Baltimore: Johns Hopkins University Press, 1979.

Wanzel, R. S. "Factors related to withdrawal from an employee fitness program." Paper presented at the meeting of the American Alliance for Health, Physical Education, and Recreation, Seattle, 1977.

Wear, R. F. "Existing health promotion programs in industry." Presented at the conference, Health Education and Promotion: Agenda for the Eighties. The Health Insurance Association of America and the American Council of Life Insurance, Atlanta, March 1980.

Whorton, M. D. and Davis, M. E. "A perspective on health intervention programs in industry." In *Managing health promotion in the workplace: Guidelines for implementation and evaluation,* R. S. Parkinson and Associates. Palo Alto, California: Mayfield, 1982.

PART TWO

GUIDELINES FOR PROGRAM DEVELOPMENT IN OCCUPATIONAL HEALTH PROMOTION

In Part One we introduced the reader to the basic foundations of occupational health promotion. The purpose was not only to provide basic definitions, but to provide the reader with a sense of the evolution of occupational health promotion and its current status.

In Part Two we will take the reader beyond the definitions and history. In this section we will provide the reader with general guidelines for program development as it relates to some of the more commonly utilized aspects of occupational health promotion. In effect, we are providing the reader with a "how-to" set of guidelines, realizing, of course, that final program development must be tailored to the specific needs and resources of the organization at hand.

CHAPTER 4

The Development of Occupational Stress Management Programs

George S. Everly, Jr.

INTRODUCTION

We as a society have gone to great extents to see that our citizens receive at least a basic education in reading, communicating, and mathematics—all in an attempt to assist them in becoming more productive members of our society. Yet we have spent considerably less time and effort in helping our citizens develop a far more basic skill—the ability to cope with the stress of everyday life, especially the pressure connected with earning a living. A statement from the office of the U.S. Surgeon General (Public Health Service, 1979a) has noted that when stress reaches excessive proportions, psychologic changes can be so dramatic as to have serious implications for mental and physical health. On the other hand, the National Conference on Health Promotion Programs in Occupational Settings (Public Health Service, 1979b) has gone so far as to conclude that stress management programs could help people to alter their life-style as to improve their health, both mental and physical. In addition, such programs may reduce absenteeism, enhance productivity, and decrease insurance and medical costs on the job. This National Conference has, therefore, included stress management as an integral part of its recommendations for promoting occupational health. This endorsement is somewhat dampened, however, by the facts that no generally accepted definition nor set of implementation guidelines currently exist for occupational stress management. In this chapter we shall examine a rationale for establishing occupational stress management programs. Furthermore, we shall provide a generic definition and flexible set of guidelines for establishing stress management programs within occupational settings.

SCOPE OF THE PROBLEM

"Can Companies Kill?" was the title of an article appearing in *Psychology Today* (Rice, 1981) which touched on the issue of "whether an employer can, in effect, by its own action or inaction, kill with stress" (p. 78). In this article the author describes a lawsuit brought by the widow of a once well-paid executive whose apparent suicide was allegedly contributed to by overwork. This same article goes on to describe the case of a secretary who was awarded $7000 for what she claimed was emotional distress caused by "her boss's continual criticism and prying questions about her family life" (p. 81). Similarly, the article describes the case of a worker employed for 20 years who one day simply walked off the job later to be hospitalized for psychiatric complaints. The worker successfully sued, claiming the psychiatric breakdown occurred due to the pressures of the job. "Can Companies Kill?" not only raises the issue of an organization's legal responsibility for the mental and physical health of its employees, but it raises a more important and fundamental issue. Can, indeed, work itself, or the work environment, cause sufficient distress as to not only affect workers' performance but their very health, as well? Can occupational stress really kill? Let us examine these issues.

Stress, Dysfunction, and Disease

At the present time, most of our society's interest in stress and stress management seems to be based upon a concern over the potentially dysfunctional (performance-inhibiting) and pathogenic (disease-causing) properties of excessive stress (often referred to as "distress").

Over the last 20 years, sufficient evidence appears to have been collected as to (1) warrant this concern for the harmful effects of stress, and (2) provide a basic rationale for the establishment of occupational stress management programs.

Evidence presented in the Report of the President's Commission on Mental Health (1978) suggests that 30%–50% of general medical practice patients (including patients of corporate medical services) are suffering from stress-related problems. Similarly, psychiatrist Roy Menninger (1978) of the Menninger Foundation estimated that up to 80% of general medical practice complaints are, in reality, physical reactions to the "problems of living." Such estimates must clearly be labeled as highly speculative, however.

Based upon more well-defined research data, Lennart Levi (1979) writes, "The causal relationship between exposure to psychosocial stressors (stress-causing stimuli) and subsequent disease is supported by numerous experimental and epidemiological studies.[1] From these studies, high-risk situations in the human environment as well as high-risk groups and high-risk reactions may be identified" (p. 11).

More relevant to the present discussion, there is considerable evidence that the *occupational* setting represents one such example of a high-risk situation in the human environment. In other words, the occupational setting represents a major potential source of human stress. For example, James Manuso (1979) reported that approximately 50% of the causes of an individual's stress can be traced back to the

[1]The reader is referred to Levi (1970), Levi (1972), Levi (in press), and Henry and Stephens (1977).

occupational setting. In a series of investigations, Levi and his colleagues examined, specifically, the occupational environment and its relationship to human stress (Levi, 1970, 1979, in press). These studies clearly demonstrated that common, generic components of the occupational, or work environment can affect potentially dysfunctional and pathogenic reactions in human beings. More specifically, for example, excessive occupational stress appears to be highly related to not only (1) decreased performance, but (2) coronary heart disease (CHD) as well.

Not all stress is dysfunctional, but *excessive* stress appears to be highly related to decreased job performance. On one hand, moderate levels of stress appear to act as a positive, motivating influence on human behavior. To some degree it has been suggested that moderately elevated stress arousal (called "eustress") can act to improve human performance (Selye, 1976b). On the other hand, however, *excessive* stress appears to hinder human performance (see Figure 4.1). Studies by Kahn et al. (1964), Rizzo, House and Lirtzman (1970), House and Rizzo (1971), French and Caplan (1973), and McLean and Jillson (1977) support the hypothesis that excessive stress can indeed result in decreased work performance. These authors typically view excessive stress as resulting from conditions such as *work overload* (work which is excessive in quantity or difficulty), *role conflict* (conflict between two occupational roles or between one's occupational role and one's personal role), and *role ambiguity* (lack of work role clarity). In a study of 5100 Swedish and

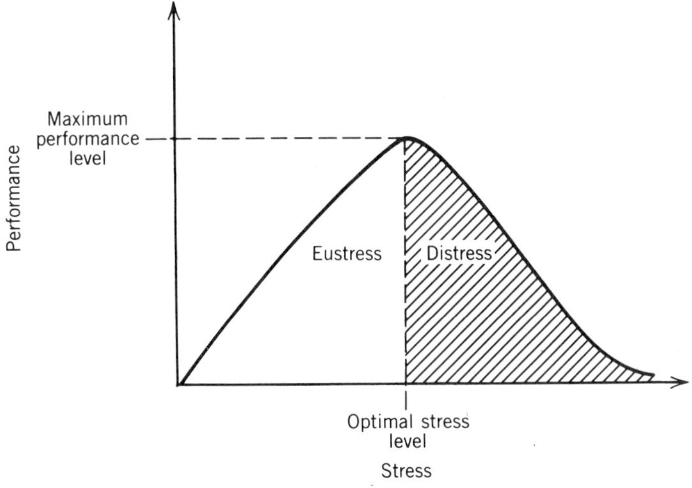

Figure 4.1. This figure graphically shows the relationship between stress arousal (horizontal axis) and performance (vertical axis). As stress increases, so does performance (eustress). At the optimal stress level, performance has reached its maximum level. If stress continues to increase into the "distress" region, performance quickly declines. Should stress levels remain excessive, health will begin to erode as well.

American workers, Karasek (1983)[2] found that jobs that possessed high levels of psychological demand combined with low levels of control were more likely to be associated with CHD.

A more direct contributor to decreased performance and decreased organizational productivity is employee absenteeism and turnover. Maslach (1976) suggests that occupational stress leading to "burnout" is a significant cause of absenteeism and job turnover. Burnout may be thought of as the mental or physical exhaustion that results from excessive occupational stress. Burnout appears to be commonly caused by work frustration and overload.

The dysfunctional effects of excessive stress upon performance certainly seem a reasonable rationale to explore stress management interventions. Perhaps more important than the effects of excessive stress upon work performance, however, is its relationship to heart disease.

Heart disease is accepted to be the leading cause of death in the United States. Any relationship between excessive occupational stress and heart disease must be considered a critically important one and further rationale to pursue occupational stress management programs.

In one of the earlier reviews of research on excessive occupational stress and coronary heart disease (CHD) Stephen Sales (1969, p. 334) concluded: "It seems, therefore, from both the literature reviewed and the laboratory investigation, that organizational factors . . . can contribute to the etiology of coronary disease." Subsequent research seemed to support Sales' conclusion. French and Caplan (1970, 1973) in their research with NASA found that occupational stress was correlated with such risk factors for CHD as elevated serum cholesterol, elevated heart rate, and excessive cigarette smoking. Sales and House (1971) found that frustration and overall job dissatisfaction were related to death from heart disease. In a later study, House (1975) found that occupational stress caused by excessive job responsibilities and overtime assignments served as a precipitator for the development of CHD.

In a series of cross-cultural worksite field studies, Levi (1972) and Timio, Pede, and Gentili (1977) found that occupational stressors were capable of inducing activation of the adrenal medullary hormones (medullary catecholamines) at levels typically seen in disease states. It has been suggested that such excessive catecholamine secretion may have pathogenic effects on the integrity of the coronary arteries and heart muscle itself (Levi, 1979).

In another worksite study of occupational stress, Weiman (1977) studied over 1500 management level personnel. He uncovered four major sources of occupational stress. They were (1) too much or too little to do, (2) role ambiguity, (3) role conflict, and (4) too much or too little responsibility. These four factors were all

[2]Columbia University Dept. of Industrial Engineering and Operations Research.

highly related to the presence of major CHD risk factors including essential hypertension, obesity, cigarette smoking, and hypercholesterolemia.

The most recently recognized risk factor for CHD is the Type A behavior pattern. This behavior pattern is characterized by an individual's chronic time urgency and impatience, high competitive drive, and tendencies for frequent anger and hostility. Richard Suinn (1978) argues that the Type A CHD risk factor may actually be perceived by the Type A individual as a short-lived coping mechanism (albeit maladaptive) employed to reduce excessive stress particularly on the job. The Type A behavior pattern appears to be significantly reinforced, if not nurtured, in many occupational settings, ultimately taking its toll not only on the individual but the organization as well.

Finally, Stanislav Kasl (1978, p. 18–19) has comprehensively reviewed the area of occupational stress and writes, "There is . . . a fair amount of suggestive evidence that job dissatisfaction and various complaints about work may be associated with CHD or CHD risk factors. . . . There is even some evidence that work satisfaction may make a modest contribution to overall longevity."

Thus we see that there is evidence to support the notion that occupational stress may not only lead to decreased work performance but, more importantly, may be related to risk factors for CHD and/or CHD itself. The concern over excessive occupational stress is such that a wave of occupation-specific publications has emerged detailing work-related stress in occupations such as education (Sparks and Hammond, 1981; Everly and Newman, 1982a), nursing (Marshall, 1980; Scully, 1980), management (Freudenberger, 1977; Ivancevich and Matteson, 1980), medicine (Walker, 1980; Gardner and Hall, 1981), social work (Minahan, 1980) and law enforcement (Kroes, 1980; Farmer and Monahan, 1980), to name just a few. If, as current evidence suggests, occupational stress has the potential to be a major problem in the areas of performance and employee health, how do these facts translate into financial terms?

Costs of Excessive Occupational Stress

Cooper and Marshall (1978, p. 81) in their review of occupational stress conclude, "Life in complex industrial organizations can be a great source of stress. . . . The mental and physical health effects of job stress are not only disruptive influences on the individual . . . , but also a 'real' cost to the organization, on whom many individuals depend: a cost which is rarely, if ever, seriously considered either in human or financial terms by the organizations, but one which they incur in their day-to-day operations."

Determining the cost impact of excessive stress on our nation is extremely difficult. However, some noteworthy attempts have been made.

Greenwood and Greenwood (1979) analyzed the "direct" costs of excessive executive stress. They concluded that such costs exceed $19 billion per year. In this estimate the Greenwoods considered such things as lost workdays, hospitalization, outpatient care, and executive mortality. In noting what the Greenwoods included in their calculations, it is even more important to note what they excluded: lost motivation, lost creativity, alcoholism, poor decision making, and no mention was made of "blue collar" stress costs whatsoever. While many of the excluded factors are extremely difficult to value in discrete economic terms, most supervisors will place a great deal of importance on such factors when it comes to getting a job done. Manuso (in press) reported that the additional cost of employing a worker with stress-related symptoms may be as high as an additional 25% of that employee's salary.

For every employee lost due to stress-related disability, the cost exceeds $200,000 according to the Health Evaluation and Longevity Planning Foundation (1981). Other estimates include the cost of replacing a top executive in a large corporation. Such costs were estimated at $1.5 million.

Clearly, estimating the dollar cost of stress is currently somewhat of a guessing game. How do we estimate lost creativity? And yet such losses may represent the greatest losses of all. Despite the difficulty in placing a dollar cost amount on excessive stress, the reality that this nation plays a very significant price for excessive stress seems beyond debate. The question is simply how much.

THE NATURE OF HUMAN STRESS

In the previous section, we described the potential problem of excessive occupational stress. In doing so a rationale for the establishment of occupational stress management programs emerged. It was shown that excessive stress represents not only a problem for the individual employee but for the organization as well. This would be particularly true in any organization where the human resource is considered the organization's most valuable resource. In order to help us understand more clearly the potential problem at hand, let us examine the nature of human stress.

Stress has been defined by endocrinologist Hans Selye (1974, p. 14) as the "nonspecific response of the body to any demand." From a biobehavioral perspective, stress may be viewed as a broad-ranging psychophysiological mechanism of mediation. By "mechanism of mediation" we mean a medium to bring about an end result, or effect (Everly and Rosenfeld, 1981). But perhaps the easiest way to conceptualize the nature of human stress is as the sum total of "wear and tear" on the person (Selye, 1976a).

In order to remove the misconception that stress is a "mental disorder" or solely an unbounded construct, let us pause here to describe the components of the stress response from a biobehavioral perspective.

The initiation of the stress response is based upon an individual's exposure to a *stressor* (i.e., some stimulus that causes a stress response). A stimulus can be transformed into a stressor via two basic mechanisms. First, any person, place, or thing can become a stressor by being viewed, or interpreted, by the individual as being threatening, challenging, or in some way undesirable. The critical point to be understood is that the vast majority of stress responses are based primarily upon the perception/interpretation of the event rather than the event itself (Arnold, 1960; Lazarus, 1966; Cassel, 1974; Meichenbaum, 1975; Selye, 1976b; Everly and Rosenfeld, 1981; and Lazarus, 1982; 1984). The corollary of this conclusion is that stress will last as long as the individual continues to ruminate upon, or worry about, the stressful event. As Epictetus, the fifth-century B.C. Greco-Roman philosopher, is credited with stating, "Men are disturbed not by things, but the views which they take of them." More recently, Hans Selye has noted "It is not what happens to you that matters, but how you take it." The second way that a stimulus may become a stressor is by direct stimulation of the sympathetic nervous system. This branch of the autonomic nervous system is the one most commonly responsible for triggering the stress response. Sympathomimetic (sympathetic nervous system stimulating) substances bypass the interpretive mechanisms of the brain and directly stimulate the sympathetic nervous system resulting in a stress response. Examples of such substances include caffeine (found in coffee, tea, chocolate, and many "soft drinks"), nicotine (found in tobacco), and high volume noise. However, it has been our experience that such substances would prove of minimal harm to most individuals if they could learn to alter their stressful perceptions.

Having examined the initiation of the stress response, let us now examine the actual nature of the stress response itself. Upon initiation of the stress response, its three major psychophysiological response mechanisms, or axes, may become activated: the neural, the neurohumoral, and the endocrine. Usually, the first axis to become activated is the *neural axis*. This axis consists of the strictly neural aspects of the sympathetic, and to a much lesser degree the parasympathetic, nervous system. The effect of this axis is thought to be the generalized arousal of the ergotropic (work-related) system in the human body (Hess, 1957). The next stress response axis to become activated is usually the *neurohumoral axis* (also called the "fight or flight" response). Stimulation of this axis results in the release of the hormones adrenalin and noradrenalin into the systemic blood circulation (Cannon, 1914). The effect will be to continue the work of the previous axis. Usually the last axis to become activated is the *endocrine axis*. Hans Selye (1976a) is primarily responsible for elucidating the characteristics of this mechanism. Stimulation of this axis results in activation of the adrenal cortex and the thyroid gland. The adrenal cortex is responsible for releasing glucocorticoid (sugar-producing) and mineralocorticoid (mineral-retaining) hormones into the systemic blood circulation. The thyroid gland releases the hormone thyroxin, which may speed up the general metabolism. Selye (1976a; 1976b) argues that long-term activation of this axis can result in significant health problems. Figure 4.2 summarizes the stress response process initiated by a psychosocial stressor.

It is clear that the stress response is a complex multidimensional psycho-

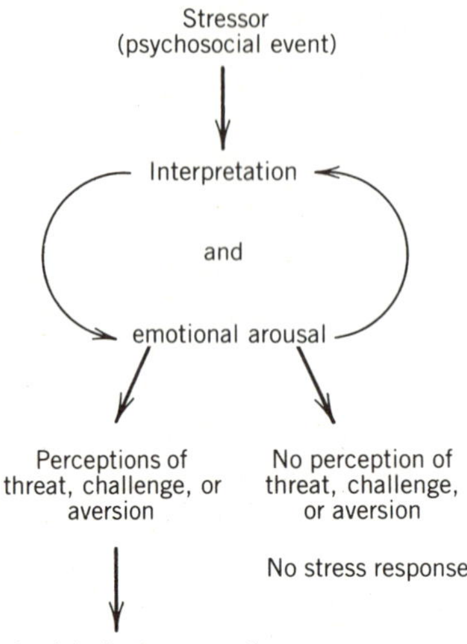

Psychophysiologic stress reactions—
The stress axes:

1. Activation of sympathetic nervous system (some parasympathetic nervous system activation).
2. Activation of neurohumoral ("fight-flight") mechanisms releasing adrenalin and noradrenalin from adrenal medulla into systemic circulation.
3. Activation of endocrine mechanisms releasing glucocorticoids and mineralocorticoids from the adrenal cortex and thyroxin from the thyroid gland.

Stress-related symptoms

Figure 4.2. The stress response. As the figure above clearly shows, the stress response is caused by an individual's interpretation of his or her environment. This is the case with the exception of the sympathomimetic stressors that directly stimulate the activation of the stress axes.

physiologically reactive mechanism, *not* a mental illness (see Everly and Rosenfeld, 1981, for a global summary of the stress response). It is also clear, as we stated earlier, that stress is not always harmful. Stress can be a positive force (eustress) that may act to improve mental and physical performance short-term.

On the other hand, stress has the potential to hinder performance and to contribute to the development of many illnesses (distress). We feel it is important to keep in mind that stress has its positive aspects as well as its negative ones. Susan Seliger (1982) has written an excellent, nontechnical, article on the positive aspects of stress.

Whether any given stress response takes on positive or negative characteristics is for the most part a matter of intensity and chronicity of activation, i.e., the more intense and/or chronic the level of activation, the more likely the result will be distress (Everly and Rosenfeld, 1981, for a review).

STRESS MANAGEMENT

It is now clear that stress can be a positive and highly desirable force when harnessed and positively controlled and directed. This is especially true in the workplace. The reader will recall the conclusion of the National Conference on Health Promotion programs in Occupational Settings that found stress management a powerful tool for not only reducing excessive stress but promoting health, productivity, and general well-being as well. But the question still remains, "What constitutes stress management?"

Stress management is *not* one specific technique or program. Rather, stress management is a generic term that may be used to define any coherent attempt at:

1. Reducing excessive debilitating stress once it has begun, thus reducing the probability of stress-related illnesses and counterproductivity.
2. Implementing strategies that can be used to prevent excessive stress or to harness and direct the stress response into a positive motivating force, thus channeling stress into improving human health and performance.

It seems useful to view stress management from both of the aforementioned perspectives, that is, not only as:

1. A *reactive* intervention used to correct a problem, but as
2. A *proactive* intervention used to prevent or improve a condition no matter what its initial status.

The U.S. Surgeon General (Public Health Service, 1979a) has called for the development of such programs designed to prevent or reduce harmful stress while improving the individual's ability to cope with stress.

The term *stress management,* therefore, simply refers to any coherent attempt at managing the elicitation and/or effects of the stress response. *Occupational stress management,* then, involves any such efforts centering around work or the work environment.

THE GOALS OF A STRESS MANAGEMENT PROGRAM

What are the realistic goals for a stress management program? Referring again to the National Conference's conclusion, goals such as reduced absenteeism, decreased medical and insurance costs, increased productivity, and even improved worker health seem within the realm of reason. Should we then expect and attempt to measure such outcome in order to assess the effectiveness of a stress management program?

What about the stress response itself? An earlier section detailed the psychophysiological nature of the stress response. Is it possible to reduce the stress response itself and consider such reduction an adequate outcome goal rather than relying on changes in more global manifestations of excessive stress such as absenteeism, productivity, etc.?

Finally, is a simple subjective evaluation of the stress management program by the participants enough to evaluate such a program? Should then subjective satisfaction with the intervention be the goal?

Clearly, the desired outcome of the program must be known and accepted by top management as well as by the designers of the program before the program can be implemented. Therefore, let us examine various options that may be used for measuring the effectiveness of a stress management intervention.

Goals such as reduced absenteeism, increased productivity, and decreased medical/insurance costs are perhaps the most desirable indicators of successful stress management interventions in the eyes of management personnel. The isolation and measurement of such variables in order to assess the efficacy of stress management could be problematic in many organizations, however. Suitability and specific strategies must be determined on a case-by-case basis.

An earlier section of this chapter detailed the psychophysiological nature of the stress response. The neuroendocrine and endocrine components of the stress response can, indeed, be measured (Levi, 1975). However, for all but the sophisticated laboratory, such measurement becomes impractical. Yet a compromise can be achieved by professionals who deem it desirable to obtain some level of measurement of stress. In such cases, the following options may be considered useful and practical.

1. Measurement of the individual's stress levels via paper and pencil self-report (Everly and Newman, 1982b; Everly et al., 1984, for a scale to measure stress; also Leavitt, 1980, for a review of anxiety scales).
2. Measurement of the individual's utilization of adaptive, proactive stress-coping strategies as well as maladaptive, health-eroding, coping strategies (Everly, 1979a). Such measurement lends itself particularly well to assessing pre- and postintervention changes in stress coping behavior.
3. Measurement of the participant's ability to relax. Commonly used techniques include electromyographic assessment of frontalis muscle tension and skin temperature (Everly, 1979b; Everly and Rosenfeld, 1981, for reviews).
4. Measurement of selected personality traits. Commonly used measures include locus of reinforcement, self-esteem, and coronary-prone behavior patterns (Leavitt, 1980; Kutash and Schlesinger, 1980).

Finally, it is our belief that the participant's subjective satisfaction with a given stress management intervention can be an acceptable short-term outcome measure, either in addition to, or when the outcome measures discussed above are not employed.

REVIEW OF RESEARCH ON STRESS MANAGEMENT

As noted earlier in this chapter, the term *stress management* is a generic one. Any given stress management program may consist of any number of interventions designed to affect stress arousal. This fact makes it impossible to test the effectiveness of *stress management* per se, in its generic sense. Rather, stress management interventions must be defined and then tested on a case-by-case basis. Let us examine some of the available research on stress management interventions.

Large-scale research results validating the efficacy of stress management programs do not currently exist, nor is there follow-up data sufficient to attest to longer-term effectiveness. However, there does appear to be preliminary support for the proposition that certain stress management interventions can lead to improved health factors and functioning.

Early research conducted at the University of Maryland's (College Park) Psychophysiological and Biofeedback Research Laboratory, under the direction of Dr. Daniel Girdano, began to discover that attitudinal changes could result from the stress management programs developed at that facility. Girdano and Girdano (1977) found that locus of control could be altered in a positive direction as a result of a 15-week (3 hours per week) multidimensional stress management program combining relaxation training, cognitive restructuring, and behavioral management techniques (Girdano and Everly, 1979, for components of the program).

In another early study, Peters and his associates found that a 12-week relaxation-based stress management program in the workplace was conducive to creating a more healthful and productive employee. Results of the study indicated that employees who practiced a noncultic meditation relaxation technique (Benson, 1975) benefited through improved psychological well-being, improved work performance, improved general health, and reduced blood pressure (Peters, Benson, and Peters, 1977).

Other occupationally oriented stress management interventions include those employed in the following studies enumerated below:

The effectiveness of a stress management program that concentrated on teaching various relaxation techniques was evaluated by Charlesworth, Murphy, and Beutler (1981). During the 5-week, 10-session program, subjects were taught visual imagery, deep muscle relaxation, autogenic training and a modified systematic desensitization technique. Results indicated that trait anxiety and state anxiety (test taking) declined over a several-month period for these subjects.

Leclerc (1980) investigated the effectiveness of biofeedback training on stress for 24 business executives. The results of the study indicated that biofeedback training assisted in improving locus of control and muscle tension levels as well as self-reported stress. These results were maintained at a one month follow-up.

The effectiveness of a "cognitive/behavioral" stress management program was tested by Forman (1981). An approach to stress management focusing upon cognitive and behavioral stress management interventions was employed in the form of one 2-hour session per week for six consecutive weeks. The data indicate that this approach was successful in decreasing self-reports of anxiety and improving job satisfaction for a group of school psychologists.

In a study comparing a cognitive approach to stress management versus a relaxation-oriented approach, Landwehr (1979) found that both approaches were effective in stress reduction but that neither one was superior to the other in doing so.

Research conducted by Everly (1980a, 1980b) tested a "non-clinical" education/training model for stress management that was especially designed for and tested with working adults. The studies utilized a "multidimensional" model developed by Daniel Girdano and George Everly that employs strategies for environmental change, attitude change, dietary change, relaxation, and exercise. (Everly and Rosenfeld, 1981, for a discussion of program structure; see Girdano and Everly, 1979; Griffin, Everly, and Fuhrmann, 1982.) Results of the studies showed that participants realized improvements in self-esteem, locus of control, and ability to relax as well as declines in Type A behavior patterns and anxiety. These studies were based upon the author's assumption that stress management programs should be flexible enough to meet a person's individual needs and preferences.

The results of these studies are important because they suggest that attitudinal and behavioral improvements can be facilitated without the use of intensive psychotherapeutic models. Attitudinal change is critically important to a stress management program because, as mentioned earlier in this chapter, most of the excessive stress in a person's life is due to the cognitive interpretations (meanings) that the person assigns to his or her environment. (Refer to Figure 4.2 to see how stress arousal is dependent upon cognitive interpretation.) Further research on the role of education and training models in health promotion must be conducted in order to assess long-term effectiveness—a major shortcoming in the research previously cited.

Finally, two studies tested the effectiveness of a "semiautomated" approach to stress management. Bhalla (1980) tested a multidimensional stress management program combining relaxation, cognitive restructuring, and environmental restructuring. The entire program was delivered solely by audiocassette (Girdano and Everly, 1979, for the components of the program). Subjects were asked to practice the techniques daily. Bhalla found that interventions that took into consideration the subject's personal preferences proved to be not only efficacious, but superior to traditional unidimensional stress management programs. Dependent measures for this study included cardiovascular, endocrine, and striate muscular variables.

A particularly noteworthy investigation in occupational stress management was conducted by Patricia Carrington (Carrington et al., 1980). In this study, conducted at New York Telephone, 154 employees were taught relaxation-based stress management techniques. Two groups were taught a noncultic form of meditation while a third group was taught a form of muscle relaxation. The results of the study indicate that the meditation relaxation groups showed significant improvement in stress and psychiatrically related symptoms after 5.5 months compared to a control group. All relaxation interventions were delivered via audio tape. Subjects were asked to practice the techniques twice a day.

Initial data on the cost-effectiveness of stress management programs comes from three sources. James Manuso (in press) tested the effectiveness of employing a multimodal stress management program in a large corporation. Results of the study indicated that employees were able to increase their ability to reduce stress (measured electromyographically), decrease by 75% their number of visits to the employee health center, and improve their work performance. Manuso calculated the return on corporate investment to be $5.52 for every dollar invested for stress management. In another program, Ed Donovan (Standke, 1979), founder of the International Law Enforcement Stress Association, states stress reduction programs for law enforcement personnel return $3 in increased productivity and community service for every $1 invested.

The most impressive data on cost-effectiveness, however, is reported by Wood (1980). In reviewing the work of Carrington and others at New York Telephone, he estimated that the semiautomated stress management intervention would result in a 10% decrease in stress-related disability absence. This 10% decrease would translate into a $268,000 savings, he estimated.

Although encouraging, *all* of the dollar estimates noted in this section must be considered highly speculative however. This caveat is based upon the fact that there exist no generally accepted guidelines for determining the financial impact of stress management programs.

In concluding this section on research, several points must be made. The term *stress management* may be applied to virtually any intervention(s) designed to affect stress arousal. While the specific interventions reviewed in this section seem promising, the findings should be considered preliminary. Before we can state that certain stress management interventions are, indeed, effective and financially prudent, certain steps may be recommended. First, the most promising studies must be replicated. Second, longer-term follow-up analyses must be conducted. Third, generally acceptable criteria should be devised to analyze cost-effectiveness of such programs. Without a firmer basis to substantiate dollar savings, such financial analyses may be deemed premature and may reduce the credibility of otherwise valuable outcome data. In a review of occupational stress management, Schwartz (1979) concludes that occupational stress management programs possess a significant potential for effectiveness, but he underscores the need for more sound empirical research to validate its overall value. In providing recommendations for the future, Neale, Singer, Schwartz, and Schwartz (1983) suggest that stress management interventions should focus upon two points: (1) the worker, and (2) the job, utilizing an integrated "holistic" format.

Before leaving this section on the research in stress management, we should make note of what may be a "natural" stress management factor: psychological hardiness. The work of Kobasa and her colleagues (Kobasa and Puccetti, 1983) has lead to the identification of three personality factors that seem to mitigate the harmful effects of stress. Kobasa has referred to these factors as *hardiness*. The three factors which constitute psychological hardiness are (1) commitment, i.e., the tendency to involve oneself in experiences in meaningful ways; (2) control, i.e., the tendency to believe and act as if one has some influence in one's own life, and (3) challenge, i.e., the belief that change is a positive and normal characteristic of living. It remains to be seen whether psychological hardiness can be taught to others in the same manner that one can teach, for example, leadership characteristics.

GUIDELINES FOR PROGRAM DEVELOPMENT

Stress management programs may come in many shapes and sizes—from one-to-one counseling interventions to group training programs. Rather than select one "ideal" stress management program, within this section we have elected to present a generalized set of developmental guidelines that are flexible enough to fit virtually any organizational setting. To date, the most popular format used for occupational stress management programs has been the group training seminar format. Here, groups of 10–50 participants are brought together for training in stress management. Such seminars may last for several hours or several days. Some in-house programs may even meet once a week for several months. The guidelines presented in this section are ideally suited for the training seminar format, but may be adapted for individual, one-to-one counseling interventions as well.

Components of a Stress Management Program

The experience of the present authors has been that participants in stress management programs appear to benefit the most when certain generic components are provided within the scope of the intervention.[3] They include:

1. An explanation of what stress is.
2. An explanation of the personal health and performance implications for eustress and distress.

[3]The work of the present authors has been greatly influenced by Levi's recommendations (1979) on the development of a preventive health program.

3. A method for identifying personal symptoms of excessive stress.
4. A method for identifying personal causes of stress.
5. An explanation of and practice in various stress management strategies, preferably organized around a cogent rationale and coherent structure (Levi, 1979; Everly and Rosenfeld, 1981).

Having listed the components of a stress management program, we shall now discuss each component in detail. Before we do, however, it is important to note that preceding all intervention will be selection and measurement, if applicable, of any outcome (dependent) variables that will be used to measure the success of the stress management intervention. Following such determination, actual intervention begins.

The word *stress* means many things to many people. The first component of a stress management intervention should entail *discovering what stress means* to the participants, noting connotations as well as denotations (both correct and incorrect) of the word. This represents a form of audience analysis. This discussion should conclude by reaching some generally acceptable and technically correct definition of stress. Thus, general agreement will be reached as to what stress really is and existing misconceptions will have been corrected by the end of this stage.

The second stage of our stress management model includes some form of *explanation of the nature of the stress response* itself. Included in this discussion should be an explanation of how stress can physiologically act to be a positive, motivating force as well as a debilitating, destructive force. While it is important to keep this discussion simple, we have found it of particular value in demystifying stress and removing it from the realm of a ''mental illness'' in the minds of the participants.

The third component in our program involves helping participants *identify the symptoms of excessive stress*. Because symptoms are idiosyncratic and are even dynamic within the same individual, this process requires personal practice on the part of each participant. It is helpful to begin with a stress symptom checklist and then instruct participants in how to recognize these symptoms in themselves. Even more importantly is the cognitive interpretation assigned to these symptoms. If the participants view these symptoms as wholely negative, a self-perpetuating feedback mechanism could be created. The result would be greater stress arousal and even panic states being initiated at the first symptom's appearance. On the other hand, recognition of these symptoms could be interpreted positively as the initiation of a eustress response that can be controlled and actually be helpful to the individual, in that it represents the mobilization of resources not previously available to the individual. Therefore, we see that not only should recognition of symptoms be taught, but that a positive interpretation of these symptoms should be encouraged when possible. We will be quick to note that certain stress symptoms could be early warning signs of significant health problems. The participants should be instructed in the recognition of such symptoms and encouraged to seek medical consultation. Such symptoms might include prolonged dizziness, prolonged gastrointestinal pain, persistent visual disturbances, persistent heart conduction abnormalities, or any severe pain or dysfunction at all.

The next component in our stress management program involves the *identification of sources of stress*—called stressors. We have found that effective stress management is often predicated upon stressor identification. To achieve that goal, it is advisable to have participants complete a "personal stressor profile." We elect to use a series of self-assessment exercises (Girdano and Everly, 1979). These self-assessment exercises yield a profile of each participant's personal sources of stress, including the job, home activities, diet, and even personality factors. When participants know what their sources of stress are, they can better begin to render those sources benign.

The fifth, most lengthy and most complex component in our stress management program involves *explanation of and practice in various stress management strategies*. The model that we have created represents a "multidimensional" or "holistic" model for intervention. By this we mean that stress management should attack the problem of excessive stress from multiple dimensions, or perspectives, in order to be comprehensive (Girdano and Everly, 1979; Levi, 1979; Everly and Rosenfeld, 1981) in the same way that the individual's total life is affected by multiple influences. Such perspectives should consider the roles of environment, personality, life-style, and biology in the cause and management of excessive stress.

Based upon a review of the literature and practical application, we have concluded that there are basically three generic perspectives, or dimensions, from which *individual* stress management may be initiated. They are

1. Helping the individual develop strategies by which to avoid/minimize/modify exposures to stressors, thus reducing the tendency to experience the stress response.
2. Helping the individual develop skills in relaxation, thus reciprocally inhibiting the stress response.
3. Helping the individual develop techniques for the healthful expression of the stress response (Everly and Rosenfeld, 1981).

A fourth dimension of stress management intervention that may be only indirectly under the control of the individual and only sometimes found in occupational stress management programs, would be affecting changes in the work environment itself. Such changes represent organizational sanctions that may range from microcosmic changes in job description or office environment to macrocosmic changes that might include alteration of the organizational structure, etc. Thus, we have a fourth intervention perspective:

4. Organizational development and redesign to reduce/eliminate occupational stressors.

We have added this last option with the understanding that some corporate climates and some jobs are inherently stressful and are unalterable. Therefore, stress management interventions must put an emphasis upon the individual rather than the job.

Table 4.1 lists the multidimensional stress management model described above, including some specific strategies and resources. In reviewing Table 4.1, it is important for the stress management trainer to be cognizant of the fact that there is no single *best* way to manage stress. Rather, effective stress management is based upon finding stress management strategies that the participant finds personally useful at a given point in time. Because all human beings are unique, what may be right for one person may not be for another. Similarly, because people are dynamic, stress management cannot be static but must be dynamic and flexible as well.

Table 4.1 A Multidimensional Model for Stress Management

I. Helping the individual develop strategies by which to avoid/minimize/modify exposures to stressors.
   A. Time management training (see Lakein, 1973)
   B. Assertiveness training (see Alberti and Emmons, 1970)
   C. Communications training (see Gordon, 1977)
   D. Hostility management training (see Layden, 1977)
   E. Management-by-objectives (see Odiorne, 1965)
   F. Practice in rational thinking (see Ellis, 1973; Ellis and Harper, 1975)
   G. Diet (see Girdano and Everly, 1979; and Everly and Rosenfeld, 1981)

II. Helping the individual develop skills in relaxation (see Everly and Rosenfeld, 1981, for a review).
   A. Deep breathing techniques (see Everly and Rosenfeld, 1981)
   B. Meditation (see Benson, 1975)
   C. Biofeedback (see Danskin and Crow, 1981)
   D. Progressive relaxation (see Jacobson, 1977)
   E. Coping with anxiety (see Kutash and Schlesinger, 1980)
   F. Mental imagery (see Lazarus, 1977)

III. Helping the individual develop techniques for healthful expression of the stress response.
   A. Physical exercise (see Cooper, 1970)
   B. Emotional catharsis (see Emmons, 1978)

IV. Organizational development (see Adams, 1980; Everly and Girdano, 1980).

We often use the analogy that stress management strategies are like golf clubs. Certain clubs should be used for best results under certain select conditions. Other clubs have a wider range of potential usage. However, clearly, there is no one "best" club. Like golf, effective stress management is based upon knowing how to select and use the most appropriate intervention in the face of widely different circumstances (stressor situations). To become proficient at this skill usually requires the individual to develop a "personal plan" for stress management selected from the options in Table 4.1. Rather than assigning interventions, the role of the trainer becomes that of facilitating the participant's creation of his or her plan.

A few more points seem worth underscoring regarding the creation of a personal plan.

First, as noted above, people are dynamic, therefore the stress management plan should be as well. Plans should be flexible, not rigid.

Second, the only way to refine the plan to its highest levels of effectiveness is through trial and error. This process may last months beyond the end of the stress management program.

Third, like golf, stress management is a skill. It takes practice to become and ultimately remain proficient. Thus, we see these latter two points act as further rationale for not immediately administering certain types of outcome measures in order to assess the effectiveness of a stress management program—it takes time and practice to become truly successful in stress management, particularly in reducing chronically elevated physiologic indicators as well as mastering a whole new set of coping skills.

Fourth, one of the most popular components of many stress management programs is the instruction of the participants in deep relaxation (trophotropic state). This can be a highly effective component in the overall program. Teaching deep relaxation is not without its potential pitfalls, however. The stress management trainer should be aware of the potential adverse side effects that may come from inducing the trophotropic state:

1. Drug reactions. Research has clearly demonstrated the ability of deep relaxation to functionally increase the dosage of many chemical substances through what appears to be increased cellular receptivity. Insulin, antihistamine, and cardiovascular medication users pose potential problems.
2. Panic states. The literature contains reports of subjects who undergo panic states in reaction to unusual somatic sensations or feelings of depersonalization, even catalepsy.
3. Excessive trophotropic states. Deep relaxation is known to be capable of inducing orthostatic hypotension and what appear to be temporary states of hypoglycemia.

Deep relaxation techniques can be extremely useful in a stress management program if the trainer is aware of the potential problems connected with their use. Refer to Luthe (1969), Basmajian (1978), Everly and Rosenfeld (1981), and Everly (1984) for further discussions of these issues.

Finally, it seems important for the trainer to realize that the goal of stress management is to have the participant learn to avoid/minimize *excessive* stress *regardless of its source*. Stress management must ultimately focus upon the *individual*. Experience has shown that it is a disservice to the individual to focus stress management solely on the external sources of stress, such as the job. To do so greatly reduces the scope of the stress management intervention and runs the risk of making stress management situation-specific (state-dependent). The rationale should be clear in that by showing the individual the tools by which he or she can reduce his or her excessive stress, the trainer has fostered self-responsibility and self-esteem in the indivudal and, of equal importance, has avoided the tendency to promote job dissatisfaction by "blaming" the working environment or the supervisor for the individual's stress problems (Everly and Girdano, 1980, for an occupationally oriented view of stress management). Furthermore, if the job situation changes, the employee can apply his or her stress management skills without having to learn a whole new set of skills or wait for organizational sanctions to correct the stress problem. This does not mean that organizational development techniques are inappropriate. Such techniques must simply be applied and integrated with the intent of fostering individual self-responsibility. In the final analysis, effective stress management occurs within the *individual* and reaches beyond the work place. If this assertion is correct, then stress management in its purest form may well be an *attitude* upon which a series of techniques may ultimately be built into an effective life-style.

Resistance to Stress Management

In concluding this chapter, it would seem useful to make the reader aware of the most common forms of resistance to the implementation of stress management programs. Major sources of resistance are as follows:

1. The lack of a large volume of research demonstrating the cost-effectiveness of occupational stress management programs. We have highlighted some of the most significant research in this chapter, and yet such research is admittedly not voluminous. Part of the problem revolves around the intrinsic difficulty of quantifying stress reduction in terms corporate executives are familiar with. This in no way reduces the merit of stress management—it just makes it harder to demonstrate.
2. The commonly held misconception that stress is "all in your head" and is therefore a "mental illness." Organizations are hesitant to delve into

areas they feel are related to mental illness. Fortunately, after a little education as to the real nature of stress and its implications for heart disease, stroke, ulcers, etc., opinions change quickly.

3. The individual who views stress management programs as a challenge to prove he or she has no stress problem. Stress problems are too often viewed as a sign of weakness. We have found that stress is often an illness not of weakness but of uncontrolled achievement where the "high achiever" pushes him or herself beyond the point of healthful reason.
4. The individual who comes to the stress management program instead of seeking therapy. Participants should understand that stress management is health education and health behavior change—not intense medical treatment or psychotherapy. Just as the participants must understand this from the onset of the program, so too must the decision makers who may not fully understand what stress management is and is not.
5. Finally, the "high achiever" who misperceives stress management as advocating passivity and underachievement. We present stress management as a program that can be used to increase performance and health at the same time.

Summary

In this chapter we have attempted to provide a rationale and guide for implementing stress management as a component within an occupational health promotion program.

While it is difficult to assess in "hard data" terms, it is generally agreed that excessive stress represents a major threat not only to the health of the individual but to the economic health of the organization. More optimistically, however, preliminary data do suggest that stress management programs can be effective in reducing excessive stress. Nonetheless, it is clear that more research needs to be conducted to validate the efficacy of stress management interventions. As we stated earlier, stress management can't be tested per se, because the term is a generic one used to define a myriad of potential interventions. Therefore, stress management programs must be defined and tested on a case-by-case basis.

Within this chapter, we have detailed one structure that may be used to formulate stress management interventions. It should be understood that the guidelines presented within this chapter are not the only such guidelines possible. The present guidelines are merely those which have been found to be effective by the present authors. Other permutations and combinations of components can be effective.

However, we do believe that in the final analysis, effective stress management occurs from within the individual and reaches beyond the workplace to all aspects of the participant's life. Therefore, in the purest sense, managing stress becomes a function of attitude and life-style.

Perhaps the best summary for the chapter is our belief that stress management should be thought of not as a way to cure an illness, but as a way to improve the quality of work and life in general. As Goldbeck and Kiefhaber (1980) conclude, it appears clear that a company cannot be truly healthy without addressing the phenomenon of occupational stress.

RESOURCE GUIDE FOR STRESS MANAGEMENT

Organizations

Biofeedback Society of America
4301 Owens Street
Wheat Ridge, CO 80033

International Stress and Tension Control Association
P. O. Box 8005
Louisville, KY 40208

National Institue of Mental Health
5600 Fishers Lane
Rockville, MD 20857

National Mental Health Association
1800 North Kent Street
Arlington, VA 22209

Society of Behavioral Medicine
P.O. Box 8530
University Station
Knoxville, TN 37996

Journals

Biofeedback and Self-Regulation

Brain/Mind Bulletin

Health Education

Health Psychology

Journal of Clinical and Consulting Psychology

Journal of Human Stress

Journal of Psychosomatic Research

Psychosomatic Medicine

Psychosomatics

REFERENCES

Adams, J., ed. *Understanding and Managing Stress: A Book of Readings.* San Diego: University Associates, 1980.

Alberti, R., and Emmons, M. *Your Perfect Right.* San Luis Obispo: Impact, 1970.

Arnold, M. *Emotion and Personality.* New York: Columbia University Press, 1960.

Basmajian, J., ed. *Biofeedback: Principles and Practices for Clinicians.* Baltimore: Wilkens and Wilkens, 1979.

Benson, H. *The Relaxation Response.* New York: Morrow, 1975.

Bhalla, V. ''Neuroendocrine, Cardiovascular, and Musculoskeletal Analyses of a Holistic Approach to Stress Management.'' Doctoral dissertation, University of Maryland, 1980.

Cannon, W. ''The emergency function of the adrenal medulla in pain and the major emotions.'' *Journal of Physiology,* 1914, **33,** 356–372.

Carrington, P. et al. "The use of meditation relaxation techniques for the management of stress in a working population." *Journal of Occupational Medicine,* 1980, **22,** 221–231.

Cassel, J. "Psychosocial processes and 'stress': Theoretical Formulation." In *The Behavioral Sciences and Preventative Medicine,* R. Kane, ed. Washington, D.C.: U.S. Government Printing Office, 1974, 53–62.

Charlesworth, E., Murphy, S., and Beutler, L. "Stress management skills for nursing students." *Journal of Clinical Psychology,* 1981, **37,** 284–290.

Cooper, C. and Marshall, J. "Sources of managerial and white collar stress." In *Stress at Work,* C. Cooper and R. Payne, eds. New York: Wiley, 1978, 81–105.

Cooper, K. *The New Aerobics.* New York: Bantam, 1970.

Danskin, D. and Crow, M. *Biofeedback: An Introduction and Guide.* Palo Alto: Mayfield, 1981.

Ellis, A. *Humanistic Psychology: A Rational-Emotive Approach.* New York: Julian, 1973.

Ellis, A. and Harper, R. *A New Guide to Rational Living.* Hollywood, Ca.: Wilshire, 1975.

Emmons, M. *The Inner Source,* San Luis Obispo: Impact, 1978.

Everly, G. *Strategies for Coping with Stress: An Assessment Scale.* Washington, D.C.: Office of Health Promotion, U.S. Department of HEW, 1979a.

Everly, G. "Psychophysiological and Biofeedback Methodologies for the Assessment of Community 'Stress Education' Programs." Proceedings of the Lifelong Learning Conference, College Park, Md.: University of Maryland, 1979b.

Everly, G. "The Development of Less Stressful Personality Traits in Adult Learners: Preliminary Findings." Proceedings of the Lifelong Learning Research Conference, College Park, Md.: U.S. Adult Education Association and the University of Maryland, 1980a.

Everly, G. "The Development of Less Stressful Personality Traits in Adults Through Educational Intervention." *Maryland Adult Educator,* 1980b, **2,** 63–66.

Everly, G. and Girdano, D. *The Stress Mess Solution: The Causes and Cures of Stress on the Job.* Englewood Cliffs, N.J.: Prentice-Hall, 1980.

Everly, G. and Newman, E. "Educational and Administrative burnout: Causes and treatment." Paper presented to the National Education Honor Society (PDK) Leadership Institute on Educator Burnout, Stress, and Renewal. Towson State University, Oct. 30, 1982a.

Everly, G. and Newman, E. "The development of a self-report scale for the measurement of stress arousal in adults." Proceedings of the Lifelong Learning Research Conference. U.S. Adult Education Association, College Park: University of Maryland, 1982b.

Everly, G. and Rosenfeld, R. *The Nature and Treatment of the Stress Response: A Practical Guide for Clinicians.* New York: Plenum, 1981.

Everly, G. et al. *Manual for the Everly Behavioral Survey.* Dept. of Psychology, Loyola College. Baltimore, Maryland, 1984.

Everly, G. "Precautions in the use of relaxation training." Paper presented at the Maryland Psychological Association Annual Convention, College Park, Md., June 8, 1984.

Farmer, R. and Monahan, L. "The Prevention Model for stress reduction." *Journal of Police Science and Administration,* 1980, **8,** 54–60.

Forman, S. "Stress management training: Evaluation of effects on school psychological services." Journal of School Psychology, 1981, **19,** 233–241.

French, J. R. P. and Caplan, R. "Psychosocial factors in coronary heart disease." Industrial Medicine, 1970, **39,** 303–397.

French, J. R. P., and Caplan, R. "Organizational Stress and Individual Strain." In *The Failure of Success,* A. J. Marrow, ed. New York: AMACOM, 1973, 30–66.

Freudenberger, H. "Burnout—The organizational menace." Training and Development Journal, July 1977, 27–29.

Gardner, E. and Hall, R. "The professional stress syndrome." *Psychosomatics,* 1981, **22,** 672–680.

Girdano, D. and Everly, G. *Controlling Stress and Tension: A Holistic Approach.* Englewood Cliffs, N.J.: Prentice-Hall, 1979.

Girdano, D. A. and Girdano, D. D. "Performance-Based Evaluation." *Health Education,* 1977, **8,** 13–15.

Goldbeck, W. and Kiefhaber, A. "Wellness . . . The new employee benefit." *Voluntary Effort Quarterly,* 1980, **2,** 1–3.

Gordon, T. *Leader Effectiveness Training.* New York: Wyden, 1977.

Greenwood, J. and Greenwood, J. *Managing Executive Stress.* New York: Wiley, 1979.

Griffin, D., Everly, G., and Fuhrmann, C. "Designing an effective stress management training program." *Training,* September 1982, 20–31.

Health Education and Longevity Planning Foundation. *Sample Health Assessment.* Scottsdale, Arizona, 1981.

Henry, S. and Stephens, P. *Stress, Health and the Social Environment.* New York: Springer-Verlag, 1977.

Hess, W. *The Functional Organization of the Diencephalon.* New York: Grune & Stratton, 1957.

House, R. and Rizzo, J. "Role conflict and role ambiguity as critical variables in a model of organizational behavior." *Organizational Behavior and Human Performance,* 1971, **7,** 467–505.

House, J. "Occupational stress as a precursor to coronary disease." In *Psychological Aspects of Myocardial Infarction and Coronary Care,* W. D. Gentry and R. B. Williams, eds. Saint Louis: Mosby, 1975.

Ivancevich, J. and Matteson, M. *Stress and Work: A Managerial Perspective.* Glenview, Ill.: Scott, Foresman, 1980.

Jacobson, E. *You Must Relax.* New York: McGraw-Hill, 1977.

Kahn, R., Wolfe, D., Quinn, R., Snoek, J., and Rosenthal, R. *Organizational Stress.* New York: Wiley, 1964.

Kasl, S. "Epidemiological Contributions to the Study of Work Stress." In *Stress At Work,* C. Cooper and R. Payne, eds. New York: Wiley, 1978, 4–48.

Kobasa, S. and Puccetti, M. "Personality and social resources in stress resistance." *Journal of Personality and Social Psychology,* 1983, **45,** 839–850.

Kroes, W. *Society's Victim—The Policeman.* Springfield, Ill.: Thomas, 1980.

Kutash, I. and Schlesinger, L. *Handbook on Stress and Anxiety.* San Francisco: Jossey-Bass, 1980.

Lakein, A. *How to Get Control of Your Time and Your Life.* New York: Signet, 1973.

Landwehr, M. "An exploratory study of teacher stress reduction programs." Doctoral dissertation. University of Cincinnati, 1979.

Layden, M. *Escaping the Hostility Trap*. Englewood Cliffs, N.J.: Prentice-Hall, 1977.

Lazarus, A. *In the Mind's Eye*. New York: Rawson, 1977.

Lazarus, R. *Psychological Stress and the Coping Process*. New York: McGraw-Hill, 1966.

Lazarus, R. S. "Thoughts on the relations between emotion and cognition." American Psychologist, 1982, **37,** 1019–1024.

Lazarus, R. S. "On the primacy of cognition." *American Psychologist,* 1984, **39,** 124–129.

Leavitt, E. E. *Psychology of Anxiety*. Hillsdale, N.J.: Lawrence Erlbaum Associates, 1980.

Leclerc, G. "Effects of biofeedback and relaxation training on stress management in business executives." Doctoral dissertation. University of Rochester, 1980.

Levi, L. "Physiologic methods of measuring industrial stress." *Occupational Mental Health Notes,* April, 1970, 3–6.

Levi, L. "Stress and distress in response to psychosocial stimuli." *Acta Medica Scandinavica,* 1972, **191,** Supplement 528.

Levi, L. *Emotions: Their Parameters and Measurement.* New York: Raven Press, 1975.

Levi, L. "Psychosocial factors in preventive medicine." The Surgeon General's Report on Health Promotion and Disease Prevention: Background Papers. Washington, D.C.: U.S. Government Printing Office, 1979.

Levi, L. ed. *Society, Stress and Disease-Working Life*. London: Oxford University Press, in press.

Luthe, W., ed. *Autogenic Therapy,* Vols. I–IV. New York: Grune & Stratton, 1969.

Manuso, J. "The Metamorphosis of a Corporate Emotional Health Program." Paper presented at the American Psychological Association, 87th Convention, New York, Sept. 1–5, 1979.

Manuso, J. "Stress Management Training in a Large Corporation." *Biofeedback and Self-Regulation,* in press.

Marshall, J. "Stress amongst nurses." In *White Collar and Professional Stress,* C. Cooper and J. Marshall, eds. New York: Wiley, 1980, 19–59.

Maslach, C. "Burned-Out." *Human Behavior,* Sept. 1976, 16–18.

McLean, P., and Jillson, R. "The Manager and Self-Respect-A Follow-up Survey." New York: AMACOM, 1977.

Meichenbaum, D. "A self-instructional approach to stress management." In *Stress and Anxiety, Vol. 1.,* C. Spielberger and I. Sarason, eds. New York: Wiley, 1975, 237–263.

Menninger, R. "Coping with life's strains." *U.S. News and World Report.* May 1, 1978, 80–83.

Minahan, A. "'Burnout' and organizational change." *Social Work,* 1980, **25,** 87.

Neale, M., Singer, J., Schwartz, J., and Schwartz, G. "Yale-NIOSH Occupational Stress Project." Paper presented to the Society of Behavioral Medicine Meeting, Baltimore, March 1983.

Odiorne, G. *Management by Objectives.* New York: Pitman, 1965.

Peters, R. K., Benson, H., and Peters, J. "Daily Relaxation Response Breaks in a Working Population: II. Effects on Blood Pressure." *American Journal of Public Health.* 1977, **67,** 954–959.

Public Health Service. *Healthy People: The Surgeon General's Report on Health Promotion and Disease Prevention.* Washington, D.C.: U.S. Government Printing Office, 1979a.

Public Health Service. *Proceedings of the National Conference on Health Promotion Pro-*

grams in Occupational Settings. Washington, D.C.: U.S. Government Printing Office, 1979b.

Public Health Service. *Promoting Health/Preventing Disease.* Washington, D.C.: U.S. Government Printing Office, 1980.

Report of the President's Commission on Mental Health, Vol. II. Appendix. Washington, D.C.: U.S. Government Printing Office, 1978, 512.

Rice, B. "Can companies kill?" *Psychology Today.* June 1981, 78–85.

Rizzo, J., House, R., and Lirtzman, S. "Role conflict and ambiguity in complex organizations." *Administrative Science Quarterly,* 1970, **15,** 150–163.

Sales, S. "Organizational role as a risk factor in coronary disease." *Administrative Science Quarterly,* 1969, **14,** 325–336.

Sales, S. and House, J. "Job dissatisfaction as a possible risk factor in coronary heart disease." *Journal of Chronic Diseases,* 1971, **23,** 861–873.

Schwartz, G. *Stress Management in Occupational Settings.* Washington, D.C.: Public Health Service, 1979.

Scully, R. "Stress in the nurse." *American Journal of Nursing,* 1980, **80,** 912–915.

Seliger, S. "Stress can be good for you." *New York.* August 2, 1982, 20–24.

Selye, H. *Stress Without Distress.* New York: Signet, 1974.

Selye, H. *The Stress of Life.* New York: McGraw-Hill, 1976a.

Selye, H. *Stress in Health and Disease.* Boston: Butterworth's, 1976b.

Sparks, D. and Hammond, J. "Managing Teacher Stress and Burnout." Washington, D.C.: *ERIC,* 1981.

Standke, L. "The advantages of training people to handle stress." *Training,* February 1979, 23–25.

Suinn, R. "Pattern A behaviors and heart disease." In *The Comprehensive Handbook of Behavioral Medicine,* Vol. 1. J. Ferguson and C. Barr Taylor, eds. New York: SP Medical and Scientific Books, 1978, 5–27.

Timio, M., Pede, S., and Gentili, S. "Urinary excretion of adrenaline, noradrenaline, and 11-hydroxycorticoids under job stress." *Giornale Italiano di Cardiologia.* 1977, **7,** 1080–1087.

Walker, J. "Prescription for the stressed physician." *Behavioral Medicine,* 1980, **7,** 12–13.

Weiman, C. "A Study of Occupational Stressors and the Incidence of Disease/Risk." *Journal of Occupational Medicine,* 1977, **19,** 119–122.

Wood, L. "Lifestyle management strategies at New York Telephone." Paper presented to the "Leadership Strategies-Health" conference. Institute for Health Policy Study, Nov. 19–21, 1980.

CHAPTER 5

Behavioral Approaches to Weight Reduction and the Treatment of Obesity

George S. Everly, Jr.

INTRODUCTION

Almost invariably, if you take a look at the New York Times "Best Sellers" list you will see a book on dieting and weight reduction appearing every few months. It almost seems as if we have a societal compulsion to diet and control weight. But this concern about becoming overweight is not restricted to the United States, nor is it anything new. In 399 B.C. Socrates noted: "Beware of foods that tempt you to eat when you are not hungry and those liquors that tempt you to drink when you are not thirsty." In this chapter we will examine the problem of obesity, review behavioral programs for weight control, and then offer a behaviorally oriented model for weight control that may be intergrated within occupational health promotion efforts.

SCOPE OF THE PROBLEM

It has been estimated that 25%–45% of all American adults are obese (Rodin, 1977), while between 10%–15% of young children are obese (Colley, 1974) and about 30% of adolescents are obese (Garn and Clark, 1976). But what is obesity?

Generally speaking from a health perspective, if a person is considered "overweight," and more specifically, obese, then there exists an excessive amount of adipose tissue in relation to the person's total body mass. Traditionally, however, obesity has been defined as being greater than 20% overweight when compared to the person's "ideal" weight using normative weight tables adjusted for height, sex, and general body structure. While such tables are useful as a rough guide, they are not sensitive enough to reflect certain subpopulation idiosyncracies. For example, muscularly hypertrophied, mesomorphic athletes are usually considered "overweight," and sometimes "obese" using such tables. As a result of such problems, other methods for evaluation have gained popularity. We will discuss these in a later section.

Is obesity a condition that warrants concern? The answer is clearly "yes." Obesity is associated with numerous health problems and medical complications.

Obesity has been shown to be a risk factor for insulin insensitivity and diabetes mellitus (Drash, 1973). The majority of diabetics have a history of obesity. It is usually found that when the diabetic loses weight, the diabetic condition improves, that is, the individual demonstrates a greater cellular receptivity to glucose.

The obese individual is more likely to suffer pulmonary dysfunction when compared to nonobese individuals (Barrera, Reidenberg, and Winters, 1967). In the case of an obese person there is greater body mass. This mass requires increased amounts of oxygen supplied by the lungs. Unfortunately the lungs may be ill-prepared to meet the oxygen requirements of such an enlarged body mass. Furthermore, adipose pockets in the abdomen may act to functionally restrict the mechanics of the breathing process. Finally, an asthmatic individual who already has difficulty breathing will find his or her breathing problems increased if he or she is also obese.

Clearly the greatest physical risk that the obese person faces is that of cardiovascular disease and dysfunction. There are several mechanisms to explain this risk. First, obesity is correlated with hyperlipidemia (a condition of high levels of serum lipids). Of special concern is the finding by Kannel, Gordon, and Castelli (1979) that obesity is correlated with elevated low density lipoprotein cholesterol (LDL) which is a risk factor for atherosclerosis and therefore coronary heart disease. Second, obesity is related to high blood pressure (Gordon and Kannel, 1973) which is one of the more generally accepted risk factors for cardiovascular disease, kidney disease, and stroke. The increased risk for cardiovascular disease appears to begin somewhere between the 20% to 30% overweight category (Kannal and Thom, 1979; Andres, 1980).

In addition to the major health risks enumerated above (insulin intolerance, pulmonary dysfunction, hyperlipidemia, and high blood pressure), obesity is also associated with medical complications during pregnancy (Peckman and Christianson, 1971) and surgery (Prem, Mensheha, and McKelvey, 1965). In general, Lew and Garfinkel (1979) have concluded that the individual who is 50% or more overweight has a 90% increased chance of premature mortality.

So far we have examined only the physical health and medical problems associated with obesity. But there are social and psychological implications as well. Brownell (1982a) argues effectively that "obesity is a social disability" (p. 821) with implications for social biases and discrimination directed against the obese individual. If, indeed, obesity is a "social disability" as some current evidence suggests, then it is reasonable to assume that there may be psychological ramifications as well. The obese individual has been shown to often suffer from a body image distortion with an increased tendency to see oneself as larger than one really is and to also see oneself as undesirable (Stunkard and Mendelson, 1967). Obese individuals are more likely to exhibit shyness, social withdrawal, social immaturity, poor self-image, and depression (Stunkard and Mendelson, 1967; Brownell and Stunkard, 1978; Monello and Mayer, 1963; Stunkard, 1976).

As a final note in this section on the problem of obesity, it is important to note that while the reduction of the overweight condition can be seen to reverse many of the health problems enumerated above, there are risks associated with the generic process of weight loss. Brownell and Venditti (1982) review these risks and conclude anxiety, weakness, irritability, fatigue, nausea, and depression are potential side effects from the generic weight loss process. Therefore it seems clear that psychological assessment should be an important and potentially ongoing aspect of weight control programs. Brownell and Venditti (1982) quote Stunkard's conclusion ''Most forms of dieting carry with them a high likelihood of emotional disturbance'' (p. 55).

FACTORS THAT CONTRIBUTE TO AND MAINTAIN THE OBESE CONDITION

Simonson (1982) has stated: ''Endocrine and organic abnormalities account for about 6% of the obese population'' (p. 2). This conclusion would lead us to believe that somewhere around 94% of the obese population suffers from an overeating problem, that is, the condition of consuming more calories than are expended. But the problem of obesity is not just a problem of self-control on the part of the obese individual. There exist several factors that have been clearly shown to contribute to and maintain the obese condition beyond simple self-control issues.

In a series of pioneering studies, Schacter and his colleagues (Schacter, 1968; Rodin, 1978) demonstrated that obese individuals are far more sensitive to external environmental cues than are individuals of normal weight. Findings indicate that obese individuals, in effect, eat in response to external cues rather than internal cues (such as hunger). Factors such as the sight of food, the smell of food, the sound of food in preparation, and environmental factors which remind one of mealtime (including the scheduled time for eating) all serve as powerful cues that can trigger the consumption of food in the obese individual. On the other hand, most normal weight individuals are far less vulnerable to environmental triggers for eating and tend to eat more in response to internal events. This externality hypothesis has come under attack, however. In an attempt to reconcile what seems to be conflicting data, Herman, Olmsted, and Polivy (1983) have concluded that the external orientation shown by obese persons toward food may be but one aspect of a greater concern on the part of these individuals ''to secure behavioral guidance from the external environment, physical or social'' (p. 926).

The cellular characteristics of the adipose tissue is another factor that can serve to maintain the obese condition. There appear to be two basic forms that obesity will take when considered from a cellular perspective: (1) hyperplastic obesity, and (2) hypertrophic obesity (Salans, Horton, and Sims, 1973). Hyperplastic obesity is characterized by an increased number of normally sized fat cells (up to five times). This hypercellular condition appears to be initiated during periods of cellular growth that typically occur before puberty. Hypertrophic obesity is a condition characterized by a normal number of fat cells that are abnormally large. This condition typically appears later in life. The clinical implications for these findings are important. Research has shown that when weight loss occurs, it occurs primarily through a process of decreasing the size of the fat cells, *not* reducing their number (Björnstrop, Carlgren, and Isaksson, 1975). "This suggests that fat-cell number may determine response to treatment and may set an upper limit on weight reduction" (Brownell and Venditti, 1982, p. 61).

"Set-point theory" (Keesey, 1978) has recently emerged from the study of metabolic processes and may play a significant role in understanding the processes of weight gain and weight loss. Very simply stated, set-point theory suggests that human beings (and other organisms) may have a biologically preset body weight. This preset body weight will be expressed in a range that the person's weight appears to naturally fluctuate within. This weight will be different for each individual and will be determined by many diverse factors, as yet undetermined. The implications for obesity and weight reduction are monumental. If valid, this theory suggests that some obese individuals are suffering from a high set-point, that is, a higher than normal predetermined natural body weight. The body's dedication to maintaining homeostasis then suggests that this obese person's body will fight in a tenacious manner as to maintain its preset body weight—even if that weight puts this person in the obese range. One mechanism the body may employ to restrict weight loss is a lowering of basal metabolism when the individual begins to lose more weight than the body deems desirable.[1] The existence of a set-point within the variable of body weight appears to be gaining empirical validation (Keesey, 1980) and may help to explain the difficulty some obese individuals have in *losing* weight and in *maintaining* the weight loss. Other factors that may in some way act to influence the obese condition are meal patterns, exercise patterns, and psychological variables.

Research by Bray (1972) and Leveille and Romsos (1974) has indicated that consuming the majority of one's daily calories in one or two meals has a greater tendency to produce adipose tissue than does consuming the same number of calories in a greater number of smaller meals. It appears that when the body is faced with large amounts of food to metabolize, it uses a different metabolic process (called adaptive lipogenesis) compared to when it is given smaller more frequent

[1]Basal metabolic rate (BMR) has been shown to be a critically important factor in body weight. It has been estimated that 70% of total calories burned during the course of an average day is a function of BMR (McArdle, Katch and Katch, 1981). It has also been observed that when a person drastically reduces caloric consumption, BMR may drop by 20% (Bray, 1969). Therefore, any mechanism that causes a reduction of BMR may actually make weight loss more difficult or may actually facilitate weight gain.

meals to metabolize. The adaptive lipogenic process tends to result in an increased fat-producing metabolism, increased serum cholesterol, and increased glycogen, all leading to potential difficulty in losing weight or to actual weight gain.

"There is a growing consensus that obese adults are less physically active than their lean counterparts" (Brownell, 1982a, p. 7). What is unclear, however, is whether the lower activity levels for obese adults is a cause of the obesity or is a result of the obesity. In either event, physical inactivity can clearly serve to maintain the obese condition. Conversely, increasing activity levels may be considered a potentially viable means of beginning to treat the obese condition.

Finally, we should consider the role of psychological variables in the initiation and maintenance of the obese condition. It certainly seems tenable, at least according to Thorndike's Law of Effect, that for many persons, eating in excess and/or the condition of obesity is in some way, reinforcing. In developing a weight reduction program, it then becomes important to discover what the reinforcements are and how they can best be substituted for or otherwise managed. Care must be exercised in altering any person's source of reinforcement and indeed eating and/or obesity may clearly possess positively as well as negatively reinforcing characteristics.

GOALS OF A WEIGHT REDUCTION PROGRAM

As noted earlier, if a person has a "weight control problem," or is considered obese, it is far too simplistic too merely say the person is overweight; rather, the real concern is that there exists an excessive amount of adipose tissue in relation to the person's total body mass. This understanding of the weight control issue is a *critical* one. Such a perspective allows us to not only develop an appropriate program but allows us to appropriately measure outcome as well. Let us review some of the outcome variables that are often used in ascertaining the success of a weight reduction program.

The most commonly used measure of success in any weight reduction program is body weight. This measure, when used alone, can be grossly inappropriate. Body weight is not only affected by adipose tissue, but by the amount of fluid in the body and by the amount of muscle mass. Many diets result in dramatic initial losses of weight that are, in reality, fluid loss and usually in no way related to the real goal of the program, that is, the loss of adipose tissue. Weight loss based on fluid loss is quickly regained and serves to disillusion the individual. On the other hand, weight loss programs that include exercise may actually contribute to somewhat of a weight gain. This weight gain results from the development of skeletal muscle mass that is heavier in relation to adipose tissue. While this is a minor issue with individuals who are extremely overweight due to excessive adipose tissue, it serves to point out the shortcomings of using body weight alone as measure of success in a weight reduction program.

Another method for determining outcome in a weight reduction program would be taking anthropometric assessments of body girth. By simply measuring the circumference of selected body areas, an individual can literally see how their body is being reshaped. It is important to emphasize that "spot reducing," that is, losing adipose tissue from one area to the exclusion of other areas, is impossible. Weight loss is a generalized somatic process.

A third method that is sometimes used to determine outcome is a cosmetic and extremely superficial method—before-and-after photography. Most participants in a weight loss program expect to improve their cosmetic appearance. Photographs taken before and after the weight loss program provide the opportunity to assess global cosmetic changes in physique. The drawbacks to such an approach are that weight loss must be dramatic enough to see noticeable differences and the weight loss process itself does take some time. This could be frustrating for persons impatient for success. In addition, if the person's body image is distorted, the photographs themselves may fail to have much of an impression upon the individual.

If the health problem of being overweight is actually based upon the existence of an excessive amount of adipose tissue in relation to the total body mass, then the best single criterion for measuring outcome in a weight reduction program must be the variable of adipose tissue present in a person's body in relation to total body mass. This may be determined in several ways. The three most popular are (1) hydrostatic weighing, (2) skinfold assessment, and (3) circumference assessment.

Hydrostatic weiging (sometimes called underwater weighing) is a method of determining the density, or "heaviness," of an object in relation to its volume. There is an inverse relationship between the overall density and the percentage of adipose tissue in one's body. Through the use of various formulae, the amount of adipose tissue can be rather reliably estimated (Katch and McArdle, 1977). The drawback in this procedure is that hydrostatic weighing requires that the subject be fully submerged in a tank of water while being weighed. The equipment needed to accomplish this task makes it impractical for utilization in most weight reduction programs.

The problems inherent in hydrostatic assessment lead to the development of another way of estimating the amount of adipose tissue in the body—skinfold assessment. In this procedure, calibers are used to measure skinfold thicknesses in selected areas. The greater the thicknesses in the selected sites, the greater the overall amount of adipose tissue in the body. While not perfect, regression analyses have identified the following points as being useful anthropometric landmarks for the assessment of adipose tissue via skinfold thickness: middle of the right tricep, just below the right scapula, and one inch to the right of the umbilicus (note: other points are sometimes used as well). Once again, by using conversion formulae the total percentage of adipose tissue in the body can be estimated (Fox, 1979; Lohman, 1981).

Finally, Katch and McArdle (1973) developed an anthropometric technique for estimating adipose percentages based upon measuring the circumference of selected body sites, (Katch and McArdle, 1977, for details). Given the difficulty that some individuals have in generating reliable information using calibers, Katch and McArdle's methods seem the most utilitarian for occupational health promotion programs.

In deciding what outcome variable to use in a weight reduction program, certainly the assessment of adipose tissue must play the central role. Yet, one might consider using some of the other variables mentioned in combination with adipose assessment. The final decision will depend on the population parameters as well as the facilities available. In the final analysis, however, it is important to keep in mind that the goal of any weight reduction program is the *long-term* management of energy intake and energy utilization so as to maintain a healthy relationship between adipose tissue and total body mass.

REVIEW OF RESEARCH ON WEIGHT CONTROL

There exist several excellent reviews of research on weight reduction programs (Stunkard, 1980; Stuart, Mitchell, and Jensen, 1981; Bellack and Williamson, 1982; Brownell, 1982a; Brownell and Venditti, 1982; Wing, 1982).

Table 5.1, adapted from Stuart, Mitchell, and Jensen (1981) will provide the reader with a general overview of how generically different programs have differed in outcome over the last few years.

Table 5.1 Summary of Treatment Outcome

			During Follow-up	
Treatment	*Duration (weeks)*	*Average Total Change (pounds)*	*Length (weeks)*	*Average Change (pounds per week)*
Fasting (0 cals.) (15 studies)	16.7	−75.01 (N = 1159)	104 (2 studies)	+.22 (N = 53)
Dieting (900 cals.) (9 studies)	23.8	−25.50 (N = 2212)	39.4 (5 studies)	+.02 (N = 47)
Behavior Modification (95 studies)	13.3	−8.99 (N = 3890)	33.5 (53 studies)	−.04 (N = 1934)
Exercise (7 studies)	19.2	−16.16 (N = 76)	None	None

Source: Adapted from Stuart, Mitchell, and Jensen (1981).

As indicated in Table 5.1, the most dramatic weight loss has been achieved using programs that employ fasting. Fasting programs (0–750 calories/day), alone, may not be well suited for many occupationally based health promotion efforts, however, such programs should be supervised by a physician due to the fact that many such programs increase the probability of ketosis, electrolyte imbalance, extreme serum glucose fluctuations, and cardiac arrhythmias (Vertes, 1978). But perhaps even more importantly it is clear that the goal of all weight control programs is the *long-term* maintenance of adipose tissue in relation to total body mass. Behaviorally based (“behavior modification” and “behavior therapy”) programs have shown the greatest potential to achieve such a goal.

The history of weight reduction programs at the worksite is sparse. Those programs which have been instituted have been behaviorally oriented. Stunkard and Brownell (1980) conducted the first worksite weight reduction program for the United Storeworkers Union member-employees of Bloomingdale's and Gimbels department stores in New York City. While weight loss ranges were encouraging (2.7–5.6 kg), the attrition rate exceeded 50% (compared to 15%–30% for such programs run in clinical settings). Subsequent research at the worksite has resulted in similar and improved outcome (compared to that initially reported by Stunkard and Brownell) and improved outcome in the attrition rate and weight loss maintenance as well (Abrams and Follick, 1983).

In the next section, we shall present a model for formulating occupational weight reduction programs.

GUIDELINES FOR PROGRAM DEVELOPMENT

Based upon the clinical and programmatic evidence available there appear to be two key elements to successful weight reduction programs: (1) adherence to the empirically demonstrated principles of behavior, and (2) adoption of a *multifaceted* format with flexibility enough for individuation of strategies (this is not to suggest that this is incongruent with the principles of learning but rather is listed separately for emphasis).

For years, perhaps since the movie *A Clockwork Orange,* many laypeople and professionals alike have conceptualized the modification of behavior based upon the principles of learning (sometimes called “behavior modification” or “behavior therapy”), as a discrete program or set of techniques. Writers such as Everly and Girdano (1980), Stuart et al. (1981), and Brownell (1982a) have argued that “behavior modification” is best viewed not as a set of techniques but rather as a set of empirically derived principles that not only offer the greatest insight into why people behave the way they do, but also give us remarkably effective opportunities for changing behavior. These principles of behavior (as discussed earlier in the text) should underlie any weight reduction program offered at the worksite or otherwise.

With such a notion in mind, multifaceted weight reduction programs have emerged and appear to be of superior efficacy when compared to single-approach treatment packages as primarily reviewed by Stuart et al., 1981). The interested reader should refer to Stunkard (1976); Everly and Girdano (1980); Stuart et al. (1981); Bellack and Williamson (1982); Brownell (1982a); Dusek (1982); Wing (1982); Perri, McAdoo, McAllister, Spevak, Lauer, and Newlin (1983). As Brownell (1982a) concludes, we are now in a "second generation" of behaviorally based intervention programs that employ multifaceted approaches to weight control. Such programs combine such things as diet, counseling, family involvement/support, nutrition education, exercise, and the more traditional behavior modification interventions, (Katch and McArdle, 1977; Everly and Girdana, 1980; Stuart et al., 1981; Bellack and Williamson, 1982; Brownell, 1982a; Dusek, 1982; Wing, 1982; Perri, McAdoo, McAllister, Spevak, Lauer, and Newlin, 1983).

Finally, it has been argued that weight reduction programs should be individually tailored as much as possible (although generally maintaining a group format) as to maximize efficacy (Stunkard and Mahoney, 1976; Stuart and Davis, 1978; Everly and Girdano, 1980; Stuart et al., 1981). As Stuart et al. (1981) conclude, "Because the sources of the urge to eat and the form that overeating takes are as individual as each person's fingerprints, some effort should be made to individualize behavioral instigations" (p. 336). Such efforts will maximize the chances that the weight control program will address the problems and meet the needs of each individual participant.

Offered below are some general guidelines for formulating a multifaceted weight reduction program. It is based upon the work of Ferguson (1976), Stunkard and Mahoney (1976), Jeffrey (1976), Brownell (1976), Everly and Girdano (1980), Wilson (1980), and Dusek (1982) and is summarized in Table 5.2.

Initial Assessments

The initial assessment phase consists of 3 steps: (1) medical history and exam, (2) psychological evaluation, and (3) collection of baseline data.

According to Powers (1980; 1982) an initial medical history and exam are necessary to eliminate "organic" causes of the obese condition as well as to achieve an understanding of the psychosocial history of the weight problem. The psychological evaluation should be used to screen for psychiatric disturbances that might be manifest in or otherwise masked by the weight problem (Powers, 1982). The Minnesota Multiphasic Personality Inventory, or the Millon Multiaxial Clinical Inventory may be used as broad-range screening tools. The Draw-A-Person projective test or a form of body cathexis scale may be used to assess body image. The Everly Behavioral Survey (see Chapter 4) may be used to assess the individual's perceived stress levels and coping mechanisms. The Millon Behavioral Health Inventory (see Chapter 16) may be used to assess personality variables as they relate to medical compliance and related issues.

Table 5.2 Guidelines for a Weight Reduction Program

I. Initial assessments
   A. Medical history
   B. Medical exam
   C. Psychological evaluation
   D. Collection of baseline data

II. Goal setting

III. Intervention (strategies should be combined and designed to meet individual preferences)
   A. Nutrition education
   B. Stimulus control modification
   C. Stress management
   D. Caloric restriction via dieting
   E. Development of social support networks
   F. Increasing physical activity
   G. Behavioral contracting of reinforcements for weight reduction and maintenance

IV. Maintenance
   A. Continued use of social support and stimulus control
   B. Continued use of maintenance behavioral contracting
   C. Exercise
   D. Booster sessions

V. Evaluation

Baseline data collection should consist of an eating diary, a physical activity diary, and baseline assessment of outcome variables. Each participant should be asked to record their general levels of physical activity for seven days so that gross approximations of caloric expenditure can be made. In addition, each individual should be asked to complete an eating diary for seven days. The diary would cover such variables as enumerated in Figure 5.1. Finally, outcome variables as discussed earlier should include some assessment of adipose tissue (usually skin-fold or circumference) as well as other appropriate variables (perhaps body weight and general circumference measures or even a photograph).

Goal Setting

Previous writers (see Everly and Girdano, 1980, for a review) have noted the importance of establishing objective goals for a weight reduction program and integrating the individual into the decision-making process. This is called "self-

Name ______________________ Date ______________________

Day of the week ______________________

Time	*Hunger Level*	*Location*	*Describe Food, Quantity, and Calories*	*Thoughts/ Feelings Before Eating*	*Activity While Eating*	*Whom Did You Eat With?*	*Minutes Spent Eating*	*Thoughts/ Feelings After Eating*
6 am								
9 am								
12:00 noon								
3 pm								
6 pm								
9 pm								
12:00 midnight								

Note: Hunger level ranges from 1 to 5: 1 = not at all hungry; 3 = moderately hungry; 5 = extremely hungry.

directed goal setting.'' The professional staff plays a nondirective facilitative role in helping the individual set *reasonable* goals. It may be necessary to break a goal of great magnitude down into subgoals in order to ''shape'' the individual's goal attainment. One mistake that is often made in the goal-setting phase is to select only short-term goals. The formulation of goals should include long-term maintenance in addition to short-term weight reduction.

Intervention

Potential intervention strategies are numerous and are most effective when combined into a multidimensional intervention paradigm. Commonly used strategies are briefly described below.

Nutrition Education is designed to educate the individual as to the caloric and nutritive values of food. The concepts of energy balance, calories, food shopping, food preparation, carbohydrates, fats, proteins, vitamins, and minerals are possible topics. These topics should be analyzed in light of each person's personal food diary. The goal of this process is to help individuals develop a well-rounded diet, selecting foods that have the highest nutritive values eliminating foods that are ''empty calories,'' that is, have little or no nutritive value (Katch and McArdle, 1977; Dusek, 1982, for guidelines).

Modification of the *Stimulus Control* of eating refers to helping the individual identify, then modify, eating that is in response to external environmental cues, as opposed to internal hunger cues. Using the eating diary, patterns of eating in response to environmental stimuli (certain social situations, TV, and other habits) will develop. By eliminating many of these cues and/or inhibiting eating behavior in such circumstances, needless calorie consumption will be greatly reduced. Stimulus control modification tactics might include restricting eating to certain times or certain places, eliminating snacking, eating small portions, using a small plate to make portions appear larger, and so on (Bellack, Schwartz, and Rozensky, 1974, for a review).

Stress Management techniques would be included in a weight reduction program because for many people losing weight is indeed stressful. Furthermore, many people eat simply as a means of coping with stress. By teaching a broad spectrum of cognitive-behavioral stress management techniques, the stress of losing weight is reduced, thereby increasing the likelihood of compliance, while at the same time eliminating the need to cope with other stressors through eating (refer to the chapter on stress management for guidelines).

Specific *Diets* are perhaps the oldest of the weight reduction tactics. There are many diets available. A useful review of 16 popular diets is provided by Dwyer (1980). Specific diets are seldom needed, however, if other tactics discussed in this section are employed. This is an important concept very well explained by Ferguson (1976) and Dusek (1982).

Social Support Networks are individuals who can lend psychosocial support to people trying to reduce their weight. In a program designed by the author, members of each participant's support network (family, neighbors, or friends) were asked to sign ''behavioral contracts'' that were nothing more than statements of commitment detailing how they would support the person in his or her attempt to lose weight. We also found that weekly meetings of active participants were useful in sharing successes and overcoming barriers to success. In addition, we found that daily phone calls from either a group leader or designated group member helped improve compliance, in addition to the weekly group meetings.

Increasing Physical Activity can be an excellent method for losing weight. Remember, weight gain occurs because more calories are being consumed than utilized as energy. Weight loss occurs when more calories are burned than consumed. Increases in physical activity will result in weight loss if calories are held constant. An even greater weight loss will be realized if exercise is combined with caloric restriction. Current literature (Everly and Girdano, 1980; Bellack and Williamson, 1982; Brownell, 1982b; Wing, 1982) argues that weight reduction programs should include an exercise component for the following reasons:

1. It is likely that a moderate increase in basal metabolic rate will result thereby increasing the resting rate of calorie utilization (which is 70% of all calorie utilization).
2. Appetite suppression has been observed following exercise.
3. An increase in energy expenditure will obviously result on days the individual exercises.
4. Exercise spares the loss of lean body mass when compared to dieting alone.
5. There is evidence that exercise may improve mood and psychological well-being.
6. Exercise improves cardiopulmonary efficiency.
7. Exercise has been shown to improve maintenance levels of weight loss.

The physical activity diary can be used to monitor physical activity levels (Katch and McArdle, 1977; Dusek, 1982, for guidelines).

Finally, *Behavioral Contracting* can be used with each individual to delineate specific rewards that individuals can give themselves for being compliant (e.g., not eating junk food, exercising, eating only at prescribed times, etc.), or for meeting specific goals, short-term and long-term goals, including the maintenance period.

Maintenance

Once the target goals have been met, the real challenge in any weight reduction program begins—maintenance of the newly attained goals. Several strategies may be useful in accomplishing this. The continued use of social support networks will provide a firm basis to prevent relapse and subsequent weight regaining. This means changing the person's environment from a long-term perspective. To have a person lose weight, then return to the unaltered environment that spawned the initial weight problem is virtually assuring that the individual will regain his or her weight. Behavioral contracting can be used to help individuals reinforce themselves for maintenance of their new weight. Exercise has been found to be effective in weight loss maintenance as noted earlier. Finally, "booster sessions" may prove of some value as well. A booster session is a brief refresher course that the individual is asked to attend on a periodic basis to sharpen the skills acquired during the weight reduction program.

Evaluation

Evaluation consists of the periodic assessment of the success of the program for each individual. The criteria used for determining success will depend on whatever outcome veriables were initially selected and recorded during baseline. Such outcome variables may vary among individuals within the same program, at the discretion of the program leader (see the "Goals" section of this chapter).

Evaluation may be an ongoing process that occurs at preselected intervals (e.g., weekly, monthly, etc.). It is important that evaluation continue into the maintenance phase in one form or another. This in itself may act to improve maintenance.

Summary

In this chapter we have discussed the problem of being overweight and provided general guidelines for designing a behaviorally based program to combat the problem as we see it. In their then "state-of-the-art" paper, Stunkard and Mahoney (1976) made recommendations for the "maximally effective" weight reduction program. Their recommendation included

1. Monitoring of behaviors associated with weight problems (e.g., eating, exercise, etc.) rather than just merely weight itself.
2. Nutrition education.
3. Exercise.
4. Stimulus control.
5. Social support.
6. Training in what we would refer to as stress management.
7. Self-directed goal setting and reinforcement.

In the time that has elapsed since the publication of their paper, empirical analyses have demonstrated the efficacy of such recommendations. The program described in this chapter not only meets these recommendations but has been empirically derived in its own right. The potential flexibility of such a program makes it well-suited for occupational health promotion settings.

As for the future, protein-sparing fasts (with approx. 800 calories/day) combined with a multifaceted behavioral program similar to that described in this chapter may hold the greatest potential for effective, sustained weight reduction. Such programs are currently under development and testing. Yet, it should be noted that to maintain maximum safeguards, such drastic calorically restricted diets should be monitored by a physician.

RESOURCE GUIDE FOR WEIGHT CONTROL

Organizations

Lean Line
151 New World Way
South Plainfield, NJ 07080
(201)757-7677

National Association to Aid Fat Americans
P.O. Box 43
Bellerose, N.Y. 11426
(516)352-3120

TOPS (Take Off Pounds Sensibly)
P.O. Box 07489
4575 S. Fifth Street
Milwaukee, WI 53207

Weight Watchers
Manhasset, NY (International Headquarters)

Journals

Addictive Behaviors

American Journal of Clinical Nutrition

American Journal of Public Health

Behavior Research and Therapy

Behavior Therapy

Health Education

Health Psychology

International Journal of Obesity

Journal of Behavioral Medicine

Journal of Consulting and Clinical Psychology

Journal of Nutrition

Metabolism

Nutrition Today

Obesity/Bariatric Medicine

REFERENCES

Abrams, D., and Follick, M. "Behavioral weight-loss intervention at the worksite." *Journal of Consulting and Clinical Psychology,* 1983, **51,** 226–233.

Andres, R. "Influence of obesity on longevity in the aged." In *Aging, Cancer, and Cell Membranes,* C. Borek, C. M. Fenoglio, and D. King, eds. Stuttgart: Thieme Verlag, 1980.

Barrera, F., Reidenberg, M., and Winters, W. "Pulmonary function in the obese patient." *American Journal of Medical Science,* 1967, **254,** 784–796.

Bellack, A., Schwartz, J., and Rozensky, R. "The contribution of external control to self-control in a weight reduction program." *Journal of Behavior Therapy and Experimental Psychiatry,* 1974, **5,** 245–249.

Bellack, A. and Williamson, D. "Obesity and anorexia nervosa." In *Behavioral Medicine,* D. Doleys, R. Meredith, and A. Ciminero. New York, Plenum, 1982, 295–316.

Björnstrop, P., Carlgren, G., and Isaksson, B. "Effect of an energy-reduced dietary regimen in relation to adipose tissure cellularity in obese women." *American Journal of Clinical Nutrition,* 1975, **28,** 445–452.

Bray, G. A. "Effect of caloric restriction on energy expenditure in obese patients." *Lancet,* 1969, **2,** 397–398.

Bray, G. A. "Lipogenesis in human adipose tissue." *Journal of Clinical Investigation,* 1972, **51,** 537–546.

Brownell, K. "Exercise and obesity." *Behavioral Medicine Update,* 1982b, **4,** 7–11.

Brownell, L. "Obesity: Understanding and treating a serious, prevalent, and refractory disorder." *Journal of Consulting and Clinical Psychology 1982a,* **50,** 820–840.

Brownell, K. *Behavior Therapy for Weight Control.* Unpublished manual. University of Pennsylvania, 1976.

Brownell, K. and Stunkard A. "Behavioral treatment of obesity in children." *American J. of Dis. Child.,* 1978, **132,** 403–412.

Brownell, K. and Venditti, E. "The etiology and treatment of obesity." In *Phenomenology and Treatment of Psychophysiological Disorders,* W. Fann, I. Karacan, A. Pokorny, and R. Williams, eds. New York: Spectrum, 1982, 51–87.

Colley, J. R. "Obesity in school children." *British J. Soc. Prev. Med.,* 1974, **28,** 221–225.

Drash, G. "Relationship between diabetes mellitus and obesity in the child." *Metabolism,* 1973, **22,** 337–344.

Dusek, D. *Thin and Fit.* Belmont, Ca., Wadsworth, 1982.

Dwyer, J. "Sixteen popular diets." In *Obesity,* A. Stunkard, ed. Philadelphia: Saunders, 1980, 276–291.

Everly, G. S. and Girdano, D. "Implementing behavior modification in a weight control program." *Health Education,* 1980, **11,** 15–17.

Ferguson, J. *Habits Not Diets.* Palo Alto: Bull Publishing, 1976.

Fox, E. *Sports Physiology.* Philadelphia: Saunders, 1979.

Garn, S. and Clark, D. "Trends in Fatness and the origins of obesity." *Pediatrics,* 1976, **57,** 443–456.

Gordon, T. and Kannel, W. "The effects of overweight on cardiovascular disease." *Geriatrics,* 1973, **28,** 80–88.

Herman, C. P., Olmsted, M., and Polivy, J. "Obesity, externality, and susceptibility to social influence: An integrated analysis." *Journal of Personality and Social Psychology,* 1983, **45,** 926–934.

Jeffrey, D. B. "Behavioral management of obesity." In *Behavior Modification,* W. E. Craighead, A. Kazdin, and M. Mahoney eds. Boston: Houghton Mifflin, 1976, 394–413.

Kannel, W., Gordon, T., and Castelli, W. "Obesity, lipids, and glucose intolerance: The Framingham Study." *Amer. Journal of Clinical Nutrition,* 1979, **32,** 238–245.

Kannel, W. and Thom, T. "Implications of the recent decline in cardiovascular mortality." *Cardiovascular Medicine,* 1979, **4,** 983–997.

Katch, F. and McArdle, W. "Prediction of body density from simple anthropometric measurements in college-age men and women." *Human Biology,* 1973, **45,** 445–454.

Katch, F. and McArdle, W. *Nutrition, Weight Control and Exercise.* Boston: Houghton Mifflin, 1977.

Keesey, R. E. "Set-points and body weight regulation." *Psychiatric Clinics of North America,* 1978, **1,** 523–544.

Keesey, R. E. "A set-point analysis of the regulation of body weight." In *Obesity,* A. Stunkard ed. Philadelphia: Saunders, 1980, 114–165.

Leveille, G. and Romsos, D. "Meal eating and obesity." *Nutrition Today,* 1974, 4–9.

Lew, E. A. and Garfinkel, L. "Variations in mortality by weight among 750,000 men and women." *Journal of Chronic Diseases,* 1979, **32,** 563–576.

Lohman, T. G. "Skinfolds and body density and their relation to body fatness: a review." *Human Biology,* 1981, **53,** 181–225.

McArdle, W., Katch, F., and Katch, V. *Exercise Physiology.* Philadelphia: Lea and Febiger, 1981.

Monello, L. and Mayer, J. "Obese adolescent girls." *Amer. J. Clin. Nutrition,* 1963, **13,** 35–40.

Peckman, C. and Christianson, R. "The relationship between pregnancy weight and certain obstetric factors." *Am. Journal of Obstetrics and Gynecology,* 1971, **111,** 1–7.

Perri, M., McAdoo, G., McAllister, D., Spevak, P., Lauer, J., and Newlin, D. "Behavioral treatment of obesity: Two studies of weight-loss maintenance of obesity: Two studies of weight-loss maintenance strategies." Paper presented at the annual Society of Behavioral Medicine Conference, Baltimore, March, 1983.

Powers, P. *Obesity.* Baltimore: Williams & Wilkins, 1980.

Powers, P. "Obesity: Psychosomatic illness review." *Psychosomatics,* 1982, **23,** 1023–1039.

Prem, K., Mensheha, N., and McKelvey, J. "Operative treatment of adenocarcinoma of the endometrium in obese women." *Am. Journal of Obstetrics and Gynecology,* 1965, **92,** 16–22.

Rodin, J. "Bidirectional influences of emotionality, stimulus responsivity, and metabolic events in obesity." In *Psychopathology,* J. D. Moser and M. E. P. Selegman, eds. San Francisco: Freeman, 1977.

Rodin, J. "Environmental factors in obesity." *Psychiatric Clinics of North America,* 1978, **1,** 581–592.

Salans, L., Horton, E., and Sims, E. "Experimental obesity in man: Cellular character of the adipose tissue." *Journal of Clinical Investigation,* 1971, **50,** 1001–1011.

Schacter, S. "Obesity and eating." *Science,* 1968, **161,** 751–756.

Simonson, M. "An overview: Advances in research and treatment of obesity." *Food and Nutrition News,* 1982, **53,** 1–4.

Stuart, R. B. and Davis, B. *Slim Chance in Fat World.* Champaign, Ill.: Research Press, 1978.

Stuart, R., Mitchell, C., and Jensen, J. "Therapeutic options in the management of obesity." In *Medical Psychology,* C. Prokop and L. Bradley, eds. New York: Academic Press, 1981, 321–353.

Stunkard, A., ed. *Obesity.* Philadelphia: Saunders, 1980.

Stunkard, A. *The Pain of Obesity.* Palo Alto: Bull Publishing, 1976.

Stunkard, A. and Brownell, K. "Worksite treatment for obesity." *Amer. Journal of Psychiatry,* 1980, **137,** 252–253.

Stunkard, A. and Mahoney, M. "Behavioral treatment of eating disorders." In *Handbook of Behavior Modification and Behavior Therapy,* H. Leitenberg, ed. Englewood Cliffs, N.J.: Prentice-Hall, 1976, 45–73.

Stunkard, A., and Mendelson, M. "Obesity and body image." *American Journal of Psychiatry,* 1967, **123,** 1296–1300.

Vertes, V. "Supplemented fasting." *Drug Therapy*. Summer 1978.

Wilson, G. T. "Behavior modification and the treatment of obesity." In *Obesity,* A. Stunkard, ed. Philadelphia: Saunders, 1980, 325–344.

Wing, R. "Obesity: The current state of the art." *Health Psychologist,* Summer/Fall 1982, 5–6.

CHAPTER 

The Development of a Smoking Cessation Program

Sharlene M. Weiss
Calvin F. Fuhrmann
George S. Everly, Jr.

INTRODUCTION

The manner in which our society views cigarette smoking has changed dramatically over the last four decades. During the 1940s and 1950s cigarette smoking was considered by many to be not only fashionable, but a sign of maturity. In the 1980s smoking is considered by a growing number to be inconsiderate of others, unclean, and a major risk factor for disease. Furthermore, it has been the target of the most concerted efforts of the U.S. Surgeon General's Office ever directed towards the reduction of a behaviorally based risk factor for disease.

Having reviewed the most recent research data available, the American Lung Association has concluded:

1. Smoking is the single most important negative health behavior in our society today.
2. Smoking should be regarded as an addictive behavior to be classified and treated accordingly.
3. Recent data indicate that aggressive health education and behavior change programs can result in a positive health effect.

In this chapter, we will examine smoking as a risk factor for disease, and we will offer guidelines for designing occupational health promotion efforts directed towards helping people reduce or eliminate their smoking behavior.

SCOPE OF THE PROBLEM

Since 1964, when the U.S. Surgeon General first issued the warning that cigarette smoking could be hazardous to human health, 33 million smokers have quit smoking. In that same time span, over 30 million smokers have tried to quit but have failed. According to recent estimates, 70–90% of the 53 million adult smokers in this country actually want to stop. Many of these smokers want to stop smoking because they believe smoking is a risk factor that is jeopardizing their health. Current evidence supports such a conclusion.

According to the World Health Organization (WHO, 1979), smokers of all ages demonstrate higher death rates, for both sexes, compared to nonsmokers. The comparisons are striking. Table 6.1 compares the death ratios of smokers to nonsmokers for selected diseases.

In general, smokers have a 70% greater overall mortality rate from disease than do nonsmokers. Furthermore, it has been estimated that almost 400,000 deaths occur each year from cigarette smoking (Public Health Service, 1979).

The problem of smoking does not appear to be confined exclusively to smokers, however. Smoking affects nonsmokers. Research by the National Fire Data Center indicates that 13% of residential fires are attributable to careless smoking. According to the American Lung Association, "second-hand smoke" (smoke released into the air as a result of tobacco smoking) is a health hazard for nonsmokers. There are in excess of 1200 potentially toxic chemicals in tobacco smoke. Substances such as carbon monoxide, cadmium, benzopyrene, nicotine, pyridine, ammonia, hydrogen cyanide, nitrogen dioxide, formaldehyde, and hydrogen sulfide are all released into the air as tobacco burns. These substances are released in sufficient quantities as to cause irritation or represent significant health risks through the burning of only several cigarettes (American Lung Association, 1974).

Research has shown that nonsmokers, working in a room in which people are actively smoking, inhale about one third the amount of smoke that the smoker receives. These dose studies have been born out by epidemiologic investigations that show that people who live with smokers show a higher morbidity and mortality than would be expected among nonsmokers (PHS, 1979).

Smoking has been shown to be a significant risk factor for many factory workers, but there is also evidence that cigarette smoke may threaten office workers as well. Carbon monoxide from cigarette smoke may combine with other gases

Table 6.1 Comparison of Death Ratios of Smokers to Nonsmokers for Major Diseases

Disease	*Smoker to Nonsmoker Ratio*
Heart disease	approx. 2 to 1
Lung cancer	approx. 11 to 1
Stroke	approx. 3 to 1
Oral cancer	approx. 4 to 1
Bronchitis and other pulmonary disease	approx. 6 to 1
Ulcers	approx. 3 to 1

Source: U.S. Department of Health and Human Services.

found in office buildings such as formaldehyde (given off by many adhesives), ozone (released through photocopier use), and radon (released from building blocks). These gases may collectively represent a serious health risk in newer, energy-efficient "tight" buildings, that is, buildings with little air exchange activity (Moramarco, 1983; Makower, 1981).

In sum, it has been argued that the annual health damage resulting from smoking is about $27 billion. This cost is considered to be a function of worker absenteeism, decreased productivity, increased medical care, and accidents (Lichtenstein, 1982). For example, recent statistics (Oster, 1984) indicate that for individuals between the ages of 40–44 years who smoke more than two packs of cigarettes a day cost their companies over $55,000 in increased health costs over their work life compared to nonsmokers. If these same individuals quit, they save their companies over $30,000 over their work life.

In consideration of the arguments that (1) smoking is a costly and significant risk factor for human disease, and (2) the majority of smokers would like to stop smoking, there appears to emerge a strong rationale for the research and development of health promotion programs designed to assist individuals in reducing or terminating smoking behavior.

PHYSIOLOGICAL MECHANISMS OF ACTION[1]

In the previous section we discussed the problems potentiated by cigarette smoking. In this section we will discuss the specific physiological mechanisms by which tobacco smoking may affect the health of an individual. The three major agents associated with smoking and known to be responsible for significant alterations in human physiological functioning are carbon monoxide (CO), nicotine, and tar.

Carbon monoxide is a gaseous agent released as the tobacco burns. Once inhaled into the lungs, it quickly enters the blood stream forming a substance known as carboxyhemoglobin (COHb). COHb acts in the blood stream by lowering the heart's ventricular fibrillation (a potentially fatal, non-functional spasm of the heart during which no blood is pumped) threshold, and by displacing oxygen-carrying hemoglobin. This latter effect severely strains the heart muscle, causing it to pump harder and faster. Such a condition may act as a precursor of myocardial infarction (heart attack). The rise in COHb has been found to be greater for women than for men.

Nicotine is a caffeinelike sympathomimetic drug which is inhaled from tobacco smoke. Nicotine, once inhaled into the lungs, begins to affect brain tissue in about 8–10 seconds. Its pharmacological action includes rapid and intense stimulation of the autonomic nervous system. Its stimulant effects include an increase in

[1]The present discussion is based upon research reviews which appear in PHS (1979), Girdano and Girdano (1976), Guyton (1976), Aronow (1975), and Astrup and Kjeldsen (1979).

heart rate, peripheral vasoconstriction leading to an increase in blood pressure, enhanced blood platelet adhesiveness thus increasing thromboembolytic tendencies as well as arteriosclerotic tendencies, and a reduction of the ventricular fibrillation threshold thus increasing the chances of heart failure. Nicotine stimulates a discharge of catecholamine hormones to create protracted stimulant effects. Unlike other stimulants, however, nicotine is somewhat unique in that it is suspected of causing a potential synaptic blockade in certain nerve branches, particularly in the neuromuscular junctions. This blockade may be responsible for the decreased muscle tone and subjective feelings of relaxation reported by smokers. The half-life (the interval within which one half of the potency of the drug decays) is only about 20–30 minutes. So, to avoid the resultant mental and physical depression that might accompany this effect, many smokers will light another cigarette—and the chain smoking cycle may have then begun. It is interesting to note that nicotine is such a powerful drug, it has been used as an animal poison.

Finally, tar is generally defined as the combination of particulate matter (the components of smoke that make it visible to the human eye) and various gases. Some of the 1200 or more substances that make up the particulate matter are thought to be carcinogenic (capable of causing cancer), while the gases such as hydrogen cyanide, formaldehyde, ammonia, hydrogen sulfide, cadmium, and nitrogen dioxide are known to be toxic. The effects of the inhalation of tar appears to be an increase in the potential for cancer of the air passages and lungs and deactivation of the protective cilia (hairlike fibers) in the pulmonary networks. Such latter conditions may lead to pulmonary inflammation and chronic obstructive lung diseases.

In summary, the physiological mechanisms initiated by the inhalation of carbon monoxide, nicotine, and tar from smoking clearly seem to set the stage for a host of health-eroding processes to begin. Nevertheless, it is impossible to predict who will suffer smoking-related diseases and who will not. This fact, alone, may be one of the major reasons why people continue to smoke.

WHY PEOPLE SMOKE

In the design of any smoking reduction program, it becomes of value to attempt to understand, in some general form, why the person smokes. The factors that surround the habit of smoking are likely to be unique for each individual. However, we do have some insight into the most common patterns which initiate and sustain smoking behavior. The evolution of the smoking habit may be analyzed in four sets of factors: (1) factors that play a role in the initiation of the smoking habit; (2) factors that sustain the smoking habit; (3) factors that facilitate cessation of the smoking habit; and (4) factors that serve to resume the smoking habit (Lichtenstein and Brown, 1980; Bone, Phillips, and Chowdhury, 1981). These four sets of factors are summarized in Table 6.2.

In reviewing Table 6.2, the reader should pay particular attention to the sec-

Table 6.2 Factors Affecting the Evolution of the Smoking Habit

I. Factors that serve to initiate smoking:
   A. Curiosity
   B. Peer influence
   C. Anticipation of adulthood
   D. Rebelliousness
   E. Modeling provided by parents, peers, siblings, or other "significant others"

II. Factors that serve to motivate sustained smoking:
   A. Psychological support—some people, sometimes called "negative-effect" smokers, smoke to reduce stress and frustration
   B. Dependence—the "addictive smoker" smokes because he or she is psychologically and/or physically dependent upon cigarettes
   C. Pleasure—the "positive-effect smoker" smokes because smoking directly provides him or her pleasure
   D. Environmental cues—the "habitual smoker" smokes in response to environmental triggers without consciously desiring to smoke

III. Factors that facilitate smoking cessation:
   A. Health concerns
   B. Monetary expense
   C. Self-mastery
   D. Social pressure and support
   E. Aesthetics
   F. To serve as an example for others

IV. Factors that contribute to resumption of the smoking habit:
   A. Withdrawal effects
   B. Stress
   C. Social pressure
   D. Alcohol consumption
   E. Frustration
   F. Weight gain

tion, "factors which serve to motivate sustained smoking." Identification of the smoker's primary motivation(s) for smoking can serve as a basis upon which to individualize the treatment intervention. The "negative-effect smoker" is one who smokes to reduce stress, anxiety, and frustration. The "addictive smoker" is one who is dependent upon the smoking habit. Nicotine addiction is the most common form of addiction. The "positive-effect smoker" is one who derives pleasure from the smoking act itself. Pleasure may be derived from manipulating the cigarette, the taste of the cigarette after eating or drinking, blowing smoke rings, or adherence to some image portrayed through smoking, that is, smoking may be perceived as fashionable or "adult." The "habitual smoker" smokes in response to environmental cues.

Despite the seemingly linear processes involved in the evolution of the smoking habit, the problem is far more complex than that, however. Reviews by Leventhal and Cleary (1980), Stepney (1980), Lichtenstein (1982), and Blaney (1983) strongly support the notion that smoking is initiated and, particularly, maintained through a complex psychosocial-physiological paradigm.

It now appears as if psychosocial factors play a prepotent role in the initiation of the smoking behavior, however, psychophysiological factors (nicotine-related phenomena) seem to combine with cognitive and environmental factors to sustain the smoking habit. Reviews by Glasgow and Bernstein (1981) and Lichtenstein (1982) support the conclusion that the smoking habit is maintained in many individuals tbrough a strong nicotine dependency interacting with the smoker's cognitive evaluation that he or she is not personally at risk from cigarette smoking. These factors combine with environmental cues, or "triggers", that serve to promulgate the smoking habit. In the case of nicotine dependency, two major styles seem to exist: the "peak" smoker who smokes in order to receive stimulation at various peak intervals, and the "maintenance" smoker who smokes in order to create a maintenance level of nicotine in the blood supply, presumably to avoid withdrawal effects. Such dependence behaviors are often supported through the smoker's belief that he or she is simply not at risk for a smoking-related illness (although some smokers simply feel the benefits of smoking far outweigh the consequences of trying to stop). Finally, environmental cues can serve to trigger the urge to smoke a cigarette. Such cues include eating, consuming alcohol, being around others that smoke, or any stressful event.

Any smoking cessation program should attempt to attend to the complete psychosocial-physiological complex that sustains the smoking habit. To fail to do so increases the chances the individual will resume smoking.

GOALS OF A SMOKING CESSATION PROGRAM

By definition, the goal of a smoking cessation program is the cessation of the smoking behavior. Some programs, however, are smoking reduction programs, that is, they don't insist on total abstinence from smoking but rather attempt to get the participants to smoke less than one-half a pack of cigarettes a day.

Some evidence does exist (PHS, 1979) that continued smoking at a rate of 1–14 cigarettes a day only slightly increases the risk of coronary heart disease. However, the long-term effects of daily smoking at virtually any level appear to be less favorable when applied to pulmonary disease, especially if other pulmonary risk factors are present.

Smoking is a behavior that can be *directly* measured by self-report and by observation. Lichtenstein (1982) questions the reliability of data generated in such manners, however.

Smoking can also be *indirectly* measured through the assessment of CO levels, nicotine levels, and thiocyanate levels. These indirect assessment procedures are only useful to check for abstinence from smoking in a smoking cessation program. They are simply not sensitive enough to detect differential rates of smoking in a smoking reduction program.

CO can be used as a means of assessing smoking behavior because as was noted earlier, CO is inhaled in tobacco smoke. This gas can be measured in two ways: (1) invasively through blood samples of carboxyhemoglobin; and (2) noninvasively through CO samples of expired air. Two problems exist with the assessment of CO levels: (1) the half-life is only two to six hours, which makes it difficult to conduct assessments often enough to accurately reflect smoking behavior, and (2) CO levels can be elevated by behavior not related to active smoking, for example, inhaling air in heavy traffic or industrial environments and by inhaling second-hand smoke.

Nicotine levels may be assessed by way of the blood or the urine. The latter offers a practical advantage over the former in that it is noninvasive and yet reliable (Paxton and Bernacca, 1979).

Thiocyanate, a metabolite of cyanide found in tobacco, can be a useful measure of smoking abstinence. It can be sampled from either the blood or saliva. The half-life of thiocyanate is about two weeks, and the potential for nonsmoking related elevations is minimal.

Penchacek (1979) recommends combining the use of CO levels and thiocyanate as measures of smoking abstinence. Such measures may be used in addition to self-report and may serve to increase compliance. In the vast majority of cases, the goal of a health promotion program targeted towards smoking behavior should be termination of the smoking behavior as opposed to mere smoking reduction. In cases where total abstinence is impossible, only then should smoking reduction be considered as the goal. Once smoking cessation is achieved, the 1-week, 3-month, and 6-month postcessation intervals appear to be particularly important. Most of the nicotine withdrawal is over at 1 week; most people resume within 3 months of stopping; and by 6 months most life-style alterations around nonsmoking have been adjusted to. Follow-up evaluations should probably occur at 3-, 6-, and 12-month intervals, if possible.

REVIEW OF RESEARCH ON SMOKING CESSATION

Historically, smoking cessation programs have been unidimensional, that is, consisting of one main intervention. We will briefly review some of the more commonly used single intervention programs.

Stimulus control programs have focused upon the environmental cues that appear to trigger the smoking behavior. Stimulus control interventions require the smoker to keep a smoking diary for one week before any attempt at reducing the smoking habit actually begins. The smoker should record when they smoke each cigarette, what the urge to smoke was, where the smoking took place, and what thoughts or feelings were being experienced at the moment before the cigarette was smoked. The smoking log may be in the form of an index card which is wrapped around the cigarette pack (see Figure 6.1). Once these cues have been identified, the smoker gradually narrows the conditions under which smoking is permitted. When used by a person whose smoking behavior is "habitual," that is, caused predominantly by environmental cues, stimulus control procedures may be effective. However, if the motivation for smoking is a different primary factor or is multifactorial,

Date: ______________________

Cigarette Number	*Rate Your Urge to Smoke*	*Where Did You Smoke*	*Time*	*Thoughts/ Feelings Before You Smoked*

Rating for "urge to smoke": 1 = Very low urge; 2 = Low urge; 3 = Medium urge; 4 = Strong urge; 5 = Very strong urge.

Figure 6.1. Smoking diary. (Use one sheet/card for each day.)

stimulus control procedures alone will most likely prove ineffective (Bernstein and Glasgow, 1979).

Hypnosis, when used as a tool for smoking education or cessation, is typically used in one of three ways:

1. As a tool for forming aversive covert associations with the smoking behavior.
2. As a tool for facilitating the positive reinforcement of nonsmoking behavior.
3. As a means of reducing stress, thereby reducing "negative-effect" smoking behavior.

To date, research on the efficacy of hypnosis for smoking cessation has yielded contradictory results (Fredericken and Simon, 1979; Bernstein and Glasgow, 1979). Why hypnosis proves effective with only some individuals and under certain conditions may well be a function of individual differences. It is commonly accepted that hypnosis lies within the patient, not the therapist. Many resistant individuals simply will not allow themselves to experience the state we often call hypnosis. The use of hypnosis within occupational health promotion programs seems unadvisable due to its as yet unproven effectiveness and the traditionally conservative view that most organizations have regarding the use of "clinical" interventions.

Controlled smoking usually consists of two strategies: (1) the use of cigarettes that are low in tar and nicotine; and (2) alteration of the "smoking topography," that is, having the smoker learn to take shorter, shallower puffs while smoking. One concern that Lichtenstein (1982) raises is that merely employing low nicotine cigarettes may inadvertently force the "addictive smoker," that is, one dependent upon nicotine, to take deeper puffs and to retain the smoke longer. Therefore, if low nicotine cigarettes are used. they should be combined with a shorter, shallower inhalation pattern. Such alterations in smoking topography have been shown to be potentially effective as a "safer" form of smoking based upon preliminary data (Fredericksen and Simon, 1978). Controlled smoking strategies may be used in smoking reduction programs for smokers who do not wish to quit or simply cannot completely stop smoking.

Positive reinforcement procedures involve providing positively reinforcing consequences for nonsmoking or substitution (doing something instead of smoking when the urge arises) behavior. Although there is little controlled research, there is some evidence that positive reinforcement paradigms can be effective in smoking cessation programs (Blaney, 1983). Such procedures may be of particular value with the "positive-effect smoker." Such paradigms seemed to be enhanced through the use of behavioral contracting and buddy systems or social support networks (Lichtenstein, 1982).

Stress management interventions have been shown to be effective for facilitating smoking cessation (Ockene et al., 1981) in many instances. Stress appears to play a dual role in the smoking habit: (1) The ''negative-effect'' smoker smokes to reduce stress and frustration; by supplying alternative methods for meeting that goal, the need to smoke will be reduced if indeed the smoker is primarily a ''negative-effect'' smoker or if that factor plays a major role in the maintenance of the behavior; and (2) Stopping smoking is usually a stressful event; the present authors have observed, under clinical conditions, the value of stress management procedures in helping people cope with the stress of stopping smoking. See the stress management chapter within this text for guidelines. Also refer to Everly and Rosenfeld (1981).

Aversion strategies entail pairing smoking behavior with some aversive stimulus with the intent of having the smoking assume some of the aversive qualities of the paired stimulus. The technique of covert sensitization involves having the person imagine cigarette smoking being associated with some highly undesirable experience. The greater the person's ability to imagine such an event, the greater the likelihood of success. For this reason hypnosis is often used. Overt sensitization involves having the person actually experience some aversive stimuus associated with the smoking act. One common technique is to mix water together with cigarettes and cigarette ashes. Then place the mixture in a jar. The smoker would take an inhalation of the mixture at periodic intervals. When finished smoking a cigarette, the person should place the remains in the jar and take another inhalation. Soon the aversion of inhaling the mixture will generalize to the smoking act itself in many individuals. This will serve to diminish the desire to smoke. Rapid smoking is another kind of aversion technique. Rapid smoking involves having the smoker inhale cigarette smoke every 6 to 8 seconds until complete satiation occurs.

While available research data indicates that aversive strategies can be effective (Lichtenstein and Rodrigues, 1977; Hall, Sachs, and Hall, 1979; Lichtenstein, 1982), they may present problems that make their use in occupational health promotion programs undesirable (Hall, Sachs, and Hall, 1979; Blaney, 1983; and Lichtenstein, 1982, for reviews). This is especially true in the case of rapid smoking. Rapid smoking is contraindicated for any individual with pulmonary disease or dysfuncion. This requires medical screening before rapid smoking can be used. In addition, even in healthy individuals, rapid smoking increases blood nicotine levels, catecholamine discharge, and carboxyhemoglobin, each of which represents a potential health risk.

In their review of smoking behavior, Bone, Phillips, and Chowdhury (1981) argue that quitting ''cold turkey'' is the best way to help the ''addictive smoker'' quit smoking. They argue that simply stopping altogether is superior to gradually reducing the number of cigarettes smoked each day. Gradual reduction runs the risk of being far more stressful and frustrating to the nicotine-dependent smoker because it stretches the nicotine deprivation withdrawal out over a longer period of time, sometimes weeks or months. Stopping "cold turkey" has the advantage of minimizing the nicotine deprivation. Most of the nicotine withdrawal will be over in about a week if one stops smoking completely. For those smokers who cannot break the nicotine habit, but don't want to gradually reduce the number of cigarettes smoked each day, nicotine injections or nicotine chewing gum may be effective. Both require medical application, however (Lichtenstein, 1982; Blaney, 1983).

Multifactorial programs are programs that consist of more than one therapeutic intervention. Traditional unidimensional programs for smoking cessation have a rather poor success record at six-month follow-up evaluations. Most programs report a 15%–40% success rate when success was defined as smoking abstinence (Lichtenstein, 1982). However, programs that have employed multiple components have reported 40 to 70% success rates after six months (Glasglow and Bernstein, 1981). Frequently used combinations of strategies include stress management, stimulus control, and positive reinforcement. Conceptually, the multifactorial program is a far superior program. Within its combinations and permutations lie flexibility enough to satisfy the individual needs that may be numerous and diversely unique to participants.

In concluding this section, the smoking habit appears to be a multifactorial one consisting of a complex constellation of psychosocial and physiological variables. The history of single intervention smoking programs has been disappointing, while multifactorial programs appear far more promising.

As noted in earlier chapters, multifactorial programs give the program coordinator the flexibility to address the individual needs of individual participants while still using a cost-effective group format. As also noted in earlier chapters, programs will maximize their effectiveness if they are based upon the principles of learning, rather than just employing a few behavioral techniques.

The section that follows describes a multifactorial program for smoking cessation developed by the present authors.

GUIDELINES FOR PROGRAM DEVELOPMENT

In this section we will discuss two multifactorial programs for smoking cessation. One program employs a method for abruptly quitting ("cold turkey"), and is summarized in Table 6.3. The other program employs a method for the gradual reduction of smoking until cessation is achieved; this program is summarized in Table 6.4.

Both multifactorial programs outlined in this section may be divided into three stages:

Stage I: Preparation

Stage II: Quitting

Stage III: Maintenance

Preparation

The preparation stage of both approaches to smoking cessation is the same. In the opinion of the present authors, the failure of many smoking cessation programs can be traced back to their failure to appreciate the individual differences among participants. This problem may be corrected by the use of a baseline assessment procedure. This procedure consists of three steps:

1. Creation of a "reasons for smoking"-"reasons for quitting" balance sheet. On the left side of a sheet of paper have participants list the reasons they smoke. On the right side of that same sheet, have them list their reasons for wanting to quit now, at this point in time. Try to discover why the motivation is stronger now than it was before, or why they think they can be successful now as opposed to previous times.
2. Assessment of environmental and cognitive/affective cues which seem to trigger the smoking behavior. This can be done through the use of a baseline collection sheet such as that in Figure 6.1, for a period of one week.
3. Assessment of factors that sustain the smoking habit (refer to Table 6.2, Section II). By identifying the primary motivation(s) for smoking, that is, what type of smoker the person is (e.g., addictive, negative effect, etc.), the factors that play the largest role in sustaining the smoking habit may be directly confronted. The reader should see Ikard, Green, and Horn (1969), American Lung Association (1980), and Bone, Phillips, and Cowdhury (1981) for assessment tools that can be used to identify smoker "types."

Once all of the baseline data have been collected, this information should be discussed with each participant, either individually or in a group format.

In addition to the baseline evaluations, each participant should make a written declaration of their intent to stop smoking. This document should be copied and publicly displayed at home and at work. Behavioral contracting procedures may be very useful here.

Finally, each participant should enlist the aid of a social support network. A group format can be used, or a buddy system, or both. For example, a buddy can call the exsmoker, daily, to see how he or she is doing and offer support and/or encouragement.

Quitting

There are two basic ways to stop smoking:

1. Quitting "cold turkey" (Table 6.3).
2. Gradually reducing the number of cigarettes smoked until cessation is achieved (Table 6.4).

Table 6.3 A Multifactorial Program for Quitting "Cold Turkey"

Stage I—Preparation for quitting
- A. Baseline evaluations (one week):
  1. Creation of a smoking balance sheet
  2. Stimulus control assessment of smoking triggers
  3. Assessment of factors that sustain the smoking habit (see Table 2, Section II)
- B. Written declaration of intent to stop smoking
- C. Creation of social support network

Stage II—Quitting
- A. Stress management
- B. Use of target date to quit

Stage III—Maintenance
- A. Use of social support
- B. Use of substitute behaviors
- C. Aerobic exercise program

Table 6.4 A Multifactorial Program for the Gradual Reduction of Smoking

Stage I—Preparation for quitting
- A. Baseline evaluations (one week)
  1. Creation of a smoking balance sheet
  2. Stimulus control assessment of smoking triggers
  3. Assessment of factors that sustain the smoking habit (see Table 2, Section II)
- B. Written declaration of intent to stop smoking
- C. Creation of a social support network

Stage II—Quitting
- A. Rate of reduction set
- B. Stress management
- C. Use of social support network
- D. Use of substitute behaviors
- E. Stimulus narrowing
- F. Controlled smoking
- G. Aversive ashtray technique

Stage III—Maintenance
- A. Continued use of social support
- B. Continued use of substitute behaviors
- C. Aerobic exercise

Before quitting "cold turkey," each participant should be exposed to stress management procedures for reasons discussed in an earlier section. This is especially true for the negative-effect and addictive smokers. It is particularly helpful if participants can be forewarned of potential adversities, particularly relevant to withdrawal.

Once participants have been prepared, they should pick a target date, then quit. Group discussions and encouragement can be useful here. Quitting "cold turkey" seems most desirable for the nicotine-dependent smoker, as noted earlier.

If quitting "cold turkey" seems unadvisable, the gradual reduction of smoking may be adopted. Numerous options are available and can be used in various combinations. First, however, it is usually advisable to decide at what pace, or rate, smoking reduction will occur. This should be open for change, however.

It has been reported (Blaney, 1983) that some individuals who try to gradually reduce their smoking habit reach a sticking point at around one-half pack per day. To help the smoker, stress management, stimulus control, and social support may be useful.

When it comes to actually cutting down the number of cigarettes smoked each day, the baseline data taken earlier will be of value. It is well known that behavior that is repeated is in some way reinforcing. One way to eliminate an undesirable behavior (e.g., smoking) is to substitute some desirable behavior in lieu of the undesirable one. Have the smoker go back to the "reasons for smoking" column on the balance sheet created earlier and also refer to the assessment of factors that sustain smoking. Then have the smoker attempt to generate a list of desirable behaviors that can substitute for the undesired smoking, that is, behaviors that can literally take the place of smoking by providing the same, or similar, reinforcements. Substitutions can be particularly effective with the "positive-effect smoker." Other strategies for reducing the number of cigarettes smoked would include stimulus narrowing, controlled smoking, and aversion techniques.

Stimulus narrowing is a technique based upon the identification of the cues that trigger smoking. Gradually, the smoker reduces the number of cues that he or she will allow to trigger smoking behavior. For example, smoking may only be allowed in one room of the house, or only at certain times, and so on. Stimulus narrowing can be particularly effective with the "habitual smoker."

Controlled smoking would include choosing low tar, low nicotine cigarettes (see Table 6.5 for approximations of tar and nicotine), as well as alterations in smoking topography (Fredericksen and Simon, 1978, for guidelines).

An aversion technique useful for occupational health promotion programs is the "aversive ashtray" technique. Ashtrays should be removed from home and work settings. They should be replaced by jars, with lids, which contain mixtures of water, cigarettes, and cigarette ashes. Each time the smoker lights a cigarette, the lid is removed, and the jar full of mixture serves as the ashtray. When finished, the cigarette butt is placed in the jar, and then the jar is resealed until the next cigarette.

Table 6.5 Approximate Tar and Nicotine Components in Cigarettes

Brand	*Type*	*Tar (mg/cig)*	*Nicotine (mg/cig)*
Now	King size, filter (hard pack)	0.5	0.05
Carlton	King size, filter (hard pack)	0.5	0.05
Now 100s	100 mm, filter (hard pack)	0.5	0.05
Cambridge	King size, filter (hard pack)	0.5	0.05
Carlton 100s	100 mm filter, menthol (hard pack)	0.5	0.1
Carlton	King size, filter menthol	0.5	0.1
Carlton 100s	100 mm, filter (hard pack)	1	0.1
Barclay	King size, filter	1	0.1
Cambridge	King size, filter	1	0.1
Tareyton Ultra Low Tar	King size, filter menthol	1	0.1
Now	King size, filter menthol	1	0.1
Benson & Hedges	Regular size, filter (hard pack)	1	0.1
Carlton	King size, filter	1	0.1
Now	King size, filter	1	0.1
Barclay	King size, filter menthol	1	0.1
Barclay	King size, filter (hard pack)	1	0.2
Kool Ultra	King size, filter menthol	2	0.2
Now 100s	100 mm, filter menthol	2	0.2
Now 100s	100 mm, filter	2	0.2
Triumph	King size, filter	2	0.3
Kent III	King size, filter	2	0.3
Iceberg 100s	100 mm, filter menthol	3	0.3
Triumph	King size, filter menthol	3	0.4
Barclay 100s	100 mm, filter	3	0.3
Lucky 100s	100 mm, filter	3	0.3
Cambridge 100s	100 mm, filter	3	0.3
Barclay 100s	100 mm, filter menthol	3	0.3
Merit Ultra Lights	King size, filter menthol	3	0.3
Merit Ultra Lights	King size, filter	4	0.3
Winston Ultra	King size, filter	4	0.4
Triumph 100s	100 mm, filter menthol	4	0.5
Carlton 100s	100 mm, filter menthol	4	0.4
Doral II	King size, filter menthol	4	0.4
Carlton 100s	100 mm, filter	4	0.4
Vantage Ultra Lights	King size, filter	4	0.4
Kool Ultra 100s	100 mm, filter menthol	4	0.4
Doral II	King size, filter	4	0.4
Vantage Ultra Lights 100s	100 mm, filter	4	0.4
Triumph 100s	100 mm, filter	4	0.5

Table 6.5 (*Continued*)

Brand	*Type*	*Tar (mg/cig)*	*Nicotine (mg/cig)*
Salem Ultra	King size, filter menthol	5	0.4
Tareyton Lights	King size, filter	5	0.4
True	King size, filter	5	0.4
Salem Ultra 100s	100 mm, filter menthol	5	0.4
Kent III 100s	100 mm, filter	5	0.5
True	King size, filter menthol	5	0.4
Winston Ultra 100s	100 mm, filter	5	0.4
Decade	King size, filter menthol	5	0.4
Decade	King size, filter	5	0.5
Carlton 120s	120 mm, filter	6	0.6
Carlton 120s	120 mm, filter menthol	6	0.6
Kool Super Lights	King size, filter menthol	6	0.5
Pall Mall Extra Light	King size, filter	6	0.6
True 100s	100 mm, filter	7	0.6
Tareyton Long Lights 100s	100 mm, filter	7	0.6
True 100s	100 mm, filter menthol	7	0.6
More Lights 100s	100 mm, filter menthol (hard pack)	8	0.6
Omni 100s	100 mm, filter menthol	8	0.7
Camel Lights	King size, filter (hard pack)	8	0.7
L & M Lights 100s	100 mm, filter menthol	8	0.8
Merit	King size, filter menthol	8	0.6
Golden Lights	King size, filter	8	0.7
Virginia Slims Lights 100s	100 mm, filter menthol (hard pack)	8	0.6
Camel Lights	King size, filter	8	0.7
Decade 100s	100 mm, filter	8	0.7
L & M Lights 100s	100 mm, filter	8	0.8
Merit	King size, filter	8	0.6
Northwind	King size, filter menthol	8	0.6
Saratoga 120s	120 mm, filter (hard pack)	15	1.0
Virginia Slims 100s	100 mm, filter	15	1.0
Marlboro	King size, filter menthol (hard pack)	15	0.9
Winston	King size, filter	15	1.0
Oasis	King size, filter, menthol	15	1.0
Chesterfield 100s	100 mm, filter	15	1.1
Viceroy Super Long 100s	100 mm, filter	15	1.0
L & M 100s	100 mm, filter	15	1.1

(*continued*)

Table 6.5 (*Continued*)

Brand	*Type*	*Tar (mg/cig)*	*Nicotine (mg/cig)*
Montclair	King size, filter menthol	15	1.0
Virginia Slims 100s	100 mm, filter menthol	15	1.0
Lark	King size, filter	15	1.1
Long Johns 120s	120 mm, filter menthol	15	1.3
Camel	King size, filter	15	1.0
Saratoga 120s	120 mm, filter menthol (hard pack)	15	1.0
Newport	King size, filter menthol (hard pack)	16	1.1
Philip Morris International 100s	100 mm, filter, menthol (hard pack)	18	1.0
More 120s	120 mm, filter menthol	16	1.2
Galaxy	King size, filter	16	1.0
Raleigh	King size, filter	16	1.0
DuMaurier	King size, filter (hard pack)	16	1.0
Benson & Hedges 100s	100 mm, filter (hard pack)	16	1.1
Benson & Hedges	King size, filter (hard pack)	16	1.2
Lark 100s	100 mm, filter	16	1.2
Kool	King size, filter menthol (hard pack)	16	1.1
Benson & Hedges 100s	100 mm, filter menthol (hard pack)	16	1.1
Pall Mall 100s	100 mm, filter	16	1.2
Winston	King size, filter (hard pack)	16	1.0
Marlboro	King size, filter (hard pack)	16	1.1
Benson & Hedges 100s	100 mm, filter menthol	17	1.1
Kool	King size, filter menthol	17	1.1
Tall 120s	120 mm, filter menthol	17	1.3
Raleigh 100s	100 mm, filter	17	1.0
Newport	King size, filter menthol	17	1.2
Benson & Hedges 100s	100 mm, filter	17	1.1
Marlboro 100s	100 mm, filter	17	1.1
Marlboro	King size, filter	17	1.1
Marlboro 100s	100 mm, filter (hard pack)	17	1.1
Old Gold Filter	King size, filter	17	1.3
More 120s	120 mm, filter	17	1.2
Philip Morris International 100s	100 mm, filter (hard pack)	18	1.1
Half & Half	King size, filter	18	1.4
Pall Mall	King size, filter	18	1.2
Picayune	Regular size, nonfilter	18	1.2

Table 6.5 (*Continued*)

Brand	*Type*	*Tar (mg/cig)*	*Nicotine (mg/cig)*
Winston International 100s	100 mm, filter (hard pack)	18	1.3
Max 120s	120 mm, filter menthol	18	1.5
Kool	Regular size, nonfilter, menthol	18	1.0
Tall 120s	120 mm, filter	19	1.5
Max 120s	120 mm, filter	19	1.5
Long John 120s	120 mm, filter	19	1.5
Old Gold Filters	100 mm, filter	20	1.5
Spring 100s	100 mm, filter menthol	20	1.1
Newport 100s	100 mm, filter menthol	20	1.4
Camel	Regular size, nonfilter	20	1.3
Chesterfield	Regular size, nonfilter	20	1.3
Philip Morris	Regular size, nonfilter	22	1.4
Raleigh	King size, nonfilter	23	1.3
English Ovals	Regular size, nonfilter (hard pack)	23	1.7
Lucky Strike	Regular size, nonfilter	24	1.5
Chesterfield	King size, nonfilter	24	1.6
Pall Mall	King size, nonfilter	24	1.6
Players	Regular size, nonfilter (hard pack)	25	1.8
Old Gold Straight	King size, nonfilter	26	1.6
Philip Morris Commander	King size, nonfilter	26	1.7
English Ovals	King size, nonfilter (hard pack)	27	2.0
Herbert Tareyton	King size, nonfilter	28	1.8
Bill Durham	King size, filter	30	2.0

The aversive smell will soon become associated with the act of smoking. It is important that smoking only be allowed when the aversive ashtray is used, so it may be necessary to have several aversive ashtrays in use.

Maintenance

While many people quit smoking, historically less than half have maintained their abstinence. During this stage, behavioral contracting, positive reinforcement of substitute behaviors, and the use of social support networks become critical.

Recently, the use of aerobic exercise has found its way into smoking cessation programs. If aerobic exercise programs are safe and available, they will do much to reduce recidivism.

Summary

In concluding this chapter on smoking cessation, several points are important and should be emphasized here.

Historically, single-factor smoking cessation programs have not proven very effective with six-month abstinence rates ranging from 15 to 40%. Multifactor programs offer a far greater potential. The reason may be that multifactor programs are broad enough to meet many of the idiosyncratic needs that individuals will bring to such a program. We believe that smoking cessation programs should be individually tailored as much as possible while still taking advantage of the social support mechanisms inherent in group processes.

The programs outlined in this chapter are merely potential guidelines. The specific program must be developed by each program coordinator based upon the needs and resources of each program. It is our belief, however, that smoking cessation should be the goal of most programs, when compared to smoking reduction. Yet smoking reduction may offer a reasonable goal when smoking cessation appears to be out of the question. In such cases, one-half pack per day should be the approximate target for participants who don't want to quit but simply want to reduce their smoking habit.

As a final note, we should mention the legalization of nicotine gum in the United States. Nicotine gum has been used in Europe for several years to assist nicotine-addicted smokers in achieving smoking cessation. In 1984, the FDA granted permission for its use in the United States for simlilar purposes. Many consider this a new ray of hope for the chronic, nicotine-addicted smoker who desires to stop smoking. Each piece of gum will contain 2 mg of nicotine. It is expected that smokers will chew from 5 to 10 pieces of gum each day. Studies conducted using the combination of nicotine gum and behavior modification strategies (such as those described in this chapter) have been found to double the success rate in smoking cessation programs.[2] Currently, nicotine gum can only be obtained by prescription, yet its proper utilization in combination with behavioral programs could offer the most powerful tool yet for achieving smoking cessation. Future data will tell.

RESOURCE GUIDE FOR SMOKING CESSATION

Organizations

Action on Smoking and Health (ASH)
2013 H Street NW
Washington, DC 20006
(202)659-4310

American Lung Association
1740 Broadway
New York, NY 10019
(212)245-8000

[2]Data submitted to the FDA by Merrill-Dow Pharmaceutical Co., 1983.

American Cancer Society (ACS)
777 Third Avenue
New York, NY 10017
(212)371-2900

Five Day Plan to Stop Smoking
6830 Laurel Street NW
Washington, DC 20012
(202)723-0800

Smokenders
Phillipsburg, NJ (National Headquarters)

Journals

Addictive Behaviors

American Journal of Public Health

Behavior Research and Therapy

Behavior Therapy

Chest

Health Education

Health Education Monographs

International Journal of Addictions

Journal of Consulting and Clinical Psychology

Journal of Respiratory Diseases

REFERENCES

American Lung Association. *Freedom From Smoking: Clinic Leader's Guide*. Washington, D.C.: American Lung Association, 1980.

American Lung Association. *Second-Hand Smoke*. Washington, D.C.: American Lung Association, 1974.

Aronow, W. S. ''Cigarette smoking, carbon monoxide, nicotine, and coronary disease.'' *Preventive Medicine,* 1975, **4,** 95–99.

Astrup, P. and Kjeldsen, K. ''Model studies linking carbon monoxide and/or nicotine to arteriosclerosis and cardiovascular disease.'' *Preventive Medicing,* 1979, **8,** 295–302.

Bernstein, D. and Glasgow, R. ''Smoking: The modification of smoking behavior.'' In *Behavioral Medicine,* O. F. Pomerleau, ed. Baltimore: Williams & Wilkins, 1979.

Blaney, N. ''Smoking: Psychophysiological Causes and Treatments.'' In *Behavioral Medicine,* N. Schneiderman and J. Tapp, eds. Hillsdale, N.J.: Lawrence Erlbaum Associates, 1983.

Bone, R., Phillips, J., and Chowdhury, P. ''The smoking habit: Physical dependence on nicotine.'' *The Journal of Respiratory Diseases,* May 1981, 10–16.

Borgatta, E. and Evans, R., eds. *Smoking, Health, and Behavior*. Chicago: Aldine, 1968.

Everly, G. and Rosenfeld, R. *The Nature and Treatment of the Stress Response*. New York: Plenum, 1981.

Fredericksen, L. and Simon, S. ''Clinical modification of smoking behavior.'' In *Modification of Pathological Behavior*. R. S. Davidson, ed. New York: Gardner Press, 1979.

Fredericksen, L. and Simon, S. ''Modification of smoking topography: A preliminary analysis.'' *Behavior Therapy,* 1978, **9,** 946–949.

Girdano, D. and Girdano, D. *Drugs: A Factual Account*. Reading, Mass.: Addison-Wesley, 1976.

Glasgow, R. and Bernstein, D. "Behavioral treatment of smoking behavior." In *Medical Psychology,* C. Prokop and L. Bradley, eds. New York: Academic Press, 1981.

Guyton, A. *A Textbook of Medical Physiology.* Philadelphia: Saunders, 1976.

Hall, R., Sachs, D., and Hall, S. "Medical risk and therapeutic effectiveness of rapid smoking." *Behavior Therapy,* 1979, **10,** 249–259.

Ikard, F., Green, D., and Horn, D. "A scale to differentiate between types of smoking as related to the management of affect." *International Journal of Addictions,* 1969, **4,** 649–659.

Levanthal, H. and Cleary, P. "The smoking problem: A review of the research and theory in behavioral risk modification." *Psychological Bulletin,* 1980, **88,** 370–405.

Lichtenstein, E. "The smoking problem: A behavioral perspective." *Journal of Consulting and Clinical Psychology,* 1982, **50,** 804–819.

Lichtenstein, E. and Brown, R. "Smoking cessation methods: Review and recommendations." In *The Addictive Behaviors,* W. R. Miller, ed. Oxford: Pergamon Press, 1980.

Lichtenstein, E. and Rodrigues, M.-R. "Long-term effects of rapid smoking treatment for dependent cigarette smokers." *Addictive Behaviors,* 1977, **2,** 109–112.

Makower, J. "Office work may be hazardous to your health." *The Washingtonian,* Sept. 1981, 223–229.

Moramarco, S. "Does your office make you sick?" *Review,* March 1983, 43–48.

Ockene, J., Nuttall, R., Benfari, R., Hurwitz, I., and Ockene, I. "A psychosocial model of smoking cessation and maintenance of cessation." *Preventive Medicine,* 1981, **10,** 623–638.

Oster, G., Colditz, G., and Kelly, N. Economic Costs of Smoking and Benefits of Quitting. Lexington, Mass.: Lexington Books, 1984.

Paxton, R. and Bernacca, G. "Urinary nicotine concentration as a function of time since last cigarette." *Behavior Therapy,* 1979, **10,** 523–528.

Penchacek, T. "Modification of smoking behavior." In *Smoking and Health.* Washington, D.C.: U.S. Government Printing Office, 1979.

Public Health Service. Report on Smoking and Health. Washington, D.C.: U.S. Government Printing Office, 1979.

Stepney, R. "Smoking Behavior: The psychology of the cigarette habit." *British Journal of Diseases of the Chest,* 1980, **74,** 325–344.

World Health Organization. "Controlling the Smoking Epidemic." Technical Report Series 636. Geneva: WHO, 1979.

CHAPTER 7

Health Behavior Strategies for the Control of High Blood Pressure at the Worksite

Donald E. Morisky

INTRODUCTION

True or false: People who have hypertension (high blood pressure) are often nervous or tense. Thin people never suffer from high blood pressure. You can tell you have high blood pressure by the way you feel. High blood pressure will go away if you take the proper medication.

All of the above statements are false, as you may have guessed. But, interestingly enough, a majority of recently surveyed individuals responded affirmatively to several or all of these statements. High blood pressure is a significant problem for working people and their employers, mainly because there are so many myths surrounding it. One of the biggest myths is that you can tell if you have it. Unfortunately, that is not true because, in the majority of cases, there are no outward signs or symptoms. Persons often can be unaware of their elevated blood pressure state for years because they do not feel sick.

SCOPE OF THE PROBLEM

According to data available from the U.S. Department of Health and Human Services, high blood pressure is a major public health problem in the United States, one that affects as many as 36 million Americans (Roberts, 1981). This amounts to more than 40% of the work force. Its significance as a major health hazard has been well documented. Studies from the National Heart, Lung, and Blood Institute have shown that individuals with high blood pressure develop approximately three times the coronary heart disease, six times the congestive heart failure, and seven times the stroke as do individuals with controlled or normal blood pressure (National Heart, Lung, and Blood Institute, 1980). This results in an estimated cost to employers of over four billion dollars a year in absenteeism, decreased productivity, disability claims, and death benefits.

What exactly is hypertension? The U.S. Public Health Service defines it as a condition in which the force exerted against the walls of the vessels by the blood flowing through them goes up too high and stays there. Normally, blood pressure goes up when the heart beats faster—during exercise or tension, for example—but it drops back to normal afterward. In a person with hypertension, the pressure may be excessive for years, causing the body's arteries to age prematurely. Eventually, this aging may lead to a sudden disruption of the blood supply going to the brain (a stroke), the heart (a myocardial infarction), or the kidneys (renal failure).

There are several theories about what causes high blood pressure, but no one knows for sure. One hypothesis is that the sympathetic nervous system, which is responsible for maintaining normal blood-vessel tone, overreacts, resulting in increased peripheral resistance (Genest et al., 1978). Another is that hypertension is caused by emotional stress—that people with high blood pressure are more excitable and tense, and tend to suppress their emotions, straining their hearts and circulatory systems (Selye, 1956; Wolf et al., 1955; Sokolow et al., 1961). Critics of this theory point out that many calm, easygoing people have hypertension, while many tense, hard-driving people have normal or even low blood pressure (Ostfeld and Lebovits, 1959). Another explanation is advanced by Laragh (1973) who did a pioneering analysis of the kidney's role in hypertension. Laragh found that a malfunction in the kidney's control mechanism sets off a chemical chain reaction that causes blood vessels to constrict and the body to retain more sodium, increasing total body fluid. Although several theories concerning the etiology of essential hypertension have been presented, the current understanding of the physiology of arterial pressure regulation indicates that the renal-body fluid volume system determines the level at which the mean pressure resides over long periods of time. The relationship between blood volume, and size and compliance of the entire vascular system, and intrinsic regulation of tissue blood-flow determine the sequence of observed changes in cardiac output and total peripheral resistance.

Finding out whether you have high blood pressure is easy and painless. A nurse or medical technician wraps a fabric-covered cuff around the person's arm just above the elbow, inflates the cuff until blood stops flowing through the artery in the arm and gradually releases the air. The average of the second and third readings generally constitute the measure. The systolic reading (the first recording) measures the increased pressure on the blood vessels when the heart muscle contracts, and the diastolic reading (the second recording) occurs when the heart muscle relaxes and the pressure drops to a minimum. Although there continues to be a great deal of controversy regarding the level at which the blood pressure is considered to be elevated, it is generally agreed that individuals with a sustained systolic/diastolic pressure 140 over 90 is considered elevated (NCHS, 1981; OHDFP, 1979).

PROBLEMS OF AWARENESS, TREATMENT, AND CONTROL

In the early 1970s, approximately 50% of the individuals who had an elevated blood pressure were aware of their condition; 50% of those individuals aware were on some form of medical treatment; and only 50% of those on treatment achieved adequate blood pressure control. This resulted in only 12% of the population with the disease having a controlled blood pressure. The nature of the problem of high blood pressure has changed substantially since that time. Today more than 80% of those afflicted are aware of their condition and more than two thirds are receiving treatment. A remaining problem, however, is the failure of the health care system to maintain long-term blood pressure control following treatment. Still approximately half of the patients who begin treatment for their high blood pressure do not remain under care or do not comply adequately to achieve blood pressure control.

The physicians have at their disposal a broad spectrum of both pharmacological and behavioral modalities that can be used to control blood pressure and improve health status. During the past 10 years, studies have demonstrated that more than 80% of persons with hypertension can have their blood pressure controlled by adhering to the medical recommendations of their health care provider (Moser, 1982). Yet, despite medical advances in the detection, treatment, and control of high blood pressure, large proportions of those with hypertension are not achieving blood pressure control.

BEHAVIORAL STRATEGIES FOR THE PATIENT SETTING

In an effort to determine the reasons for these low rates of control among diagnosed hypertensive outpatients, the National Heart, Lung, and Blood Institute sponsored several studies to identify cost-effective strategies for improving patients compliance (McGill, 1978). Although not all the studies have been published, several studies have identified common characteristics that support the contention that short-term improvement in blood pressure control can be achieved by almost any educational strategy based on a well-founded needs assessment of both patients and providers of care. Alderman and his colleagues (1982) have summarized the research findings from these studies, all of which could be easily implemented and tested in an occupational setting. Briefly, those educational strategies found to be effective in improving blood pressure control were as follows:

1. Increasing the amount of contact time for discussion and provider-patient interaction.
2. Actively involving the patient in the management of high blood pressure control, for example, goal setting for blood pressure control or behavioral change.

3. Involving a significant other person to provide ongoing social support and reinforcement in the home environment.
4. Self-monitoring of blood pressure.

Each of these educational strategies have been found to be effective in improving and maintaining adherence to the drug regimen, the major problem in the prolonged control of blood pressure. Successful methods to achieve adherence over long periods of time are not yet certain, but the importance of ongoing social support in the home environment has received considerable attention. In fact, in one of the previously mentioned reviews, the involvement of a significant other in the management of hypertension outside the physician's office accounted for a significant proportion of the variability for having one's blood pressure in control (Morisky et al., 1983). Hypertensive patients in an outpatient clinic were assigned to either experimental groups in which they identified a person (usually a spouse of a family member) who would provide social support in the home or to a control group in which they continued to receive standard medical care but no social support. The social support person received an educational counseling session in which specific behavioral objectives were identified and written as to how he or she would assist the hypertensive patient in treating their high blood pressure. The main message of the educational intervention was, "The patient needs your assistance," and this was reflected in several ways. Specific behaviors were suggested to the family member and he or she was encouraged to make a commitment to carry them out. For example, some family members lacked information concerning the need for a daily therapeutic regimen in the absence of any obvious symptoms. Once they understood the nature and sequellae of high blood pressure, they began to identify ways in which they would help remind the hypertensive patient to take his or her daily medication. Individuals assigned to the family support intervention were significantly more likely to have their blood pressure under control (77%) compared to individuals not assigned to this intervention (51%; $p < .001$).

Follow-up data on blood pressure control levels were obtained on the patient population three years following the conclusion of the educational program. Study patients assigned to any of the educational interventions demonstrated a statistically significant 65% increase in blood pressure control (from 40% to 66%) over the five-year period. Patients who were not assigned to any of the educational interventions but who continued to receive standard medical care displayed a nonsignificant 22% increase (from 41% to 50%) in the proportion having their blood pressure under control. Statistically significant differentials in all-cause as well as hypertension-related mortality using a five-year life table analysis were also noted for the group of patients assigned to the educational program. A 53% reduction in hypertension-related mortality was found in the group assigned to the educational program at the five-year follow-up (Morisky et al., 1983).

BEHAVIORAL STRATEGIES FOR THE PROVIDER OF HEALTH CARE

The hypertensive patient has not been the only target to which educational strategies have been directed. The Working Group to Define Critical Patient Behaviors in High Blood Pressure Control (1979) has also developed guidelines for professionals to incorporate into their practices. The health professional who is involved in the initial diagnosis and treatment of high blood pressure sets the stage for subsequent interaction. Several studies (Reeder, 1972; Francis, Korsch and Morris, 1969; Svarstad, 1976; Wolinsky and Steiber, 1982; Inui et al., 1982) have documented the importance of the provider-patient interaction and its impact on health behavior and medical outcomes. The construct is based on the hypothesis that active participation by the patient will favor successful management of high blood pressure. Each patient must make the decision at this critical point in time to begin and continue the proposed treatment recommendations. If patients view themselves as active partners in treatment rather than passive recipients, then continued control of blood pressure should result through an increased sense of responsibility, interaction, and self-worth. It is incumbent upon the provider to clearly define what is expected of both parties at the outset and to ensure that continued reinforcement through physicians, nurses, pharmacists, employees, family, and others is offered to help maintain adequate blood pressure control. When the patient is involved in the decision as to how he or she will keep his or her blood pressure under control, (1) the patient knows that the ultimate responsibility lies within himself or herself, and (2) the health care provider has the responsibility of advising and assisting the patient in assuming this role in the decision-making process.

In assisting the patient to adhere to the medical recommendations, it is important for the provider to communicate why it is important for the patient to follow the treatment plan, to emphasize the chronic nature of the disease, and to correct any misunderstandings before the patient leaves the office. In a recent survey, it was noted that approximately one third of all individuals who were ever on antihypertensive medication had discontinued treatment. When probed as to why they had stopped, over 70% responded that either my doctor told me to stop (49%), I no longer have high blood pressure (12%), or I no longer need to take medication. The majority of such individuals had elevated blood pressure readings (Morisky et al., 1980). These results indicate that educational efforts should also be directed toward providers aimed at decreasing misunderstanding and misconceptions through improved provider-patient interaction and education. In this same survey, approximately 39% of those patients currently taking antihypertensive medication indicated some problem with noncompliance, with 15% complying poorly. The greater variability with respect to high compliance is seen to be in age, with the older hypertensive population having significantly higher levels compared to their younger counterparts. This finding would lead us to recommend that patient education approaches in regard to long-term compliance behavior should be directed primarily at reinforcing high compliance behaviors in the older groups and increasing compliance behavior in younger individuals.

Monitoring progress toward blood pressure control is essential for both the patient and the provider. For the patient each subsequent visit represents an opportunity to measure progress and to clarify his or her understanding of the therapy and his or her response. The provider is encouraged to use each visit as a time not only to assess control and response to a specific therapy, but as an opportunity to praise explicitly and reinforce patient competence and progress. A simple method of doing this is to ask the patient at each visit how successful he or she has been in following the medical recommendations, what problems have occurred, and what questions or suggestions he or she may have about the regimen. A four-item scale that measures patient self-reported compliance behavior has been developed to address this specific concern. The scale has been found to be internationally consistent (Cronbach's alpha = 0.61) in both a patient and statewide survey. Furthermore, it has been found to adequately predict blood pressure control ($r = 0.45$; $p < .01$). A total of 74% of patients scoring high on the compliance scale were found to have their blood pressure under adequate control compared to only 50% under control for patients scoring low (Morisky et al., 1982).

Providing feedback to the patient in an atmosphere of openness to questions and willingness to explain and clarify the instructions rewards the patient for his participation and progress. A study recently investigated the effectiveness of a health education intervention designed to increase patient question asking during the patient's medical visit (Roter, 1977). Findings indicated that the intervention group asked more direct questions and fewer indirect questions than did a nonintervention group; and secondly, the experimental group demonstrated higher appointment-keeping ratios than did the nonintervention group. Keeping patients aware of their blood pressure trends by charting the systolic and diastolic measures over time or by teaching the patient to monitor and record his or her blood pressure at home, places emphasis on therapeutic goals and makes both patient and provider focus their attention on outcomes as well as barriers at each visit.

RATIONALE FOR HYPERTENSION CONTROL PROGRAMS AT THE WORKSITE

A great deal of information with respect to patient and provider health behavior strategies to increase compliance and hopefully lead to better control of blood pressure has been gained in several educational settings in outpatient clinics. Many of these strategies have in turn been successfully applied in various community, school, and worksite settings.

Several companies have begun to set up high blood pressure programs and encourage these kinds of behavioral strategies because of the following factors:

1. Increased costs of medical benefits paid by employers (health care benefit costs to employers have increased more than 800% over the past 25 years).
2. High rates of premature deaths among valuable employers.
3. Growing employee interest in health, fitness, and illness prevention.
4. Lagging American productivity.
5. The convenience and social support advantage of worksite wellness programs.

Many health professionals have realized that the worksite is an ideal location for both health promotion and preventive medicine. The worksite offers a relatively stable population, the opportunity of meaningful incentives, potential peer-group support for change and reinforcement, a captive audience for health education and a built-in organizational structure for implementation. Many companies have set up hypertension control programs because employees are more likely to take their medication, cut down on salt, and exercise more if their co-workers and friends are following a similar routine.

A number of models have been advanced to implement different types of high blood pressure control programs in the workplace (Weinstein and Stason, 1976; Alderman and Davis, 1976; Foote and Erfurt, 1977; Alderman and Davis, 1980). Basically these programs differentiate according to whether onsite treatment is provided or services are limited to screening and referral. Many companies have reported success in identifying employees who were unaware that they had high blood pressure, but without a built-in follow-up effort, the proportion of employees who are subsequently diagnosed, given a treatment regimen, and achieve blood pressure control is quite low. For example, a voluntary onsite screening, referral, treatment, and follow-up program at the home office of Massachusetts Mutual Life Insurance Company led to an increase in the percentage under control from 36% to 82% following one year of operation (NHBPEP, 1980). No indication of the number of dropouts, however, in the various stages that lead to control (screening, referral, diagnosis, treatment, and control) was indicated. In a three-site industrial high blood pressure control program that offered all the components of care, 92% of 120 auto workers, 138 sanitation workers, and 106 postal workers identified as having an elevated blood pressure saw their physician and 93% of those seeing a physician had treatment initiated (Foote and Erfurt, 1977). Of those initiating treatment, about 84% achieved adequate blood pressure control over a 16-month period.

Other behavioral methods to lower blood pressure, particularly those afflicted with mild hypertension (diastolic pressure in the 90–105-mm Hg range), include nonpharmacological modalities such as biofeedback, relaxation, psychotherapy, suggestion and placebo, and environmental modification (Shapiro et al., 1977). The rationale for a nonpharmacological approach stems from the fact that events in the emotional life of the patient influence the lability of the blood pressure and affect the progress of hypertension. Most of these methods have been used as adjunctive and not alternative approaches. Further studies regarding their long-term effects, and careful comparison through cooperative studies is warranted. Like any treatment for which one wishes to define clinical affectiveness, these methods should be scrutinized to determine rationale, origins, and physiologic effects; magnitude and duration of effects on blood pressure; methods and frequency; potential toxicity and side effects; and therapeutic indications, including cost in relation to other types of therapy. In addition to the positive benefits achieved by employers for initiating high blood pressure control programs in the workplace, employees also benefit from their successful participation. For example, the life insurance industry has provided financial incentives for effective blood pressure control by reducing or eliminating the extra premiums hypertensives have to pay if they are under effective control. At least one major insurer, Lincoln National Life, accepts treated blood pressure readings as the basis for calculating premiums if the insured has been under treatment for more than five years and if his or her blood pressure has remained under control (NHLI, 1975).

In general, employee high blood pressure control programs that include a mechanism for insuring continuing control, through onsite continuing treatment and aggressive follow-up, are the most cost-effective in reducing cardiovascular risks. Applying the behavioral strategies identified in patient educational settings to the workplace has proven to be an effective modality in the management of high blood pressure. Other studies are also providing evidence that such programs involving socially supportive and self-reinforcing behaviors have a broader effect in simultaneously inducing positive changes in other behavioral determinants detrimental to cardiovascular function, for example, increased exercise, smoking cessation, and stress reduction (Bertera, 1981). Therefore, health promotion programs that incorporate an integrative approach toward reducing those risks significantly associated with cardiovascular morbidity and mortality will continue to demonstrate both success and cost effectiveness.

GUIDELINES FOR PROGRAM DEVELOPMENT

The type of high blood pressure control program one implements into the work setting would depend mainly on the resources available, both in terms of personnel and financial constraints. As previously mentioned, hypertension control programs in occupational settings fall into two general categories: (1) those programs only concerned with the screening, detection, verification, and referral of individuals having an elevated blood pressure; and (2) those programs that add a diagnosis, treatment, and management component.

Previous studies have adequately documented the low success rates found in worksite hypertension control programs that only address those stages in the care process identified in category one. These low rates, particularly in the areas of verification and referral, can be greatly improved, however, by incorporating various follow-up efforts at each stage. Recent studies have shown that the accessibility of medical treatment for hypertensive patients in the workplace is not the essential ingredient responsible for achieving blood pressure control (Alderman and Davis, 1980). For example, a large proportion of individuals who are initially detected as having an elevated blood pressure at the screening site will not take the next step to verify that their blood pressure is truly elevated. Reasons for the delay at this early stage in the screening process may stem from the fact that individuals generally do not seek medical care unless they are experiencing some obvious physical ailment. Hochbaum (1956) originally identified this fact with regard to individuals obtaining a chest X-ray for the detection of tuberculosis. Perceived susceptibility to tuberculosis contained two elements: (1) belief about whether tuberculosis was a real possibility; and (2) strength of belief that one may have tuberculosis and no symptoms. He found that among those exhibiting both beliefs, 82% had obtained at least one voluntary chest X-ray during a specified period preceding the interview, as opposed to only 21% of those exhibiting neither belief. In a similar vein, taking the necessary steps to verify if the blood pressure is truly abnormally high may be aggravated by the asymptomatic nature of the disease. Other elements of the Health Belief Model (Rosenstock, 1974) may also affect an individual's preventive-care behavior, namely, (1) the individual's readiness to take action, determined by both perceived likelihood of susceptibility to the particular illness and probable severity of its consequences; (2) the individual's evaluation of the feasibility and efficaciousness (benefits) of the advocated health behavior weighted against other barriers (physical, psychological, or financial) involved in the proposed action; and (3) a cue to action triggering the appropriate health behavior (this stimulus can be either internal—e.g., symptoms, or external—e.g., interpersonal interactions, mass-media communications). But of all the factors, the perceived susceptibility of the disease, together with its concomitant asymptomatic nature, is probably the most significant for promoting delay in seeking medical treatment.

In order to address this problem, worksite health programs should initiate internal follow-up mechanisms within the work setting. Persons responsible for the screening of employees would normally handle this function. Category one type programs would identify at the time of detection the name of an individual's health care provider and obtain the employee's permission to send information to their physician regarding the results of the screening. Efforts should be made to set up a tickler file in which names of employees with elevated blood pressures could be contacted in a reasonable period of time. This contact would determine if a verification measure was obtained and whether or not the employee was diagnosed as having high blood pressure. For individuals who do not have a usual source of medical care, names and addresses of physicians who have agreed to accept referrals from the worksite should be provided to the employee. Furthermore, the fees that are charged by the physician should also be indicated. Previous experiences have documented the importance of specificity of provider and timeliness of referral to be major contributors to whether or not a contact is made. Often, the health coordinator at the worksite will assist the employee in setting up the initial appointment with the physician. In any case, not more than two weeks should elapse before setting up the appointment; earlier referrals should be made in cases of critically elevated blood pressure levels.

Once an employee has been diagnosed and placed under treatment, the health worksite coordinator should encourage the employee to discuss with them problems they may be having in regard to their treatment regimen. The health coordinator should provide opportunities for other diagnosed hypertensive patients in the workplace to discuss how they are coping with and solving similar problems. Small group discussions also provide a format for making more visible the differences and similarities in the management requirements of a shared problem, a beginning point for the discussion of compliance issues, as well as a way to stimulate group cohesiveness in an attempt to bring their blood pressure under control.

If employees are having difficulty achieving family support and understanding, booklets should be prepared that would explain the nature and the problems of uncontrolled high blood pressure. Employees would then share and discuss ways in which family members could become more actively involved in the care of the hypertensive patient. For example, most individuals with high blood pressure are placed on some form of a salt-restricted diet. Individual family members concerned with household food preparation must be knowledgeable as to why salt should be avoided, as well as plan for substitute meals. A certain proportion of hypertensive employees will be advised to lose weight and nutritional menus could also be included in the family-member booklet. These issues could also be topics for discussion in the small group sessions (Bowler and Morisky, 1983).

Worksite high blood pressure control programs that include a treatment component generally consist of a health team approach, with a nurse, supervised by a physician, providing care according to a systematic protocol. A major advantage of this approach is that there is no direct cost to the employee for visits, medication, or laboratory tests. Furthermore, patient adherence rates to the treatment regimen have been quite high, with an annual attrition rate amounting to less than 10%. In one worksite, satisfactory blood pressure control has been achieved and maintained by 80% of active employees, with a decline in both absenteeism and hospitalization for treated employees (Alderman et al., 1980). However, it is not the mere provision of medication and treatment at the worksite that results in the improved rates of blood pressure control. Rather, it is the educational activities that address those areas of confusion and misunderstanding regarding the medical regimen. If employees do not perceive the benefits of treatment, do not understand the chronic nature of high blood pressure, or receive little support and understanding from family members, they will not comply with the treatment plan. The employee requires this added dimension of an ongoing support and reinforcement, and this can be provided in any employee health program, irrespective of a treatment component.

In summary, the model high blood pressure control program would include a team of health professionals who would be involved in each stage of the blood pressure control process, that is, screening, detection, verification, diagnosis, treatment, and control. Special efforts must be given to insure referral and follow-up are done in an expeditious manner at each of these stages of care. Information should be readily available for employees to read and take home. Brochures that outline how family members can become involved should be provided as well. Finally, the hypertensive employee should be encouraged to discuss problems he or she may be encountering with respect to their treatment program with the health care team. This requires that convenient hours be arranged so that employees can avail of these services. When these elements are in place, the high blood pressure control program in the workplace should be able to achieve a high success rate as well as operate both efficiently and effectively.

Summary

This chapter addressed several of the behavioral problems that are commonly found in hypertension control programs in an occupational health setting. Factors that affect and often lead to the behavioral problem were identified and diagnosed. Emphasis was placed on the needs assessment stage because the entire educational program is founded on the premise of a well-researched baseline diagnosis. The health planner equipped with this information can then begin to assemble the components of the actual educational strategies used to address each behavioral problem.

In order to further increase the number of hypertensive patients under effective therapy, increased attention must be directed at reducing early dropout and promoting continuity of hypertension management. Adherence to the medical regimen must be regarded as an interactive process involving the health care provider, the patient, and the social support network. Management must continue to remain as uncomplicated and as inexpensive as possible. Patients with mild hypertension should be treated, if at all possible, by nonpharmacologic methods, taking as much time as possible to help the patient effect the behavioral changes. All health care providers should be involved in this process. The patient's progress should be monitored carefully throughout the initial treatment period, for it is this stage where noncompliance will begin to evidence itself.

Utilizing these simple techniques with adequate educational efforts, more than 80% of hypertensive patients should have their blood pressure under control, regardless of the initial severity of their blood pressure. The time has come for the employer to take a closer look at the effect of a hypertension control program and its impact on long-range cost and productivity.

RESOURCE GUIDE FOR HYPERTENSION CONTROL PROGRAMS

Organizations

High Blood Pressure Information Center
120/80 National Institute of Health
Bethesda, MD 20205

Blue Cross and Blue Shield Associations
840 North Lake Shore Drive
Chicago, IL 60611

Xerox Corporation
P.O. Box 7000
Leesburg, VA 22075

Prudential Insurance Company
213 Washington Street
Newark, NJ 07101

American Heart Association
7320 Greenville Avenue
Dallas, TX 75231

Chase-Manhattan Bank
1 Chase-Manhattan Plaza
New York, NY 10081

Macalloy Corporation
P.O. Box 130
Charleston, SC 29402

Massachusetts Mutual Life Insurance Company
Springfield, MA 01111

United Storeworkers
101 West Thirty-first Street
New York, NY 10001

Workers Health Program
Institute of Labor and Industrial Relations
University of Michigan
401 Fourth Avenue
Ann Arbor, MI 48103

Campbell Soup Company
Camden, NJ 08101

Ford Motor Company
900 Parklane Towers West
1 Parklane Boulevard
Dearborn, MI 48126

Metropolitan Life
One Madison Avenue
New York, NY 10010

Kimberly-Clark Corporation
Neehah, WI 54956

Journals

American Journal of Public Health

Chest

Circulation

Health Education

Journal of Occupational Medicine

Journal of the American Medical Association

New England Journal of Medicine

REFERENCES

Alderman, M. H. and Davis, T. K. "Hypertension control at the work site." *J. Occup. Med.*, 1976, **18,** 793–796.

Alderman, M. H. and Davis, T. K. "Blood pressure control programs on and off the worksite." *J. Occup. Med.*, 1980; 22, 167–170.

Alderman, M., Green, L. W., and Flynn, B. S. "Hypertension Control Programs in Occupational Settings.'' In *Managing Health Promotion in the Workplace: Guidelines for Implementation and Evaluation,* R. S. Parkinson and Associates, eds. Mayfield, 1982.

Bertera, R. L. ''The effects of blood pressure self-monitoring in the workplace using automated blood pressure measurement.'' Doctoral dissertation. Johns Hopkins University School of Hygiene and Public Health, Baltimore, 1981.

Bowler, M. H., and Morisky, D. E. ''A small group strategy for improving compliance behavior and blood pressure control.'' *Health Educ. Quarterly,* Spring 1983.

Foote, A. and Erfurt, J. C. ''Controlling hypertension: A cost-effective model.'' Prev. Med., 1977, **6,** 319–343.

Francis, V., Korsch, B. M., and Morris, M. J. ''Gaps in doctor-patient communication: Patients' response to medical advice.'' *N. Engl. J. Med.,* 1969, **280,** 535–540.

Genest, J., Nowaczynski, W., Boucher, R., Kuchel, O. Role of the adrenal cortex and sodium in the pathogenesis of human hypertension, *Canadian Med. Assoc. J.,* 1978, **118,** 538–549.

Hochbaum, G. M. ''Why people seek diagnostic X-rays.'' *Pub. Health Rep.,* 1956, 71, 377–380.

Hypertension Detection and Follow-up Program Cooperative Group. ''Five-year findings of the hypertension detection and follow-up program: I. Reduction in mortality of persons with high blood pressure, including mild hypertension.'' *JAMA,* 1979, **242,** 2562–2571.

Inui, T. S., Carter, W. B., Kukull, W. A., and Haigh, V. H. ''Outcome-based doctor-patient interaction analysis. I. Comparison of techniques.'' *Med. Care,* 1982, **20,** 535–549.

Laragh, J. ''Vasoconstriction-volume analysis for understanding and treating hypertension: the use of renin and aldosterone profiles.'' *Am. J. Med.,* 1973, **55,** 261–274.

McGill, A. M. ''A National High Blood Pressure Education Research Program.'' Abstracts of Papers presented at the First International Congress on Patient Counseling. *Patient Counseling and Health Education,* 1978, **1,** 35–38.

Morisky, D. E., Ward, W. B., Levine, D. M., and Bone, L. R. ''Using educational diagnosis results to target specific hypertension control strategies for population subgroups.'' Paper presented at the annual meeting of the American Public Health Association. Detroit, 1980.

Morisky, D. E., Levine, D. M., Green, L. W., and Smith, C. R. ''Health education program effects on the management of hypertension in the elderly.'' *Arch. Intern. Med.,* 1982, **142,** 1835–1838.

Morisky, D. E., Levine, D. M., and Green, L. W., et al. ''Five-year blood pressure control and mortality following health education for hypertensive patients.'' *Am. J. Public Health,* 1983, **73,** 153–162.

Moser, M. ''Physician and patient adherence in the management of hypertension.'' *Urban Health,* May 1982, 41–43.

National Center for Health Statistics. *1980 Report of the Joint National Committee on Detection, Evaluation, and Treatment of High Blood Pressure.* DHEW Publication No. (NIH) 81-1088, 1981.

National Heart, and Lung Institute. ''The Underwriting Significance of Hypertension for the Life Insurance Industry.'' DHEW Publication No. (NIH) 75-426, 1975.

National High Blood Pressure Education Program. "Massachusetts Mutual: Off Site Care and Good Monitoring Reduce Medical Costs." *High Blood Pressure Control in the Worksetting,* Winter 1980.

Ostfeld, A. M. and Levovits, B. Z. "Personality factors and pressor mechanisms in renal and essential hypertension." *Arch. Intern. Med.,* 1959, **104,** 43–52.

Reeder, L. G. "The patient-client as a consumer: Some observations on the changing professional-client relationship." *J. Health Soc. Behavior,* 1972, **13,** 406–412.

Roberts, J. "Hypertension in adults 25–74 years of age, United States, 1971–1975." (Vital and health statistics: Series 11, Data from the National Health Survey; No. 221) DHHS publication No. (PHS) 81-1671.

Rosenstock, I. M. "Historical origins of the health belief model." *Health Educ. Mongr.* 1974, **2,** 328–335.

Roter, D. L. "Patient participation in the patient-provider interaction: The effects of patient question asking on the quality of interaction, satisfaction and compliance." *Health Educ. Monog.,* **5,** 281–315, 1977.

Selye, H. *The Stress of Life.* New York: McGraw-Hill, 1956.

Shapiro, A. P., Schwartz, G. E., Ferguson, D. C., Redmond, D. P., and Weiss, S. M. "Behavioral methods in the treatment of hypertension: A review of their clinical status." *Ann. Intern. Med.,* 1977, **86,** 625–636.

Sokolow, M., Kalis, B. L., Harris, R. E., Bennett, L. F. "Personality and predisposition to essential hypertension." Proceedings of the Symposium on Pathogenesis of Essential Hypertension, p. 14. Prague: State Medical Publication House, 1961.

Svarstad, B. "Patient-physician communication and patient conformity with medical advice." In *The growth of bureaucratic medicine: An inquiry into the dynamics of patient behavior and the organization of care.* Mechanic, D., ed. New York: Wiley, 1976.

Weinstein, M. C. and Stason, W. B. *Hypertension: A Policy Perspective.* Cambridge: Harvard University Press, 1976.

Wolf, S., Cardon, P. V., Shepard, E. M., and Wolff, H. G. *Life Stress and Essential Hypertension: A Study of Circulatory Adjustments in Man.* Baltimore: Williams & Wilkins, 1955.

Wolinsky, F. D. and Steiber, S. R. "Salient issues in choosing a new doctor." *Soc. Sci. Med.,* 1982, **16,** 759–767.

Working Group to Define Critical Patient Behaviors in High Blood Pressure Control. "Patient behaviors for blood pressure control: Guidelines for professionals." *JAMA,* 1979, **241,** 2534–2537.

CHAPTER 8

Occupational Health Promotion through Physical Fitness Programming

Donald Weller
George S. Everly, Jr.

INTRODUCTION

Did you ever wonder why it may be hard for you to sit still when you are anxious, or why people who wait impatiently can often be seen pacing up and down or frequently moving about? It has been suggested by numerous writers (Kraus and Raab, 1961; also Everly and Rosenfeld, 1981, for reviews) that physical activity is the body's natural means of expressing, or ventilating, the challenges, threats, or frustrations that may confront an individual. Indeed, it has been suggested that it is the "wisdom of the body" that such challenges or threats should be ventilated; failure to do so increases the risk of psychological and physical dysfunction or disease (Kraus and Raab, 1961).

It is reasonable to assume that thousands of years ago, the highly active lifestyle that was led by primitive human beings afforded them sufficient opportunity for the physical and psychological expression of the dangers and frustrations they faced. In addition, there was probably ample opportunity for those humans to derive the physiological advantages we now know can be derived from regular physical exercise. However, as humans progressed from "physical beings" to "thinking beings," we provided ourselves with fewer and fewer opportunities to ventilate our fears and challenges or to develop our minds and bodies through exercise. Organized physical fitness programs serve as a means of providing modern humans with just such opportunities. As an extension of the overall physical fitness movement, occupational physical fitness programs were founded and promulgated on the belief that physically fit workers are happier and more productive workers. In this chapter we will examine physical exercise as an occupational health promotion strategy, and we will provide general guidelines for program development.

SCOPE OF THE PROBLEM

In their treatise on the role physical exercise plays in health status, Kraus and Raab (1961) argue that little over a hundred years ago, hard, physical labor was a way of life that may have served as a form of protective mechanism against a host of diseases, commonly referred to today as the "diseases of civilization." Kraus and Raab now argue that our modern sedentary life-styles may have put that protective mechanism "all but out of commission." They state:

> The system that has been put all but out of commission, the striated musculature . . . has an important role which exceeds the mere function of locomotion. Action of the striated muscle influences directly and indirectly circulation, metabolism, and endocrine balance. . . . Last but not least the striated muscle serves as an outlet for our emotions and nervous responses, the means by which we react and respond to stimuli and emotional stresses. Obliteration of an important safety valve . . . without opening of vicarious outlets, might well upset the original balance to which the bodies of primitive man have been adapted (p. 4).

Thus they coined the term "hypokinetic disease" to refer to physical and mental dysfunction or disease induced by inactivity.

The notion that the lack of physical activity can be a predisposing condition for dysfunction and disease has been supported by the World Health Organization (Chavat et al., 1964). It was concluded that suppression of somatomotor behavior in response to physical arousal (e.g., stress) is likely to lead to increased cardiovascular strain.

Finally, based upon epidemiologic data from the Framington heart studies, it has been concluded that a sedentary life-style is, indeed, a risk factor for coronary heart disease (Kannel, 1966).

Surveys of Americans have indicated that as many as two thirds of Americans do not regularly exercise (Harris, 1978). Therefore, if there is merit in the previous conclusions, much of our population is at risk from a sedentary life-style.

While the problem of a sedentary life-style as a risk factor for disease may be rationale enough for establishing physical fitness programs, there exists another, more subtle, rationale. Given the fact that participation in fitness-related activities is steadily increasing (Presidents Council on Physical Fitness and Sport, 1974), there exists the problem that most of the adults who are "rediscovering" exercise are simply not aware of the appropriate types of exercises to perform, nor of the frequency, duration, and most importantly, the intensity levels that are safe for their present physical condition. This lack of education most frequently ends up in a negative exercise experience yielding little, if any, success, frequent dropouts, and sometimes injury and even sudden death (Steodefalke, 1977).

Well-planned and structured physical fitness programs have the potential to attack the problems of the hypokinetic disease phenomenon while reducing the problems associated with physical exertion as a health promotion strategy.

REVIEW OF RESEARCH ON PHYSICAL ACTIVITY AS A HEALTH PROMOTION STRATEGY

In this section we will review findings relevant to the use of physical exercise as a health promotion strategy.

In preface, it becomes necessary to delineate just what form of exercise we will be addressing. Most basically, there are two forms of exercise: anaerobic and aerobic. Anaerobic exercise consists of any physical activity in which energy is derived through a metabolic process that does not require the presence of oxygen. Such exercises create what is called an oxygen debt and may be maintained for only about two minutes of constant exertion. Anaerobic exercises are typically thought of as strength exercises rather than endurance exercises. Aerobic exercises, on the other hand, are exercises which require a constant flow of oxygen to maintain the metabolic process. They are most commonly rhythmic, endurance exercises such as jogging, swimming, running, and brisk walking. Because the majority of occupational health promotion programs employ aerobic exercise predominantly, or exclusively, we will limit our discussion to aerobic exercise.

Physiological Benefits of Exercise

The physiological benefits of regular, aerobic exercise are most frequently noted in the cardiovascular system. It is well known that programs of aerobic exercise can lead to increased cardiac efficiency (Scheuer and Tipton, 1977). The increased cardiac efficiency that results from aerobic training seems to be a function of several mechanisms. One of the most important is an increase in the amount of collateral coronary blood flow. This increased blood flow appears to be a function of a sustained increase in the coronary arterial lumen (two to three times the normal size) as well as an increased number of collateral coronary arteries. It is interesting to note, in addition, that systemic arterial lumen also seem to increase in size as a response to aerobic fitness training. Another indication of increased cardiopulmonary efficiency is the decrease in resting heart rate that results from aerobic exercise. This phenomenon appears to be a function of a decrease in the intrinsic firing of the S-A node, an increase in the tonic vagal cardioinhibitory activity, and a decreased cardioaccelerator effect that might be experienced from respiration. A final mechanism accounting for increased cardiopulmonary efficiency may well be a healthful hypertrophy of the cardiac muscle. Such hypertrophy would be demonstrated through improved cardiac output (Clarke, 1975; also Fox, 1979).

There is even evidence that aerobic exercise may be useful in lowering the blood pressure of hypertensive individuals independent of fluid loss and/or weight reduction (Boyer and Kasch, 1970; Horton, 1981). This effect may be a function of some of the mechanisms just described.

It has been shown that regular aerobic exercise may modify cardiovascular risk profiles (Kannel and Sorlie, 1979) through several mechanisms. Epstein and Wing (1980) suggest that physical exercise can be a useful method for reducing weight and for weight loss maintenance. Soman et al. (1979) concluded that moderate physical exercise is useful in diabetic conditions due to its apparent ability to reduce plasma insulin, improve glucose tolerance, and provide increased insulin effectiveness at cellular receptors. Criqui (1980) found that another cardiovascular risk factor, cigarette smoking, decreased as exercise levels increased. Finally, during exercise, mobilized lipid stores are hydrolized to free fatty acids for energy production (Zieler, Maseri, Klassen, Rabinowitz, and Burgess, 1968). This process results in a more rapid clearance of fatty acids from the bloodstream, thus preventing their conversion to circulatory low density lipoproteins (LDL). The LDLs are the most common lipid factor in arteriosclerotic conditions (Miller, 1980).

So we see physiologically, that regular aerobic exercise has been shown to not only improve general cardiac fitness but to favorably affect many of the cardiovascular risk factors as well.

Psychological Benefits of Exercise

Reviews by Layman (1974), Folkins and Sime (1981) and Martin and Dubbert (1982) suggest that regular aerobic exercise can have beneficial effects on psychological functioning. More specifically, in the studies reviewed by these authors, regular physical exercise was seen to be correlated with improvements in depressive conditions, self-esteem, sense of control, anxiety, cognitive functioning, and general mood (Sachs and Buffone, 1984).

Recently, there has been speculation that the observed increase in circulating beta-endorphin observed in humans after physical exercise (Gambert et al., 1981) may be the cause of the antidepressive, anxiolytic (antianxiety) and even analgesic effects reported during and following physical exercise at sustained levels. However, this hypothesis has yet to be proven (Sacks and Sachs, 1981; and Sachs and Buffone, 1984, for discussions of the therapeutic effects of running).

Occupational Benefits of Exercise

While clearly it is most difficult to reach confident conclusions concerning the effects of physical exercise on work-related behavior, some studies in this area do exist. Papers by Brennan (1981), Cunningham (1982), and the excellent review by Donoghue (1977) offer evidence that regular exercise does improve work-related performance. Workers who exercise report that they hold a more positive attitude toward their work, they have less turnover, lower absenteeism, and generally have less errors. The results of these studies may be suspect due to methodological problems, however. We are still awaiting well-controlled studies on this important area of exercise and work-related behavior.

One final area that holds considerable potential for occupational health promotion programming is the area of cardiac rehabilitation. Aerobic exercise has clearly demonstrated its value in cardiac rehabilitation applications (Wilson, Fardy, and Froelicher, 1981). Cardiac rehabilitation efforts may be organized and delivered through the medical departments of mid- to large-sized organizations, thus potentiating a reduction of insurance and outside medical expenditures that would otherwise be accrued without such in-house facilities.

GOALS OF AN EXERCISE PROGRAM

It has been stated that in order for aerobic exercise programs to be cardiovascularly effective, participant compliance must meet three criteria:

1. Intensity level.
2. Duration level.
3. Frequency level.[1]

According to the American College of Sports Medicine (1980):

minimum intensity = 60% of maximum heart rate regardless of the specific exercise engaged in

minimum duration = 15 minutes of continuous exercise so that the attained heart rate does not drop below 60% of maximum

minimum frequency = 3 times per week

Should sustained participant compliance be achieved as described above, tbe goals of improving the participants' cardiovascular and overall fitness levels will be attained. More specifically, the goals of improved cardiovascular and overall fitness can be operationally assessed through the use of outcome measures such as

1. Exercise tolerance (workload capacity).
2. Body composition.
3. Lung capacity.
4. Body flexibility.
5. General body strength.
6. Resting heart rate.
7. Resting blood pressures.

[1]All of these factors have been combined in an aerobic point system developed by Cooper (1968, 1977).

All of these variables may be combined into a fitness profile that should be created in a baseline condition, before the physical training intervention, as well as at selected points during and after (follow-up) the physical training intervention. These variables, and the composite fitness profile, may then serve as outcome measures to assess the effectiveness of the occupational fitness program.

GUIDELINES FOR PROGRAM DEVELOPMENT

Just providing a fitness program is not enough. Provision of the most modern facilities and even time to exercise does not significantly increase the fitness status of employees. The newest, most advanced exercise facilities available will simply rust from lack of use unless employees can see exercise as personally rewarding!

If asked, most people will indeed recognize the "need" to improve their personal fitness levels, yet within the first year of any program, "dropout."

rates exceed 50% (Stone, 1980). Based on reviews by Dishman (1982), Oldridge (1982), and Martin and Dubbert (1982), the factors which seem to affect participant compliance are listed in Table 8.1.

One of the most promising models for predicting exercise compliance may be a model that integrates psychosocial and biologic factors as predictors of participant compliance (Dishman and Gettman, 1980; Dishman, 1982).

Table 8.1 Factors Affecting Compliance in An Exercise Program

I. Program factors that improve compliance
  A. Convenient exercise facilities
  B. Exercise intensity levels that are perceived as being moderate
  C. Social support
  D. Reinforcement procedures

II. Characteristics of the program "dropout"
  A. Smoking
  B. Low self-motivation
  C. Inactive leisure time
  D. Spouse neutral or negative towards exercise
  E. Poor credit rating
  F. Obesity

III. Reasons for noncompliance
  A. Inconvenient location of program
  B. Inconvenient time of program
  C. Lack of support
  D. Lack of reinforcement
  E. Exercise levels perceived as being too intense

Obtaining compliance in an exercise program is usually dependent upon achieving two goals: participant education and participant motivation. Participants must be educated as to the personal value of exercise and a healthy, fit body. This process includes instruction in how to exercise in a safe and productive manner. Participants must also be motivated to exercise. Initially, this motivation may be assisted through reinforcement procedures, stimulus control, behavioral contracting procedures, and social support procedures. After a month or so of activity, there must be a shift to more subtle forms of internal motivations. It must be recognized that the exercise program or regimen is worthwhile and necessary as a life-long pursuit (Wilmore, 1977).

The motivation of success and maintenance is normally dependent upon the design of any program. Generally, successful programs contain the following elements: a specific design, a degree of individuality and a specific reinforcement/support system for gains (Hall and Hall, 1974).

Overview of Program Design

The worksite is an ideal setting for all types of health-related programs, especially a fitness type program that requires some extraordinary commitment in time and effort to succeed. There are over 100 million persons in the workforce, and 75% are in small units of 100 workers or less, in some two million corporations (Pearson, 1980). Since the work is normally in some central location, and for approximately 40 hours per week, aspects of program control, organization, and the all important follow-up are simplified. Normal barriers to success, such as travel time, travel cost, finding a facility, providing equipment, and making extra time can be eliminated. Participation is much more attractive, especially if a specific time allowance is made available. It has been reported that participation rates are as high as 95% in worksite multiphasic health screening; yet similar free community-based programs reflect only a 30% participation rate (Alderman, 1980; Haskell, 1980; Marcotte and Price, 1983).

As stated earlier, initially, participant motivation is extrinsically controlled and dependent upon environmental conditions and guidance by the program coordinator. The program design must be very specific with easily identifiable and attainable goals and outcome. A structure for entrance into participating programs, a method for identifying appropriate exercise levels, a specific method of attaining the goals/outcomes (the exercise prescription), plus ways of objectively measuring progress must be identified. Once an intermediate goal is met, further attainable goals must be set and further motivation must be generated (Cooper, 1968, 1977; Weller, 1981; American College of Sports Medicine, 1980).

A structured system must allow for some individual variations for age, sex, present physical fitness status, plus individual preferences of time and activity. Some form of evaluation is not only essential to gain the knowledge to modify

Table 8.2 Program Guidelines

- I Initial information session
- II Baseline/screening
  - A. Exercise tolerance (creation of exercise prescription)
  - B. Body composition
  - C. Lung capacity
  - D. Body flexibility
  - E. General body strength
  - F. Resting heart rate
  - G. Resting blood pressure
- III Educational sessions
- IV Implementation of exercise program
- V Periodic assessments of baseline variables
- VI Post-program evaluation
- VII Follow-up and Maintenance (at periodic intervals)

program elements, but is irreplaceable as a motivational tool for goal setting and reinforcement. Table 8.2 describes an occupational fitness program.

Parameters of Fitness Assessments: Baseline

Each person has a specific maximum potential as well as a specific existing level of physical proficiency. To promote physical fitness via the means of regular exercise, it is critical to determine the preprogram level of capacity that will be valid, reliable and cost-effective. Since every individual responds physically in a unique fashion, measures of existing levels are critical in the promotion of a specific, safe and beneficial program. We shall delineate those parameters which are necessary to compile a composite picture of existing levels, as well as procedures for increasing one's exercise tolerance and, thus, one's general fitness level.

Determination of Exercise Tolerance and Exercise Prescription

The most important evaluation done during physiological research is that involving the cardiopulmonary aspect of an evaluation—that being the way one's heart and lungs respond to a specific work load, called exercise tolerance.

There are several standard methods of assessing one's cardiovascular exercise tolerance, each varying in its degree of sophistication, accuracy, and safety. All tests are similar in one aspect in that each tries to measure how much oxygen can be utilized per pound of body weight, per minimum of specific work loads. This is called the Maximum Oxygen Consumption or VO_2 Max.

There are four basic tests utilized widely in the world by physiologists and physical educators that have been developed as tools for which individual test results can be measured. They utilize different apparatus in different environments, but essentially yield comparable results (Strauss, 1979).

These tests are (1) Kenneth Cooper's 12 minute run-walk, (2) the Harvard Step Test, (3) the Astrand Ergometer Test, and (4) the Treadmill Electrocardiogram.

Fitness programs have traditionally utilized maximum exertion exercise evaluations, such as the Harvard Step Test or Cooper's 12 minute run-walk, as evaluation tools for program entrance and exit fitness levels. However, traditional techniques were developed to evaluate a young, asymptomatic population. In persons with known health risk factors, over the age of 35, or sedentary for any length of time, such maximum exertion tests are very risky to say the least.

Although each of these tests yields basically the same results, each has a factor of safety that varies greatly. In every instance, persons especially untrained or deconditioned, run a risk of some physiological malfunction without properly monitored supervision. Only one of the procedures listed above utilizes physiological monitoring equipment to generate immediate feedback regarding subject responses. This technique, the treadmill electrocardiogram, is a superior method of evaluation, for it individually and accurately measures vital heart and lung responses to increasing work loads by sphygmomanometer readings of blood pressure and heart-rate responses during the entire procedure. It can also be used as a diagnostic tool for measuring asymptomatic heart disease. As a diagnostic tool, it has proven to be about 88% accurate in screening out asymptomatic heart disease (Chin, Messinger et al., 1979). Thus, a tolerance test can be used as a screening device as well as an accurate, productive, and ultimately safe assessment of fitness levels in the diagnostic process (Weller, 1981).

In the fitness program, a standard exercise tolerance test should be given to each participant as both a preprogram entrance evaluation and a postprogram evaluation as a comparative basis for

1. Determining existing work capacities (fitness levels).
2. Determining productive individual heart rates to regulate work-load intensity during exercise sessions (creation of target heart rates for exercise prescription), see Weller (1981) and American College of Sports Medicine (1980) for exercise prescription guidelines.
3. Educating participants about their particular physiological responses to exercise (educational function).
4. Screening latent, undetected asymptomatic abnormalities evident only under physical exertion (safety factor) (Weller, 1981).

The evaluation is done on a motorized treadmill at various degrees of incline and speed with a constant monitoring of blood pressure and heart rate, as well as use of the electrocardiogram to monitor heart rhythm. Under some conditions, a stationary bicycle ergometer may be substituted (Johnson, 1983). Such a procedure is recommended for anyone who is over age 35, overweight, or inactive for a period of time. It has been used for years as a medical assessment of cardiac function and diagnosis. Although controversial, there is a need for nondiagnostic testing for the purpose of fitness assessment.

According to the American Medical Association and American College of Sports Medicine Guidelines for stress testing, asymptomatic persons under age 35 may be tested without a physician present. But symptomatic persons, or persons over age 35, should only be tested in the presence of a physician (American College of Sports Medicine, 1980; American Heart Association, 1972; 1975). See Figure 8.1 for a flow diagram for participation clearance.

These guidelines are feasible in a clinical situation involving a small number of persons. But in a mass evaluation program, such guidelines are cost and time prohibitive. Therefore, local physicians and hospital staff should be consulted as to the necessity for testing given the impracticality of existing guidelines when applied to mass evaluation exercise programs, before a conscious decision should be made

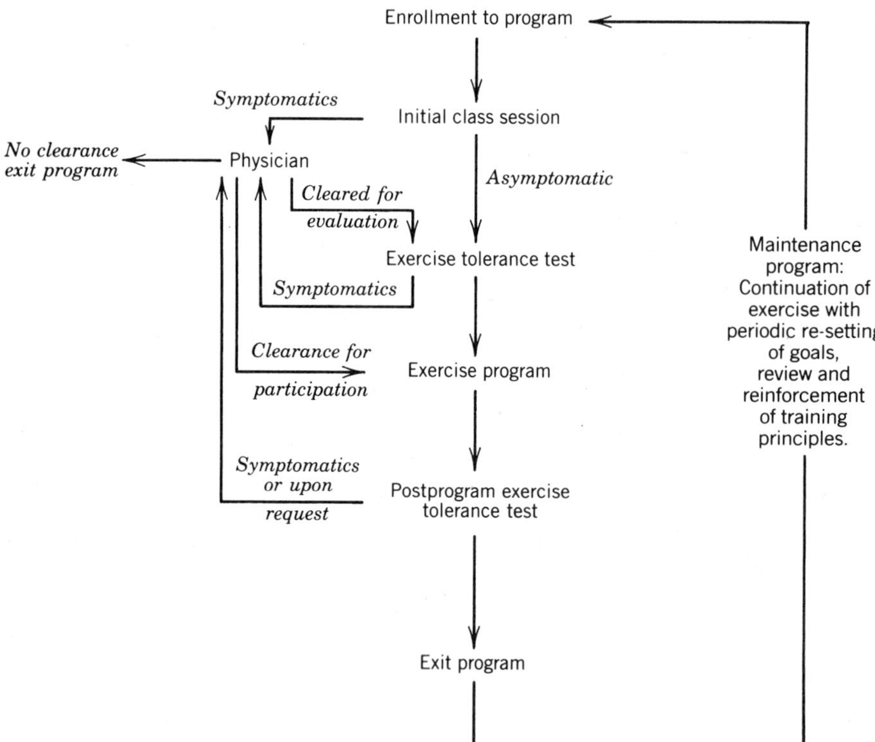

Figure 8.1. Clearance for participation.

to use a form of exercise tolerance test as a safe and effective means of gaining baseline entrance and exit levels (Weller, 1981).

Testing criteria should be developed and strictly maintained close to the established guidelines in order to reduce the risk of maximum exertion tests on symptomatic persons or people over age 35. People who are over age 35, or those who have any established symptoms, should be given a conservative submaximal test (approximately 80–85% of predicted maximum heart rate or a predetermined rate indicated by their physician).

Specific guidelines should be established to screen out symptomatic people prior to testing. In the first orientation session, each participant should fill out an extensive medical history, which upon review will be used as a tool to screen out all those persons with a known history of heart, or some related, disease. To participate in the exercise program, such persons would be required to have a medical permission slip from their personal physicians. Without a physician's permission, a symptomatic person would not be given the exercise tolerance test nor be allowed to participate in the exercise sessions.

It should be noted, at this point, that a critical factor in inducing involvement in any type of program is removal of barriers (Noland, 1981). The indiscriminate requirement for medical clearance for all participants may create a barrier that may distract greatly from the goal of full participation (Weller, 1983).

Besides the actual functional work capacity tests, additional motivational forms of evaluations are easily completed and can be instrumental in demonstrating program worth and accomplishments.

Body Composition. The easiest to attain and yet the most misunderstood measure of body mass is gross body weight. Most people rely on bathroom scales to regulate diet and exercise in an effort to maintain acceptable body composition. The use of this measure as a reference of actual lean body mass versus body fat can fluctuate drastically. Since a muscle tissue is a more dense and thicker material than adipose or body fat, a person may maintain an identical body weight for years, but have a drastic (upwards of twenty pounds) redistribution of usable muscle mass to fat for energy storage (see Figure 8.2 below).

A more effective method of checking on body distribution is the use of a simple tape measure. Girth measurements at specific body landmarks will give a

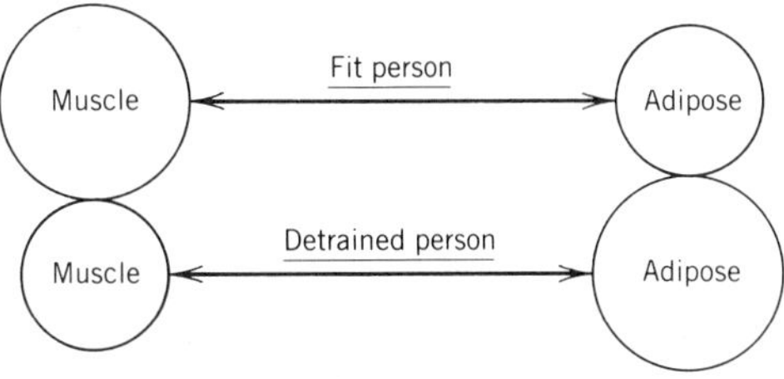

Figure 8.2. Diagram of two people with same body weight but different lean body/fat ratio.

more accurate picture of body composition changes. As a person begins an exercise and a diet program, the hypertrophy and atrophy process will be documanted by changes in body landmark measurement. Although gross body weight may increase, positive distribution changes may be occurring. The changes may take the form of reduction of waist girth, thigh girth, or arm girth. Typical landmark checks are chest, waist, right upper arm and right thigh. Further measurements may be taken but these are sufficient to detect a trend of redistribution (Himes, Roche, and Stiernogel, 1977).

One of the greatest concerns of persons entering a fitness or diet program is determining what their weight or ''ideal'' weight should actually be. The standard method of determining ideal body weight for an individual was derived from insurance charts which give ''normal'' ranges from ''small, medium, and large frame'' persons of specific age, sex, and height. The problem with this method is that these are norms for relatively active individuals (approximately 66% of the total population). Those persons at the extremes, either extremely sedentary or extremely active, are given a false ideal weight. Those who are inactive have a false sense of security by being within the ''norm,'' while active, highly muscular individuals are penalized due to the density of their muscle mass and low body fat, being deemed overweight or even obese (Lohman, 1981; Mayer, 1968).

Over the past decade, other methods have been utilized to eliminate these biases and to simply and accurately determine lean body mass and fat distribution on an individualized basis. These methods, which were discussed in the weight control chapter, include hydrostatic weighing, skinfold assessment and anthropometric assessments.

Lung Capacity. Another simple test that can be given to assess fitness is the pulmonary function test, called vital capacity. Measurement of vital capacity is achieved by having the subject exhale maximally after a maximum inspiration. Expiration should be slow, forced, and complete. As an absolute reading, variations must be at least 20% below predicted normal to be regarded as subnormal. Changes, however, are important regardless of absolute value. Serial reduction in the vital capacity may quantitate the degree of functional impairment that may ensue with progressive pulmonary disease. An increase, on the other hand, may quantify the degree of improvement achieved by either acute or long-term therapeutic measures.

The measurement of vital capacity simply indicates how much usable air space is present in the lung capsule. A reduction or an increase in vital capacity will impact on the cardiopulmonary adjustment to exercise. Positive changes in vital capacity may be noticed by sedentary people who begin to exercise or who simply stop smoking. Both cause the lungs' air sacs, the alveoli, to expand or cleanse themselves so that more air exchange is noted (Wilmore, 1982).

Once again, the instrumentation necessary for measurement of vital capacity can vary tremendously. The degree of accuracy versus the cost of instrumentation, plus the need for interpretation, are the major considerations. In a fitness assessment, as opposed to a medical pulmonary function evaluation, a hand spirometer will give sufficient accuracy to make some cursory evaluations of "normalcy" and improvement (Weller, 1981).

Body Flexibility. Range of motion, the natural motion of joints, is dependent upon the ability of the muscles to stretch as well as to contract. Exercise tends to shorten muscles, thus diminishing the stretching function. Because two major benefits of an exercise program are to make the person more mobile and effective, flexibility must be considered.

The greatest problem of middle-aged and older persons is lower back problems. This is primarily due to an imbalance of flexibility and strength. Weak abdominal muscles tend to cause the pelvis to tilt forward. Lack of motion and dynamic exercise with stretching both allow muscle groups to adaptively shorten. In the legs, the back of the legs (or hamstrings) shorten and further pull the pelvis forward. The total result is a weak and short back muscle. Any quick or dynamic motion strains or tears the muscle body.

In an effort to monitor this occurrence, a simple flexibility reach test can be employed. During an exercise program, the tendency is to reduce flexibility because of gains in muscle strength. Therefore, particular attention must be given to stretching muscles as well as strengthening them (Wilmore, 1982).

General Body Strength. In general, it is impractical to test each person on each muscle group to establish ratings as to their well-being. Therefore, researchers have identified specific tests in which generalizations about individuals can be made.

An indicator of power or strength (single rep. max. exertion) is identified from a grip strength exercise called a dynameter test. Hand dynameters have been used to secure an index of general bodily strength, to secure an index of right-handedness versus left-handedness, and for comparative purposes generally. They may also be used to obtain an index of endurance or fatigue, or combined with other forms of strength measurement.

For the purpose of fitness assessment, the dynameter can given an easily attained baseline measure of general body strength. After a program of training, a retest will indicate improvement or validate the exercise regimen.

Motivational Factors

Program success, as measured by participation and increases in fitness proficiency, is not a matter of choice. It requires not only a specific, scientifically based design, but a group of incentives that will increase the integration of exercise into the employee's life-style (Martin and Dubbert, 1982; Oldridge, 1982, for reviews of behavioral approaches to increasing compliance). Similarly, Brown (1981) has developed an effective incentive-based employee fitness program for occupational health promotion. Commonly used incentives for exercise compliance might include vacation time, a shortened workday for exercise purposes, trips, and even monetary incentives. Exercise groups may be formed that not only provide support but sponsor fitness awards, plaques, banquets, and so on. Token economics, however, where tokens can be earned through exercise participation and then applied towards purchases of desired commodities, may be the most powerful of the highly structured incentive programs.

According to Haskell and Blair (1981) the program leader or coordinator can be an effective motivational tool for participants. The program leader is likely to be perceived as a role model and therefore should be perceived as energetic, enthusiastic, and dedicated to helping people improve their fitness and health. They conclude that an effective leader is more important to the success of a fitness program than are the facilities available.

The Educational Component

Once the baseline measures are taken, and the motivational components have been attended to, participants should enter the educational component of the fitness program. The educational component can be scheduled before or along with the actual exercise classes. The goals of the educational component are to instruct participants in

1. What fitness is.
2. How exercise increases fitness.
3. How to exercise safely.
4. How to recognize and reduce the problems associated with exercise.

Implementation of the Exercise Program

The exercise prescription represents the desired levels of exercise to be engaged in by each participant. The prescription was generated in the work-load capacity evaluation conducted at baseline. It will be expressed for each participant along three criteria:

1. Heart rate range during exercise.
2. Duration of exercise.
3. Frequency of exercise.

The prescription may also delineate the type of exercise. Once these factors have been ascertained, it remains only for the program prescription to be explained during the educational component and then to be implemented. The present authors prefer exercise groups as the implementation format.

Evaluation/Follow-up/Maintenance

The cornerstone of any worthwhile program is an evaluation procedure. Not only will this give accurate preprogram entrance levels which will dictate methodology of a fitness regimen, but also act as a motivational tool.

Demonstrated changes in parameters such as work-load endurance, strength, body composition, girth, and lung capacity give critical reinforcement to internalize the motivation for continuing in the program.

As a program continues, evaluation follow-ups offer the needed feedback to the program leaders that enables them to actually modify activity, intensity, and/or other elements of the program (Haskell and Blair, 1981), thus a maintenance function is performed.

Summary

It has been significantly determined that an active life-style, which includes the two components of health behavior modification and a realistic dose of regular exercise, will benefit the quality of one's life. These data are sufficient to warrant implementation of, and the continuation of, a fitness facility or program in the industrial setting (Barnes, 1983; Fox and Boyer, 1972).

Although costly, the physiological benefits of having an increase in the general work force fitness proficiency has been translated into reduction in general health care costs, rehabilitation costs, and time lost from the job. Thus, a by-product of employing a physical fitness program is increased productivity and, generally, better morale.

It is not enough to offer a beautiful facility. A planned program must accompany the actual physical surroundings. It must be based on a sound design that will educate and motivate the participants. It must be scientifically bound to utilize the best information available to get the most success with the least amount of effort. It must be individually responsive to the participants' particular needs and desires. Program variety must reflect individuality in preference, sex, age, and fitness status.

Critical to this motivational and modification aspect of the program are fitness assessments. There are numerous assessment techniques that will quantify program success. Each will give to the program directors the necessary input as to the entrance levels of participants. These assessments will also give participants accurate appraisals of present levels and aid in the formation of realistic short-term and long-term goals. Fitness assessments will automatically evaluate the success of any program. If performed in a realistic fashion, assessments will objectively reinforce the program quality and the success of program design. Such data will be of tremendous value in recruitment of new participants, continuation of those already enrolled, and act as a specific rationale for program costs.

The key element in any program is leadership. It provides direction, evaluation, motivation, and modification. Of all components, a strong, enthusiastic, and knowledgeable leader may well be the single best commodity that will determine program success. A professional health educator or other health professional with a background in physiological indicators and contraindicators of exercise, as well as in behavior modification techniques, is essential.

With all aspects considered, a successful, long-term program can be planned. Personnel, costs, population, and the setting of the program must be blended into a workable unit. To make sure planning is accurate and goals are realistic, investigation into other programs deemed successful should be supported. Those persons that are to be served should be interviewed and program growth should be slow and critically evaluated and modified as necessary. A pilot program should be instituted with carefully selected participants that are willing to evaluate program strengths and weaknesses. Once all aspects have been tested and evaluated, then, implementation with a better chance of success and a minimal amount of problems can be expected.

Maintenance of the employer's valuable machinery is always a top priority of successful businesses. Since employees are the most valuable resource available for industry, their physical status must likewise be a top priority, and maintained in some regular and organized fashion (Barnes, 1983).

RESOURCE GUIDE FOR FITNESS PROGRAMS

Organizations

American Alliance of Health, Physical Education, Recreation and Dance
1900 Association Drive
Reston, VA 22091

American Association of Fitness Directors in Business and Industry
Corporate Park Two
3 Gannett Drive
White Plains, NY 10604

American College of Sports Medicine
1440 Monroe Street
Madison, WI 53703

President's Council on Physical Fitness and Sports
Department of Health and Human Services
Washington, DC 20201

Equipment/Supplies

EKG/Treadmills/Bicycle Ergometers/General Supplies

Blood Pressure Kits
Owl Biomedical
Lewis Associates
Carolina Biological Supply Co.
2700 York Road
Burlington, NC 27215
(800)334-5551

Grip Dynometer
Smedley Grip Dynometer
Owl Biomedical
Lewis Associates

Lafayette Instruments
P.O. Box 1279
Sagamore Parkway
Lafayette, IN 47902
(317)423-1505

Lewis Associates
10605 Concord Street, Suite 302
Kensington, MD 20895
(301)942-1112

Monark Corp.
c/o Quinton Instruments
2121 Terry Avenue
Seattle, WA 98121
(206)223-7873

Owl Biomedical
455 Cayuga Park
Buffalo, NY 14225
(716)634-0458

Propper Manufacturing Co., Inc.
Diagnostic Instrument Division
36-04 Skillman Avenue
Long Island City, NY 11101

Stoelting Co.
1350 S. Kostner Avenue
Chicago, IL 60623
(312)522-4500

Student Spirometer
Phipps & Bird, Inc.
P.O. Box 27324
Richmond, VA 23219

The Burdick Corp.
Milton, WI 53563

Journals

Corporate Fitness and Recreation

Health Education

Journal of Sports Psychology

Medicine and Science in Sports and Exercise

Research Quarterly of Exercise and Sport

Sports Medicine Digest

The Physician and Sports Medicine

REFERENCES

Alderman, M. ''Hypertension control program in occupational settings.'' *Public Health Reports,* 1980, **95,** 158–163.

American College of Sports Medicine. *Guidelines for Graded Exercise Testing and Exercise Prescription.* Philadelphia: Lea and Febiger, 1980.

American Heart Association. *Exercise Testing and Training of Apparently Healthy Individuals: A Handbook for Physicians.* New York: American Heart Association, 1972.

American Heart Association, The Committee on Exercise. *Exercise Testing and Training of Individuals with Heart Disease or at Risk for its Development: A Handbook for Physicians.* Dallas, TX.: American Heart Association, 1975.

Barnes, L. ''AAFDBI: Bringing fitness to corporate America.'' The Physicans and Sports Medicine, January, 1983, **2,** 127–133.

Boyer, J. and Kasch, F. ''Exercise therapy in hypertensive men.'' *Journal of the American Medical Association,* 1970, **211,** 1668–1671.

Brennan, A. ''Health promotion in business: Caveats for success.'' *Journal of Occupational Medicine,* 1981, **23,** 639–642.

Brown, J. ''An incentive-based employee fitness program.'' *Health Education,* March/April 1981, **12,** 23–24.

Chavat, J. Dell, P., and Folkow, B. ''Mental factors and cardiovascular disorders.'' *Cardiologia,* 1964, **44,** 124–141.

Chin, C., Messinger, J., Greenberg, P., and Ellestand, M. "Chronotropic incompetence in exercise testing." *Clinical Cardiology,* 1979, **2,** 12–18.

Clarke, D. *Exercise Physiology.* Englewood Cliffs, N.J.: Prentice-Hall, 1975.

Cooper, K., *Aerobics.* New York: Bantam Books, 1968.

Cooper, K. *The Aerobics Way.* New York: Bantam, 1977.

Criqui, M. "Cigarette smoking and plasma high density lipoprotein cholesterol." *Circulation,* 1980, **31,** 97–108.

Cunningham, R. *Wellness at work.* Blue Cross Association, Chicago, IL, 1982, 22–27, 108–114.

Dishman, R. "Compliance/Adherence in health-related exercise." *Health Psychology,* 1982, **1,** 237–267.

Dishman, R. and Gettman, L. "Psychobiologic influences on exercise adherence." *Journal of Sports Psychology,* 1980, **2,** 295–310.

Donoghue, S. "The correlation between physical fitness, absenteeism, and work performance." *Canadian Journal of Public Health,* 1977, **68,** 201–203.

Epstein, L. and Wing, R. "Aerobic exercise and weight." *Addictive Behaviors,* 1980, **5,** 371–388.

Everly, G. S. and Rosenfeld, R. *The Nature and Treatment of the Stress Response: A Practical Guide for Clinicians.* New York: Plenum, 1981.

Folkins, C. and Sime, W. "Physical fitness training and mental health." *American Psychologist,* 1981, **36,** 373–389.

Fox, E. *Sports Physiology.* Philadelphia: Saunders, 1979.

Fox, S. and Boyer, J. "Physical activity and coronary heart disease." *Physical Fitness Research Digest,* April 1972, **2,** 1–13.

Froelicher, V. et al. "Physical activity and coronary heart disease." *Cardiology,* 1980, **65,** 153–190.

Gambert, S., Hagen, T., Garthwaite, T., Duthie, E., and McCarty, D. "Exercise and the endogenous opioids." *New England Journal of Medicine,* 1981, **305,** 1590–1591.

Hall, S. and Hall, R. "Outcome and methodological considerations in behavioral treatment of obesity." *Behavior Therapy,* 1974, **5,** 59–68.

Harris, L. *Fitness in America.* Washington, D.C.: National Center for Health Statistics, 1978.

Haskell, W., "The physical activity component of health promotion in occupational settings." *Public Health Reports,* 1980, 109–118.

Haskell, W. and Blair, S. "The Physical Activity Component of Health Promotion in the Occupational Setting." In *Managing Health Promotion in the Workplace,* R. Parkinson et al., eds. New York: Mayfield, 1981.

Himes, J., Roche, A., and Stiernogel, R. "Compressibility of skinfolds and the measurement of subcutaneous fatness." *American Journal of Clinical Nutrition,* 1979, **34,** 1734–1740.

Horton, E. "The role of exercise in the treatment of hypertension in obesity." *International Journal of Obesity,* 1981, **5,** 165–171 (Supp.).

Johnson, L. "Fitness testing: The basic ingredients." *Athletic Purchasing and Facilities,* March 1983, **7,** 36–43.

Kannel, W. *The Framingham Heart Study: Habits and Coronary Heart Disease*. Washington, D.C.: U.S. Government Printing Office, 1966.

Kannel, W. and Sorlie, P. "Some health benefits of physical activity: The Framingham Study." *Archives of Internal Medicine*, 1979, **139,** 857–861.

Kraus, H. and Raab, W. *Hypokinetic Disease*. Springfield, Ill.: Thomas, 1961.

Layman, E. "Psychological effects of physical activity." In *Exercise and Sports Sciences Reviews*, J. H. Wilmore, ed. New York: Academic Press, 1974.

Lohman, T. "Skinfolds and body density and their relation to body fatness: A review." *Human Biology*, 1981, **53,** 181–225.

Marcotte, B. and Price, J. "The status of health promotion programs at the worksite." *Health Education*, July/August 1983, 4–9.

Martin, J. and Dubbert, P. "Exercise applications and promotion in behavioral medicine." *Journal of Consulting and Clinical Psychology*, 1982, **50,** 1004–1017.

Mayer, J. *Overweight: Cause, Cost and Control*. Englewood Cliffs, N.J.: Prentice-Hall, Inc., 1968.

Miller, G. "High density lipoproteins and atherosclerosis." *Review of Medicine*, 1980, **31,** 97–108.

Noland, M. "The Efficacy of a Model to Explain Leisure Exercise Behavior. Dissertation, University of Maryland, 1981.

Oldridge, N. "Compliance and exercise in primary and secondary prevention of coronary heart disease: A review." *Preventive Medicine*, 1982, **11,** 56–70.

Pearson, C. "The emerging role of the occupational physician in preventive medicine, health promotion, and health education." *Journal of Occupational Medicine*, 1980, **22,** 104–106.

Physical Fitness Research Digest, President's Council on Physical Fitness and Sports. April 1974, 1–27.

Sacks, M. H. and Sachs, M. L., eds. *Psychology of Running*. Champaign, Ill.: Human Kinetics Publishers, 1981.

Sachs, M. L. and Buffone, G., eds. *Running As Therapy*. Lincoln, Neb.: University of Nebraska Press, 1984.

Scheuer, J. and Tipton, C. "Cardiovascular adaptations to physical training." *Annual Reviews of Physiology*, 1977, **39,** 221–251.

Soman, V., Koivisto, V., Deibert, D., Felig, P., and DeFronzo, R. "Increased insulin sensitivity and insulin binding to monocytes after physical training." *New England Journal of Medicine*, 1979, **301,** 1200–1204.

Stoedefalke, D. "Physical Fitness Programs for Adults." In *Exercise in Cardiovascular Health and Disease*, K. Amsterdam et al., eds. New York: Yorke Medical Books, 1977.

Stone, W. "Motivation for fitness." *Arizona Journal of HPRR*, Fall 1980, **24,** 5–7.

Strauss, R., ed. *Sports Medicine and Physiology*. Philadelphia: Saunders, 1979.

Weller, D. "Dundalk Community College Fitness Center Procedures Manual." Unpublished Document, 1981.

Weller, D. "Elements of Self-Perception in the Maintenance of an Attained Fitness Proficiency and Body Composition." Dissertation, University of Maryland, 1983.

Wilmore, J. *Training for Sport and Activity: The Physiological Basis of the Conditioning Process*, 2nd ed. Boston: Allyn and Bacon, 1982.

Wilmore, J. "Acute and Chronic Physiological Responses to Exercise." In *Exercise in Cardiovascular Health and Disease*. K. Amsterdam et al. New York: Yorke Medical Books, 1977.

Wilson, P., Fardy, P., and Froelicher, V. *Cardiac Rehabilitation, Adult Fitness, and Exercise Testing*. Philadelphia: Lea and Febiger, 1981.

Zieler, K., Maseri, A., Klassen, D., Rabinowitz, D., and Burgess, J. "Muscle metabolism during exercise in man." Transcripts of the Association of American Physicians, 1968, **81,** 266–268.

CHAPTER 9

Employee Assistance Programs—Alcohol Abuse and Related Concerns

Melvin Sandler
Steven A. Sobelman

INTRODUCTION

When an employee is troubled by personal problems involving marital or family strife, financial concerns and worries, depression or lowered self-esteem, alcohol abuse, and so on, it is less likely that the employee will contact an employer as a comforting listening ear for assistance in the mitigation or resolution of these problems. More specifically, employees who suffer from alcohol abuse and its related problems more consistently than not shy away from personal contact with a supervisor or employer lest the employee jeopardize his or her employment. In this regard, employees will typically deny that alcohol interferes with work performance or productivity on the job. The employee's fear of job loss may cause the masking of personal problems. As can quickly be seen, an adversarial position may begin to exist between the employer and the employee when personal problems begin to surface. For example, would union officials suggest that employees "run" to employers if alcohol-related problems begin to occur in the employee's life? The inherent fear may be that by going to the employer, the focus of attention may be placed on the employee for the problem and this may hide the fact that working conditions may be a strong factor in the worker's response (through alcohol abuse) to poor work conditions or stress in the workplace. Therefore, where is the employee to seek assistance? Does the employer wish to "salvage" an alcoholic worker whose mistakes and accidents may have already been costly to the company? These questions and questions involving alliance and advocacy roles for employer/employee relations rather than adversarial roles need to be addressed when employees suffer from alcohol abuse and its related concerns.

Answers to questions involving advocacy roles for employer/employee relations that focus on alcohol-related concerns need to begin with an awareness and acceptance that the "problem" has a basis in reality. With this notion in mind, local, state, and federal agencies have begun to recognize the negative effects attributed to alcohol-related concerns on individuals in the workplace. Employers have begun to feel the financial burden and impact that alcoholism has on the American worker's productivity. According to Donald Goodwin, director of Occupational Branch of the National Institute of Alcohol Abuse and Alcoholism, more than 5000 companies and government agencies are instituting programs to assist individuals who have succumbed to the ill effects of alcohol abuse (Wrich, 1983). During the last 40 years, U.S. companies and unions have established workplace assistance programs. Such programs have commonly been referred to as employee assistance programs (EAPs). Leo Perlis, director of Community Services for the AFL–CIO, believes that the employer and union should be in concert in assisting the employee with his or her alcohol problems (Perlis, 1977). Although EAPs tend to be alcohol-focused, more recently they are encompassing all aspects of employee concerns: financial, psychological, and health.

What follows is a discussion of the origin of employee assistance programs, their goals, and an example of a model program.

SCOPE OF THE PROBLEM

"If the nation were confronted with an epidemic that left 26,000 people dead every year, there would be a ground swell of public outrage that would not end until a halt was put to the carnage. Drunk driving alone takes 26,000 lives a year but no outcry is heard." (Senator Dan Quayle, U.S. Senate, R-Indiana, 1983).

Over the past two decades, research has supported the theories that alcoholism is taking its toll in the American work force. The Cooperative Commission on the Study of Alcoholism estimated that close to 50% of all problem drinkers in America are currently employed (Plant, 1967). In the early 1970s, Bertram Brown, director of the National Institute of Mental Health, indicated that emotional problems were responsible for approximately 20–30% of employee absenteeism and from 15–30% of the work force were seriously handicapped by emotional problems (Weiner, Akabos, and Sommer, 1973). It is estimated that lost production caused by problem drinking amounts to $9.35 billion annually (U.S. Department of Health, Education, and Welfare, 1974). In a report to the Florida Office of Mental Health, Jarvis (1975) concluded that employees who were problem drinkers were responsible for 15% of insurance claims to the industry annually. Rates for alcoholics were found to be significantly higher when related to absence and medical costs due to on- and off-the-job accidents, digestive and musculoskeletal disorders and respiratory infections. Trice and Roman (1978) indicated that declining work performance and disruption of activites of co-workers and supervisors are major results of deviant drinking. They further found that blue-collar and service employees are those individuals with the most alcohol-related problems. However, they clarified their findings by stating that all levels of employees suffer decreased work efficiency and that problem drinking may be more frequent in "high-status" occupations with no particular industries having a higher concentration.

Mannello (1979) studied seven large railroads and found that drinking among male workers employed by the railroads was equal to the national average. However, it was also shown that 44,000 of the 234,000 railroad workers were problem drinkers. Mannello's research supported estimates by the National Council on Alcoholism (1976), which stated that employers lose approximately 25 cents on every dollar paid in wages to alcoholic employees. The council further estimated that, compared to the average nonalcoholic employee, alcoholic employees:

1. Have an accident rate 3.6 times higher.
2. Have 2.5 times more absences of 8 days or longer.
3. File 5 times more compensation claims.
4. Receive 3 times the sickness benefits.
5. Are subject to garnishment proceedings 7 times more often.

Mannello (1979) went on to indicate that the railroad industry spent more money to fire a problem drinker ($1050) than they paid to rehabilitate the problem drinker ($840). He estimated that the company costs of employee drinking in 1978 were

Item	Cost
Absenteeism	$3.1 million
Lost Productivity	$25 to $100 million
Injuries	$0.58 million
Accidents/damage	$0.65 million
Insurance premiums	$2.30 million
Grievance process	$0.41 million

These above figures seem consistent with lost monies associated with alcohol abuse in many major companies. Jack Rose, employee assistance counselor at Lockheed of California, Burbank, estimated that alcohol-related problems cost his company $932,000 in 1982 in grievance procedures, $547,000 for treatment of illness and accidents by plant medical personnel, and $6.6 million for sick leave. In 1982, Lockheed paid more than $8 million for alcohol-related problems (O'Connor, 1984).

From a medical standpoint, a causal relationship between heavy alcohol intake and cancer, heart disease, liver disorders, and central nervous system dysfunction has also been found (U.S. Department of Health, Education, and Welfare, 1974). See Table 9.1 for an illustration of the health-eroding complications from long-term alcohol abuse. Certainly, these illnesses bring with them a related employer cost.

Behaviorally, it comes as no surprise that the employee with alcohol problems falls down in productivity. Furthermore, the employee begins to have unpredictable mood swings, alternates in his or her feelings of guilt, hostility, depression, lone-

Table 9.1 Health-Related Problems Associated with Long-Term Alcohol Abuse

Disease	*Etiological Mechanisms*
Cirrhosis of the liver	Unclear at this time, although fatty infiltration of liver cells and malnutrition suspected
Pancreatitis	Unclear at this time, however toxic and malnutritive processes suspected
Gastritis	Toxic, inflammatory processes
Cardiomyopathy	Fatty infiltration of myocardial tissue
Peripheral neuropathy	Damage to peripheral nerves through malnutrition
Wernicke-Korsakoff syndrome	Destruction of brain cells through chronic malnutrition

liness, and many times feels powerless or helpless to change. The person may begin to have personal or marital, financial, and/or legal difficulties. The alcoholic withdraws from support systems and generally will deny that there are any problems. Basically, this individual points a blaming finger at work, society, or others as having the ''problem'' and asks to be left alone (O'Donnell and Ainsworth, 1984).

In 1982, the Employment and Productivity Subcommittee of the United States Senate Committee on Labor and Human Resources estimated that there are approximately 100 million people employed in the United States. Approximately 10% of these individuals are afflicted by alcoholism. An additional 35 million people are indirectly (families, relatives, neighbors, co-workers) affected by a problem drinker. It is further estimated that there is a cost in excess of $30 billion in lost productivity due to alcohol-related problems. The Committee's findings were supportive of past research concerning employee absenteeism, accidents, compensation claims, and industrial fatalities due to alcoholism.

Thus, it becomes quite obvious that employees' personal problems and specifically those problems caused by alcoholism exact a great toll from the workplace. A rationale for the EAP begins to emerge.

OCCUPATIONAL ALCOHOLISM PROGRAMS (OAP): PREDECESSOR TO THE EMPLOYEE ASSISTANCE PROGRAM

Attempts on the part of employers and unions to assist employees with their personal problems originated with a focus on one major problem: alcohol. Occupational alcoholism programs had their roots in the early 1940s in order to assist employees whose jobs were negatively affected by their use of alcohol. In 1960 the National Council on Alcoholism developed an industrial department to encourage and assist companies in developing alcoholism programs. This service is currently very active.

Recognition and identification of the employee with alcohol problems is a seemingly frustrating task for supervisors. Today, training of supervisors can enhance their "early detection" of potential problem drinkers and make proper intervention. E. M. Jellinek, a Yale researcher, identified five major categories of alcoholics:

The "alpha" type. This individual needs a psychological lift or boost to increase self-esteem in order to reduce inhibitions or gain courage.

The "beta" type. This individual drinks in order to reduce physical pain. The irony here lies in the fact that drinking may indeed cause more pain than it may tend to relieve.

The "gamma" type. Approximately 90% of all Americans fall into this category. These individuals have a combination of physical and psychological dependence and tissue tolerance to alcohol. As the tolerance increases, more alcohol is needed to achieve the desired effect.

The "delta" type. These individuals are physiologically but not psychologically dependent on alcohol.

The "epsilon" type. These are the "binge" drinkers who alternate between abstinence and "benders."

However, alcoholics tend to deny that they have a problem with drinking and go to great lengths to defend their behavior (Kellerman, 1980). Supervisors are

Table 9.2 Immediate Effects of Alcohol Consumption—Blood Alcohol Levels

Number of Drinks[a]	*Blood Alcohol*[b,c] *Concentration*	*Effects*
1	.02–.03%	Relaxation, mood elevation, disinhibition
2	.05–.06%	Increased disinhibition, impaired muscular control
3	.08–.09%	Slight impairment of balance, speech, and hearing
4	.11–.12%	General decrease in mental functioning, significant balance and motor control problems
5	.14–.15%	Slurred speech, blurred vision, loss of muscular control
10	.30%	Severe intoxication, stuperous
15–20	.40–.50%	Unconscious, near coma
>20	.55%	Death from respiratory failure

[a]= one drink equals one beer (12 oz.) or mixed drink with 1 oz. whiskey
[b]= for 150-pound person with an empty stomach
[c]= for each one-hour time lapse, subtract about one drink

learning, though, that repeated or chronic absenteeism, tardiness, unexplained "disappearances," and the like may warrant consideration for making a referral to a "counseling" resource. Table 9.2 lists the immediate effects of alcohol use, while Table 9.3 lists the general diagnostic criteria for alcholism. A word of caution is needed, however. As we know, a little bit of knowledge can be dangerous. It is very important for supervisors to be trained and educated in the diagnostic procedures associated with possible alcohol detection. Roman (1981) believed that participation in alcoholism programs by the employee who has a drinking problem should be offered as an alternative to a job action. Currently, occupational alcoholism programs are in operation and endorsed by many groups, yet have taken on a more "broad brush" approach and recognize the wide range of social, emotional, family, legal, medical, and financial problems that have an impact upon employees. Thus, the Employees Assistance Program, with its broader scope has emerged.

Table 9.3 General Diagnostic Criteria for Alcoholism (National Council on Alcoholism)

Behavioral
Drinking in gulps
Morning drinking
Blackouts
Continued drinking despite erosions of health and/or job performance
Several unsuccessful attempts to stop drinking
Psychological
Drinking to relieve stress, anger, fear, resentment, or related emotions
Feelings of loss of control
Denial of obvious drinking problem
Severe mood changes while drinking
Physical
Pancreatitis
Gastritis
Wernicke-Korsakoff syndrome
Cirrhosis
Peripheral neuropathy
Cardiomyopathy
Dependence as evidenced by signs of withdrawal:
Tremors
Hallucinations
Seizures

EMPLOYEE ASSISTANCE PROGRAMS (EAP)

The focus of the Employee Assistance Program (EAP) is to offer assistance to employees with social, emotional, family, legal, medical, and financial concerns. This focus points toward the improvement of the employees' on-the-job performance and productivity by offering confidential counseling or treatment for a myriad of employee concerns. Current estimates indicate that approximately 10% of all employees have serious or severe emotional problems; 25% of productivity is lost at all ages, but especially from employees between ages 30 to 49; approximately 50% of employee problems are alcohol-related; and, the use of health insurance by alcoholics is three times that of nonalcoholics (Berry, 1981).

Programs have been established within business and industry to combat the alarming statistics associated with lost productivity. The following are selected examples of companies that have utilized EAPs:

1. Illinois Bell Telephone began its alcohol assistance program in 1951. A direct savings of $459,000 from reduced absences associated with alcohol problems was seen in a study of the 650 employees participating in an alcohol rehabilitation program. "Poor performance" ratings dropped from 28 to 12% within this group. Additionally, 58% of those in the program had recovered from alcoholism and another 19% had shown vast improvement (Asma et al., 1975).
2. In 1975, the Oldsmobile Division of General Motors revealed an estimated savings of $226,334 after individuals had begun to utilize the EAP. Decreases in the following areas were also noted: lost work hours (49%), sickness and accidents benefits paid (30%), leaves of absence (56%), grievances (78%), disciplinary action (63%), and on-the-job accidents (32%). These statistics are quite impressive (Jones and Vischi, 1979). General Motors now boasts that there is an estimated 3 to 1 return on dollars invested in its EAP (Berry, 1981).
3. Kimberly Clark employees participating in an EAP had a 70% reduction in on-the-job accidents (Goldbeck and Kiefhaber, 1980).
4. Kennecott Copper Corporation of Salt Lake City reported that they were receiving a $6 return for every $1 invested in their EAP. Approximately 9000 employees have utilized the EAP since its inception in 1970. Absenteeism was reduced by 53%, sickness and accident costs dropped by 75%, and health, medical, and surgical costs were cut by 55% (Sehnert and Tilloston, 1978).
5. Equitable Life Assurance Society indicated a $5 to $1 ratio when looking at the rate of return for individuals involved in the company EAP (Berry, 1981).
6. E. I. duPont de Nemours & Co., a pioneer in alcohol-abuse programs since 1942, saves $419,200/year from decreased absenteeism, since an

alcoholic employee averages 13 disability days/year when compared to only 5.8 for a control group (Shenert and Tilloston, 1978).

Certainly, the list of successful uses of the EAP can be shown and documented. However, the obvious advantages far outweigh any potential "downside" effects of the EAP. It is important to reiterate the statement that the EAP is developed to respond to employee needs. The scope, extent, and style of service vary from setting to setting. All of these services are keeping in line with the basic goal of the EAP—to provide personnel with alternatives for overcoming problems that interfere with productivity.

Vehlow and Kropp (1983) have developed a number of objectives that are directed toward achieving the goal of the EAP:

I. Services to the Employee
- **A.** Prevention of work problems through early detection of problems.
- **B.** Motivation of employees to seek help when necessary.
- **C.** Assessment of employee problems.
- **D.** Referral of employees to the best sources of assistance.
- **E.** Monitoring of treatment employees are receiving.

II. Services to the Company
- **A.** Assistance in development of a policy statement that will clearly outline the company's EAP in relation to personnel problems.
- **B.** Training supervisors to recognize problems, effectively and efficiently confront inadequate work performance, and make referrals for assistance.
- **C.** Promoting the EAP within the company through dissemination of written material.
- **D.** Ensure the confidentiality of services to employees.
- **E.** Problem assessment and treatment of the employment data.
- **F.** Keeping the company informed of EAP use through reports of nonconfidential statistical data.
- **G.** Provide effective follow-up services for employees who enter the EAP.

It is important to note here that only through educating the employee about the mechanics of the EAP will he or she properly utilize its services.

GUIDELINES FOR PROGRAM DEVELOPMENT

Several factors need to be considered for the effective utilization of the EAP. Clearly, the success of the EAP is based upon the communication between management, unions (if applicable), and employees. The employee should view the EAP as constructive and in his or her best interest, as opposed to a destructive, "witch hunt" perspective. An EAP in its earlier developmental stages needs to address the following issues:

1. Location.
2. Policy statements.
3. Confidentiality.
4. Insurance issues.
5. Goals and measurement issues.
6. Staffing.
7. Resource development.

Foundation Building

Location. Initial issues of access and program location are very important. In order to service employees at various levels a program must have access to supervisors at that level. It must also be able to develop priorities for functions needed to keep the program operating effectively (Erfurt and Foote, 1977). Thus the need for the EAP to have support from and access to decision makers (Yasser and Sommer, 1974). Certainly it must be located in a department that is the most compatible to its goals. Roman notes that in programs for alcoholic employees in corporations studies, an equal percentage of programs is based in personnel or labor relations departments and shares responsibility with the above and medical departments. A slightly lesser number is located primarily in medical departments.

Erfurt and Foote (1977) describe criteria for assessing which departments in a particular company are best able to house the program and offer program satisfaction.

These are

1. The "helping" atmosphere associated with the organizational unit.
2. The degree to which the unit was a center of communication about employees.
3. The degree of autonomy and independence granted to the program (Erfurt and Foote, 1974).

Unions also have several possible auspices. A union-based program can be located in the local structure, in the benefit or health and welfare department. The benefit and health and welfare departments are often seen as service arms of the union and administer benefits and medically related services for the member locals. The programs sponsored by the International Ladies' Garment Workers' Union and District Council 37 of the American Federation of State, Country and Municipal Employees are located in this manner. Both connect with the locals as a means of reaching the memberships (Yasser and Sommer, 1974; Bailey, 1979).

Policy Statement. A policy statement is a critical component in the foundation-building process. It needs to have the support of representatives of all parties. It should be drafted as a collaborative venture with the EAP administrator facilitating the process. Thus a program in a company or agency needs inclusion of relevant departmental leaders as well as leadership of all representative unions (Byers, 1976). Similarly, a union that represents employees of various small companies or agencies needs internal support of the policy and representation of the employers. (Weiner et al., 1973). Developing such a policy in itself can be seen as an initial effort in pulling all of the parties together in a mutual effort related to the EAP.

The policy statement governing the United Airlines employee assistance program was developed by the administrator in collaboration with top management and leadership from the representative unions. It includes the protections indicated above, separates engagement in the EAP from all industrial reactions issues and outlines the manner in which the policy is to be implemented. The roles and obligations of each of the parties are included. Sample policy statements are outlined by Wrich and Byers (Wrich, 1980; Byers, 1976). See Appendix (page 164).

Confidentiality. The issue of confidentiality exists each time we share private information with someone in a professional capacity. Thus, if we were to confide in a lawyer, doctor or clergyperson we would hope that what we said would not go further. There are laws governing confidentiality. These include federal laws for federal agencies and state laws governing state-licensed agencies and practitioners. Clearly, aware people would not share openly with a professional if they knew that what they were about to say could become public knowledge.

An EAP must be concerned about this concept as well. An employee must be aware of the conditions under which his or her contact with the program will be shared by people inside and outside the company and/or union. Clearly, the confidentiality practices must conform with state or federal statutes (Akabas, Bellinger et al., 1981).

Employees should be able to control who will be aware that they are using the program and specifically the nature of what will be shared, why, with whom, and when. They should also have the right to revoke this permission. This informed consent should be by a written release signed by the employee (see page 164). This protects both the employee and the program. If the rules governing confidentiality are part of the policy statement, the program is able to give it the importance merited.

It is important to be clear under what circumstances information will be shared without consent. Such disclosure is governed by state and federal statutes. One example might be knowledge that an employee can be a danger to self or others.

Location of the program is also an important element here. If the office is in a heavily trafficked area or one exclusively used for it, confidentiality is violated as soon as someone is seen walking in or out of the door. Thus the need for a separate location or one in which the entrance is not obvious. In some instances both in-house and contracted programs see employees off-premises. (Business Week, 3/27/79). However, this might limit convenience and offer less availability (Akabas, Bellinger et al., 1981).

It should also be noted that in making a referral, the referrer is giving up some control and risking exposure of his or her department or function through the employee's open sharing of his or her perception of conditions to the EAP counselor. If the EAP were to share this information with supervisors who might violate the rights of the referrer, it might also discourage supervisors from referring employees. Confidentiality is also an important safeguard for the referrer.

Insurance. Clearly, as consumers we can only purchase what we can pay for. This principle applies as well to purchases of mental health and alcoholism services. A program that advertises assistance to employees with emotional or alcohol problems is obliged to have the tools to carry out its mandate. If a counselor assesses that an employee needs in-patient or out-patient care he or she can only recommend that the employee seek an affordable service.

Thus, in addition to knowing their economic status, it is important that the company or union's insurance plan by analyzed. Knowledge of which services are covered and the extent of such coverage will enable the counselor to develop a realistic pool of community resources for the employee population. In some instances the implementation of such a program has influenced the insurance coverage (Eggum, et al., 1980).

Goals and Measurement. Goals of the EAP are related to needs of the employee, the employer, and where there is representation, the union. Hopefully, implementation will be consistent with all of the above.

In order to continue functioning, a program must consistently have the support of decision makers both from management and union spheres. This is seen as the basis for an effective program (Miller, 1981; Benedict, 1973). Leo Perlis, former director of Community Services for the AFL-CIO asks that the adversarial relationship between labor and management be put aside when assisting employees with personal problems (Perlis, 1977).

Job performance and maintenance are concerns that affect all in the workplace. Erfurt and Foote in a study of 21 occupational programs indicate that a goal that separates them from community treatment agencies is that "an occupational program . . . has concrete incentives for insuring successful treatment for its client . . . improving poor work performance. . . . Employee Assistance Program personnel are in a position to monitor the client's work activities while the treatment agency is not." (Erfurt and Foote, 1977). Someone who goes to a community service for perceived needs can discontinue for a number of reasons, including motivational issues of the client or appropriateness or adequacy of the service in meeting those needs. He or she may also continue without experiencing a change in functioning. The assurance that an employee gets adequate and appropriate care that functionally relates to his or her needs is one major goal of the employee assistance program.

A program that is effectively implemented will be accepted and utilized by the employee population. Supervisors and union officials will endorse and refer personnel as well as utilizing it themselves. Employees will recommend that co-workers utilize it. Outreach efforts will also effectively reach employees who are in need of help. Thus, one measure of a program's goal attainment is the utilization rate. The counseling program of Polaroid services 5% of the employee population annually (Miller, 1981). According to the United Airlines Medical Department's Employee Assistance Program, the utilization rate is 3.9%. However, it was first implemented three years ago in its executive office and services 5.5% of the company's upper management. According to Wrich, the annual rate should approximate 1.5–2% of the total work force. It is difficult to compare utilization rates. Thus a program with group services might have a different utilization criterion than one that only services individuals. Issues related to geographic accessibility, closeness of supervision, and attitudes toward mental health services must also be considered.

The counselor can monitor care only in instances related to job performance or expand his or her role as suggested by Wrich: ". . . to assist employees who either have job performance problems or are likely to as a result of personal problems." (Wrich, 1982). Miller and Perlis perceive the monitoring role extending to all types of personal problems (Miller 1981; Perlis 1977). Labor's concern about job performance and maintenance is illustrated by the union-based program described by Yasser and Sommer (Yasser and Sommer, 1974). Instruments developed by Wrich and Eggum are applicable here (Wrich, 1982; Eggum et al. 1980). They both elicit reports of supervisors regarding the performance assessment over time. Eggum utilizes quality reporting while Wrich has a quantitive scale.

Given the destructive and progressive nature of alcoholism, it is hoped that an industrial program will be able to identify and reach alcoholics in representative numbers. The position of the National Institute of Alcohol Abuse and Alcoholism in endorsing the employee assistance rather than the occupational alcoholism approach is that it is more likely to identify and treat alcoholic employees and do so at an earlier stage (U.S. Department of Health, Education, and Welfare, 1973). Roman relates concern that this goal and its measurement can be intruded upon by the employee assistance focus (Roman, 1981). It would be of interest to see a comparison of the utilization by alcoholics of an OAP and EAP. Differences in setting and counselor perception in these respective programs may limit the reliability of such data. In general, it would be hoped that due to the nature of the program in an occupational setting, alcoholics would have a greater chance of recovery. Thus, the measure of recovery rates would be important in evaluating this goal. The reported rates of recovery for United Airlines employees is 92% for pilots, 82% for flight attendants, and 73% for ground personnel. The Kelsey-Hayes OAP estimates a recovery rate of between 65% and 70% (Byers, 1976). One must be aware that criteria for recovery and its measure can vary. Furthermore, if recovery implies not drinking, can we be sure that job-site observers can reliably agree if someone is drinking? Does recovery imply job performance improvement? What period of time is being considered?

A measure that some programs use is cost-effectiveness. It would seem that if utilization is high and people's functioning and dependability improve with the proper help, the company will benefit financially. Bryons calculates cost-effectiveness based upon decreased absence-related costs at $1627 per employee serviced. Eggum includes savings of disability and medical costs in her calculation (Eggum et al., 1980). Foote has developed a comprehensive approach to calculating cost effectiveness (Foote, 1978). It would seem that such data could be an aid in presenting the program's effectiveness to the decision makers of the company. They often need to justify both allowing and limiting various expenditures (to a great extent) in relation to the bottom line. The consideration is probably more important during difficult economic circumstances when resources are limited. It would appear though, as suggested by Benedict and Miller, that in order to commit the time and resources needed to establish a program, there must be a fundamental concern for the employees' well-being. (Benedict, 1973; Miller, 1981).

Staffing. The staff of an EAP must be chosen carefully. It must be able to accurately and quickly assess a broad range of problem areas, develop a plan for problem resolution, assess and help people utilize community resources, intervene in crises, interface appropriately with various job-site representatives. Immediate functional and interpersonal problems on the job site are often lingering in the forefront and require a staff that comes with strong skills.

According to Wrich (1982) a company with 3000 or more employees in one area merits a full-time permanent employee to coordinate the EAP. In their study Erfurt and Foote discovered that the mean average of one EAP staff member serviced 2400 employees, with substantial variation between programs. According to James Francek, who served as the director of Ford Motor Company's EAP, a company would justify one mental health professional for every 2500 employees (Business Week, 1979). It can be expected that utilization will vary and can be influenced by programmatic actions.

It is important that the criteria for staffing be clearly thought through. Erfurt and Foote (1977) noted in their survey that the majority of EAP staff have masters-level training in social work, education, psychology or counseling. A lesser number have bachelor-level social-work training and some have nursing training. Guida (1967) and Eggum (1980) suggested that given their special contact with employees, occupational health nurses are in a good position to assist alcoholic employees. In addition to criteria related to skill and empathy, Wrich (1982) puts strong emphasis on the staff's knowledge of alcoholism and chemical dependence. As cited by Roman (1981), staffing by recovering alcoholics with no other credentials did not prove adequate to meet the broad spectrum of employee needs. A clinician whose training is in long-term psychotherapy or who is locked into any one approach for providing care will be less likely to assess openly a broad spectrum of resources and less likely to have the flexibility to match an individual with an available resource. Similarly, someone who is biased in favor of management or labor will be less likely to assist an employee in his or her interactions with either side. There is also value in having someone who can relate well to groups of employees in an outreach effort. Given the nature of alcoholism and other types of chemical dependence and their likelihood to manifest in overt disruption, expertise in crisis intervention is essential. Given the complexity of the above, consideration of both psychological and medical consultation for the EAP staff would also appear to be of value.

Companies with too few employees to justify a full-time staff person for an EAP need to view other alternatives. A company can contract with a provider who will offer services similar to an in-house program. There are various groups offering these services. Hale (1974) described the efforts of a community mental health center to approach industry and develop programs with full policy and outreach and service delivered in the center (Hale, 1974). Leeman describes a program in which a hospital assigns a counselor to a bank to service its employees on a contracted basis. Family Service Association of America provides a primer for member agencies on contract with industries (Leeman, 1974). McQuirk (1980) discusses a consortium model in which the chief executives of local companies appoint representatives to a board of directors that establishes policies, procedures, and monitors the program. In this instance, a college counseling service provides the technical and service delivery expertise. Additionally, it is estimated that over 100 independent mental health organizations have been formed solely to service corporate clients. Their service is provided on a fee-per-covered-employee basis. Some provide assessment and referral and some full counseling services (Business Week, 1979). Given this range of services and the aggressive marketing that is done, a company is wise to use consumerism to assess these providers. They should consult with similar companies to assess their experiences. They should also maintain a liaison through a department that is best equipped to monitor the process. Labor unions also have similar arrangements. In certain locations, labor unions, in concert with local community services, provide assistance for their members as well (Perlis, 1977).

Resource Development/Networking. Clearly, if the EAP offers assessment and referral, a major responsibility of the counselor is to connect the employee with the best resource for his or her particular problem. This requires that a pool of community-service providers be developed that are able to help with the resolution of specific problems. If job function is being affected, the help must provide relatively quick behavior change.

Weissman (1977) suggests a linkage technology that includes keeping up with and maintaining open access to relevant service providers and follow-up with clients in problem resolution. In a program that is offered in a setting for workers whose income is limited and are without or with limited ambulatory mental-health and/or alcoholism-treatment insurance, the counselor must consider sliding-scale, low-cost, and quick availability with required skills. For example, the Personal Service Unit of District Council 37 of the American Federation of State, County, and Municipal Employees is faced with these constraints with many of its members. They found that community agencies that they relied on did not often meet the needs of employees due to lack of quick appropriate response, lack of availability during nonworking hours, and prohibitive cost (Yasser and Sommer, 1974). In addition to linkages, this program provides counseling and crises intervention in-house.

An employee population with more limited resources may also have more direct involvement with government systems such as housing enforcement and subsidy, consumer affairs, and entitlements such as unemployment, food insurance, food stamps, public assistance, social security, and large health delivery systems. It is also noteworthy that in times of economic retrenchment employees who had independence in the marketplace can suddenly find themselves dependent upon these systems. In such circumstances the resource pool must include first-hand knowledge of and liaison with the above systems.

It is clear that given the responsibility for linkage, staff will be spending time in the community assessing and updating resources and forming networks offering employees optimal service.

Implementation

Outreach. The outreach effort has begun in the context of the collaboration involved in foundation building. This must now be translated into implementation of the service directly to the employees. Outreach accompanies efforts to make employees aware of the service and utilize it.

Initially, the program needs to be presented to key employees. These include top management, union leadership, managers, supervisors, union representatives, and stewards. It is important that the orders of presentation and grouping be thought out organizationally. Wrich includes union and management employees in the same groupings (Wrich 1982). The purpose is to gain sanction, support, and work-related referrals from people who are in touch with employees around work function issues. Erfurt and Foote describe a training format that is composed of presenting general program information and its purposes (Erfurt and Foote 1977). Some education about how personal problems can impact functioning and legitimating the help-seeking process could also have values. The role of the attendants needs to be discussed in the context of their experiences. Functions include evaluation of job performance, counseling or confronting, and referral to the EAP. Identifying and approaching troubled employees more from the union official's peer role should also be included.

It is important that in this orientation the trainers be as keyed in as possible to potential political and functional issues of the group. Many such issues would have been identified in the foundation-building phase. However, suggesting that supervisors and union officials engage workers around their personal problems has many implications. Some may see it as potentially enhancing their ability to carry out their role. In making this referral a referrer is also giving up some control. Supervisors are also strongly affected by their alcoholic employees and a great deal goes into confronting them (Trice and Roman, 1978).

There would seem to be some wisdom in awaiting to respond to an outreach effort before attempting further program exposure. Initial serving delivery is a process of quick mutual learning and reading on the part of the program staff, supervisors, union officials, and employees. One does not want to hazard a barrage of intakes initially (Yasser and Summer 1974; Kurtz, 1980). One can also expect that, initially, troubled employees who have had the greatest impact on the supervisors and union officials will be the first to be referred. Thus the toughest cases both clinically and organizationally will probably be the first to be seen.

With organizational support, continued outreach can take place depending upon the ability of the program to respond to the increased request for help. Articles about the program and related topics can be placed in the company or union newsletter. Program brochures or posters can be distributed and posted. Counselors can meet with or make presentations to employee groups about the program or relevant topics. Direct mailings about the program can also be done to the employees' families. (Erfurt and Foote 1977; Wrich, 1982; Byers, 1976).

Benedict's outreach to trucking company employees required small meetings with employee groups at diverse locations (Benedict, 1973). For example, in order to provide a service to union members in numerous small government groups, counselors can travel to the employees' shops with union officials to publicize the program.

Service Delivery. The counselor's role varies depending upon the setting, population, and services provided. In a direct clinical service role, his or her task is to focus on the presenting problem which may or may not be job related. A skilled counselor will then be able to make an assessment that will uncover issues and patterns that are less obvious but necessary in order to develop a proper plan. Thus, someone with family, financial, or job problems might be dependent on alcohol or other chemicals. He or she will likely deny this dependence to him- or herself and others. However, a plan that does not include this factor will have little chance of being effective. A counselor without knowledge of these factors will be less likely to secure the appropriate help.

In a job-related referral the employee needs to be helped to integrate the feedback from the supervisor and/or union representative and explore realistic options. Among others they could be shown to accept it and work to explore behavior, or reject it, and continue as before. If the choice is the former, the EAP role continues. This data can then be used as an important part of the assessment since it includes behavior/functional information. Thus an employee who is reported to have increasing latenesses and temper tantrums at work could be influenced by various factors ranging from personal or family problems, impulse disorders, organicity or chemical dependence. Thus the need is to make a thorough assessment with work dysfunction as the presenting problem. This assessment must be followed by helping the employee conceptualize a strategy for working toward problem resolution.

The approach toward problem resolution can vary depending upon program definition. In a referral program, assessment is followed by identifying and referring to the appropriate helping resource and monitoring for effectiveness. Monitoring needs to include contact with the employee in job jeopardy with the supervisor and union official involved in his or her work life.

In some programs, counselors engage in problem resolution. Programs with limited ambulatory mental health or alcoholism insurance might have a greater need for providing in-house counseling for problem resolution. Employees with limited financial resources might require advocacy services with concrete issues: e.g., housing. Thus the need for the development of advocacy skills integrated with counseling skills becomes apparent. Miller and Bailey describe provision of numerous group services. Some are discussion groups, training, and personal problem-solving groups. Whether the above are conducted by staff or invited specialists, this advisory role requires numerous skills—organizational, programmatic, and group leadership.

The above services are all potentially of value. However, it is the role of the administrator to make sure that the nature and scope of the service is consistent with and supported by the company and union. This requires relating with the decision makers, interpreting the service to them, evaluating the program for programmatic and organizational knowledge, and implementing a program philosophy and direction. Program management, budget, and staffing are also functions requiring ongoing attention. If a program is not large enough for a full time administrator, a consultant can be considered. (Wrich, 1980).

Summary

Programs to assist employees are being implemented in companies, government agencies, and unions. Unassisted personal problems interfere with employees' job performance in numerous ways and are very costly to employers and of great concern to unions. Given the special nature of the disease alcoholism, many programs either focus only on outreach and assistance to alcoholic employees or put strong emphasis on them while helping with all types of problems.

In developing a model for an employee assistance program, it is important to recognize that they will vary depending upon the employee's and/or union's needs. The optimal program is one that is jointly sponsored by management and the union. Other factors are insurance for mental health and alcoholism, economic status of employees, and geographic dispersion. Thus some programs offer information and referral, some short-term counseling and others advocacy for concrete needs. The role of the counselor in the above is discussed. In some instances, programs have input into company policy. In all instances job functioning and maintenance are a major focus of such programs.

The administrative function is also seen as an important one in offering program definition and direction. ''The future is bright for employee assistance programs. The process is practical, and it works.'' (O'Donnell and Ainsworth, 1984, p. 506).

Appendix United Airlines EAP (Employee Assistance Program) Structure*

1. INTRODUCTION

Dependency on alcohol or other mood-altering chemicals, as well as other serious personal problems, are conditions that can interfere with health and job performance. When a serious personal problem of any nature exists, delayed action can result in the loss of the employee and in severe physical deterioration or death. Experience shows that the vast majority of personal problems are treatable.

2. PROGRAM OBJECTIVE

United Airlines, AFA, ALPA, and the IAMAW representing United's employees recognize that personal problems not directly associated with one's job function can have an effect on an employee's job performance and health. Usually, the employee will overcome such problems independently and the effect on health or job performance is negligible. In other instances, normal supervisory assistance serves either as motivation or guidance by which such problems can be resolved so the employee's job performance will return to an acceptable level. In some cases, however, neither the efforts of the employee nor the supervisor has the desired effect of resolving the employee's problems and unsatisfactory performance persists over a period of time, either constantly or intermittently. United Airlines, AFA, ALPA, and the IAMAW believe it is in the interest of the employee, the employee's family, the company and the union to provide an employee service that deals with such persistent problems. The objective of this program is to assist employees in a manner consistent with good therapeutic and business practice.

*Printed with permission of Employee Assistance Program, Medical Dept., United Airlines.

3. PROGRAM FRAMEWORK

Without altering or amending any of the rights or responsibilities of the employee or company, it is the policy of United Airlines to handle such problems within the following framework:

A. *United recognizes that almost any human problem can be successfully treated* provided it has been identified in its early stages and referral is made to appropriate care. This applies whether the problem is one of alcoholism, other forms of chemical dependency, or other personal problems. Therefore, employees having a problem that they feel could adversely affect their job performance or personal well-being are encouraged to voluntarily seek assistance on a confidential basis by utilizing the professional assistance of the Employee Assistance representative or the flight surgeon in the medical department.

B. *When an employee's job performance or attendance is unsatisfactory* and the employee is unable or unwilling to correct the situation either alone or with normal supervisory assistance, a personal problem may be the cause of the job performance difficulties. In such instances the employee's immediate supervisor has a responsibility to recommend that the employee utilize the Employee Assistance Program. Employees referred to the program by their supervisors will receive recommendations appropriate to their needs to secure services necessary to resolve the problem.

C. *It is the responsibility of the employee* to comply with the referral for assistance and follow the recommendations of the Employee Assistance Program.

D. *Under the provisions of this policy,* employees are assured that if personal problems are the cause of unsatisfactory job performance, they will receive careful consideration and an offer of appropriate assistance to help resolve such problems in an effective and confidential manner. Employees are assured that their jobs and future with the company will not be jeopardized by utilizing this employee assistance service and that program records will be preserved in the highest degree of confidence.

E. *In instances where it is necessary,* time-off may be granted for treatment or rehabilitation consistent with the applicable benefit package.

F. *Since employee work performance and attendance* can be adversely affected by the problem of an employee's spouse or other dependents, the employee can receive consultation from the Employee Assistance representative regarding problems of family members as well.

G. *It is emphasized that the Employee Assistance Program* and the disciplinary system each has its appropriate use. An employee's refusal to use the Employee Assistance Program is not in itself a cause for disciplinary action.

However, the Employee Assistance Program is not to be used as a substitute for appropriate discipline or as a basis to compromise applicable rules, regulations, or working agreements, and neither is discipline to be used as retribution for refusal to use the program.

RESOURCE GUIDE FOR EAPs

Organizations

Alcohol and Drug Problems Association of North America
1101-15th Street NW, Washington, DC

Association of Labor-Management Administrators and Consultants on Alcoholism
1800 N. Kent Street Suite 907
Arlington, VA 22209.

Columbia University Graduate School of Social Work
Industrial Social Welfare Center
225 West 113 Street
New York, NY 10025.

National Association of State Alcohol and Drug Abuse Directors
918 F Street NW
Suite 400, Washington, DC 20004.

National Council on Alcoholism
Labor Management Departments
1133 Third Avenue
New York, NY 10017.

National Institue of Alcohol Abuse and Alcoholism
Occupational Branch
5600 Fishers Lane
Rockville, MD 20857.

Journals

Addictive Behaviors

E.A.P. Digest

Labor-Management Alcoholism Journal

National Clearing House for Alcohol Information

International Journal of Addictions

American Journal of Public Health

REFERENCES

Akabas, S., Bellinger, S., Fine, M., and Woodrow, R. "Confidentiality issues in workplace settings: A working paper." Unpublished manuscripts, Industrial Social Welfare Center, Columbia University School of Social Work, NYC, 1981. NIMH Grant: 5-T21-MH-14462.

Akabas, S., Kurzman, P., and Kolben, N., eds. *Labor and industrial settings: Sites for social work practice*. Columbia University School of Social Work, Hunter College, School of Social Work, Council on Social Work Education, NYC. 1979.

Asma, F., Hilker, R., Shevlin, J. et al. "Twenty-five years of rehabilitation of employees with drinking problems." *Journal of Occupational Medicine,* 1980. **22,** (4).

Benedict, D. S. "A generalist counselor in industry." *Personnel and Guidance Journal,* 1973, **51**(10), 717–722.

Berry, C. A. *Good Health for Employees.* Washington, D.C.: Health Insurance Institute, 1981.

Business Week ''More help for emotionally troubled employees.'' March 12, 1979, 97–99.

Byers, W. ''The Kelsey Hayes Experience.'' *Labor and Management Alcoholism Journal,* 1976, **5**(5), 3–19, 26.

Cooper, B. S., Worthington, N. L. ''National health expenditures 1929–72.'' *Social Security Bulletin,* 1973, **36,** 3–19.

Dana, A. H. ''Problem drinking in industry, research report.'' *Social Science 6* 1963, **(1),** Florida State University, Institute for Social Research.

Eggum, P., Keller, P., and Burton, W. ''Nurse/health counseling model for a successful alcoholism assistance program.'' *Journal of Occupational Medicine* **22,** 1980, 545–548.

Erfurt, J. and Foote, A. *Occupational employee assistance programs for substance abuse and mental health problems,* 1977, University of Michigan-Wayne State Institute of Labor and Industrial Relation, Ann Arbor.

Foote, A., *Cost-effectiveness of occupational employee assistance programs.* 1978, University of Michigan-Wayne State Institute of Labor and Industrial Relations, Ann Arbor.

Goldbeck, W. B. and Kiefhaber, A. K. ''Wellness: the new employee benefit.'' *Voluntary Effort Quarterly,* 1980, **2,** 1–3.

Gruida, M. ''The occupational health nurse's role in corporate alcoholism programs.'' *Occupational Health Nursing,* 1967, **24**(3), 22–24.

Hale, W. E. ''The process of establishing occupational programs in industry.'' 1974, Proceedings of the Third Annual Alcoholism Conference, Nutritional Institute of Alcohol Abuse and Alcoholism, 239–247.

Jarvis, K. ''Insurance cost savings due to an adequate alcoholism health benefit.'' Prepared by insurance consultant for the State of Florida Department of Health and Rehabilitation Services, Program Planning and Development Office of Mental Health, Alcoholism and Rehabilitation Programs, 1975.

Jones, K. and Vischi, T. ''Impact of alcohol, drug abuse and mental health treatment on medical care utilization: A review of the research literature. *Medical Care,* 1979, **17**(12), 61–68.

Keller, M., McCormick, M., Efron, V. *A dictionary of words about alcohol.* Publications Division, Rutgers Center of Alcohol Studies, New Brunswick N.J., 1982.

Kellerman, J. L. *Alcoholism a merry-go-round named denial.* Minn: Hazeldon Educational Services, 1980.

Kuertz, N. and Williams, W. (1980). ''Supervisors' of an occupational alcoholism program.'' *Alcohol Health Research World,* **4**(3), 44–49.

Leeman, C. ''Contracting for an employee counseling service.'' *Harvard Business Review,* March–April 1974, 20–24.

Mannello, T. A. *Problem drinking among railroad workers: Extent, impact and solutions.* Washington, D. C.: University Research Corporation, 1979.

McQuirk, T. ''Evaluation and development of employee assistance programs.'' *Alcohol Health Research World* 1980, (3), 17–21.

Miller, L. Giving a helping hand with personal problems. *Advanced Management Journal,* Spring 1981.

New York State Task-Force on Alcohol Problems "Action against alcohol problems." Report on the status to planning effort on alcohol problems, 1975.

O'Connor, R. "Employee alcoholism programs: Sobering up the corporation." *Corporate Finesse & Recreation,* February/March 1984, 16–20.

O'Donnell, M. P. and Ainsworth, T. H. Health promotion in the workplace. New York: Wiley, 1984.

Pelf, S. and D'Alonzo, C. "A five-year mortality study of alcoholics. *Journal of Occupational Medicine,* 1973, **15,** 120–125.

Perlis, L. *The human contract in the organized workplace.* Washington, D.C.: National Conference of Catholic Charities, 1977.

Plant, T. "Alcohol problems: A report to the nation." *Cooperative Commission on the Study of Alcoholism.* New York: Oxford University Press, 1967.

Quayle, D. "American productivity: The devastating effect of alcoholism and drug abuse." *The American Psychologist,* **38**(4), 454–458.

Roman, P. "From employee alcoholism to employee assistance." *Journal of Studies on Alcohol,* March 1981.

Sehnert, K. W. and Tilloston, J. K. *How business can promote good health for employees and their families.* Washington, D.C.: National Chamber Foundation, 1978.

Taylor. "Counseling in a service industry." In *Task Centered Practice,* Reidell and Epstein, L., eds. New York: Columbia University Press, 1977.

Trice, H. and Roman, P. *Spirits and demons at work. Alcohol and other drugs on the job,* 2nd ed. New York State School of Industrial and Labor Relations, Ithaca: Cornell University, 1978.

United States Department of Health, Education, and Welfare, Public Health Service. *Alcohol and Health New Knowledge.* Second special report to the U.S. Congress, National Institute of Alcohol Abuse and Alcoholism, Rockville, Maryland, 1974.

United States National Institute on Alcohol Abuse and Alcoholism "Occupational alcoholism, some problems and some solutions." DHEW Pub.-No.HSM-73-9060. Rockville, Md., 1973.

Vehlow, C. M. and Kropp, C. L. "The development of an employee assistance program." *Innovations in Clinical practice: A source book.* Florida: Professional Resource Exchange, 1983.

Weiner, H., Akabas, S., and Somer, J. *Mental Health Care in World of Work.* Research demonstration supported in part by Grant no. RO-1-MH-14890. National Institute of Mental Health and Grant No. RD-1453-G, Rehabilitation Service Administration, Dept of Health, Education, and Welfare. N.Y.: Association Press, 1973.

Weissman, A. "Counseling in the steel industry." *Task Centered Practice,* Ingid, W. and Epstein, L., eds. New York: Columbia University Press, 1977.

Wrich, J. *The Employee assistance program: Updated for the 1980's.* Minn.: Hazeldon Publications, 1980.

Wrich, J. *Guidelines for developing an employee assistance program.* AMA Membership Publications Division, American Management Associations, N.Y., 1982.

Yasser, R. and Sommer, J. "One union's social service program." *The Social Welfare Forum.* National Conference on Social Welfare, N.Y.: Columbia University Press, 1974.

CHAPTER 10

Health Promotion for the Working Couple: A Health Promotion Program for Dual-Career Couples

Eileen C. Newman

INTRODUCTION

The historical precedents of employee assistance programs and stress management, as described in previous chapters, provide rationale as well as motivation for corporations to assist employees with challenges to their personal and professional health. In this chapter we will examine one of the newest and potentially complex challenges to the emotional well-being of a growing number of working families—the "dual-career" relationship.

Employment patterns for the sexes have undergone tremendous changes in the last 30 years as women began to enter the workplace in unprecedented numbers. In 1950 for instance, two out of every three middle-income families had one wage earner, primarily because the husband was the sole breadwinner (Smith, 1979). "Since 1947, the number of working women has increased 205% and the number of women enrolled in previously male-dominated professional programs has tripled" (Gurtin, 1980, p. 29). Thus, it is predicted that by 1990 "two thirds of all mothers, and more than one half of all mothers with children under six, will be working" (McCroskey, 1982, p. 32).

In the wake of this sociologic change, a new phenomenon, the "dual-career" marriage has evolved. Furthermore, the incidence of these marriages has increased 7% per year over the past 10 years so that in the United States they now number over three million (Rice, 1979, cited in Parker et al., 1981). What are the implications of this phenomenon for the health of the person, family and organization? This chapter will focus on the impact of the dual-career marriage on the ability of men and women to live and work together. We will examine the new genre of conflicts and problems that have arisen for the members of dual-career families and the respective organizations that employ them. Finally, we shall explore some tactics for meeting these new challenges to occupational and personal health.

Dual-Career Marriages Defined

Traditionally, the wife has been viewed as a satellite of her husband with little knowledge of, or interest in, affairs outside the domestic sphere. The husband was the breadwinner and the wife provided him and the children with homemaking and emotional support. By 1975, women held 42% of the jobs, and at this rate, will likely hold the majority of jobs by the end of the century (Bird, 1979). The dual-career marriage appears as if it is here to stay.

Precisely what is a *dual-career* marriage? Rapaport and Rapaport (1976) first coined the term ''dual-career families'' in the late 1960s. They defined such a family as a ''type of family structure in which both heads of household—the husband and wife—pursue active careers and family lives'' (p. 9). Although the word ''career'' is typically defined as any sequence of jobs, in the true sense it means a sequence of jobs demanding considerable commitment and characterized by *continuous* development. Catalyst (1981) is a national, non-profit organization that works to foster the full participation of women in corporate and professional life and to help employers with the problems generated by two-career families. They conclude that a career involves an attitude whereby the individual views work as a lifelong commitment, characterized by increasing levels of responsibility and personal growth. Therefore, the dual-career marriage may be thought of as a situation in which both husband and wife pursue not just employment, but careers designed to provide continuing personal and professional development.

In contrast to the dual-career marriage are the *two-paycheck* (or dual-worker) marriage, and the *two-person* career. The two-paycheck marriage differs from the dual-career marriage in that the former lacks the long-term commitment to professional development that is present in the dual-career marriage. Usually, this type of marriage involves the husband as the main breadwinner while the wife works to supplement his income or to simply ''get out of the house.'' Hendrick and Hendrick (1983) describe another family type, the ''two-person career,'' and distinguish it from the dual-career couple. Here, one spouse, typically the husband, has a job that demands an enormous commitment on the part of the other spouse, typically the wife. For example, ''a man may be hired to administer a community hospital: this involves management of hospital systems, hiring and firing of personnel, and frequent contact with governing board members and perhaps the community at large. This man's wife may be expected to support the hospital through volunteer work, participation in a number of related community activities, and entertainment of administrative staff, medical staff, board members, and community leaders. Thus, if both persons perform their designated roles, we have a two-person career or two for the price of one!'' (p. 254). Although each of these family models have similarities, it is the dual-career marriage that holds the greatest implications for both realizing and threatening personal and familial health.

The impetus for the dual-career marriage appears in many forms:

1. More women than ever before are receiving postgraduate and professional education.
2. Economic conditions have begun to necessitate dual incomes.
3. Many couples now prefer to share their lives more fully, thereby encompassing not just personal but professional activities.
4. Women are becoming more conscious of their ability to enter and succeed in occupational domains once exclusively male.
5. There is a general sociologic thrust for equality among the sexes (although such equality has yet to be truly realized).

The advantages that can accrue from the dual-career marriage are numerous. They include such things as increased economic viability and flexibility for the family; more equal participation in childrearing; greater potential for self-actualization on the part of both husband and wife; an increased feeling of interpersonal partnership, both personally and professionally; and, finally, a more equal distribution of familial responsibilities. All of these things can be seen as literally promoting personal and familial health.

While such advantages clearly seem attractive, the dual-career marriage can be fraught with pressures, conflicts, and stress for the husband and wife personally, for the family unit, and for the organizations that employ members of a dual-career marriage.

SCOPE OF THE PROBLEM

The dual-career marriage is far from an ideal condition. Even though its advantages can far outweigh its disadvantages, the dual-career marriage is laden with potential pitfalls for the husband and wife personally, for the family unit, and for the organizations that employ the members of a dual-career marriage. Problems endemic to the dual-career marriage primarily involve interpersonal and personal issues, family issues, and organizational issues.

Personal and Interpersonal Issues

Research by Thomas et al. (1982) uncovered some of the key perceptions of husbands and wives vis-à-vis the dual career marriage. Listed below are some of the perceived problem areas:

1. Husbands tend to underestimate the amount of stress their wives are under.

2. Wives see their husbands' support for their careers slipping when the pursuit of a career conflicts with traditional female responsibilities such as childrearing and household chores.
3. Dual-career marriages may inadvertently set up competitive situations between husband and wife to see who can be more successful professionally. This can be especially troublesome if both husband and wife are in similar occupations.
4. Arguments may develop when the demands of one spouse's career conflict with the demands of the other's career. There is usually difficulty deciding whose career will be pre-eminent.

A personal issue that often affects mothers in the dual-career marriage is the issue of being a mother *and* having a career. Mothers often express self-doubt and guilt about not being a "traditional" mother. This issue may be a problem at work as well. There is a stigma attached to being a mother while simultaneously being employed full-time. Some employers still practice overt discrimination against mothers who want to work.

Also, studies show that being married is negatively related to income for women while positively related to income for men (Perrucci, 1978). Furthermore, when data are adjusted for the increased number of women in the labor force, the percentage of employed women in managerial positions increased by only 0.4%, and in professional positions by only 0.5% in the 1965–1974 decade (Hiller and Philliber, 1980). Bryson and Bryson (1978) claim that women tend to be paid less and, when they work with their husbands, antinepotism policies are disproportionately applied to the detriment of women.

Finally, both husband and wife in the dual-career couple may be perceived as somewhat deviant by more traditional, married couples. Traditionalists might question the woman's femininity in her pursuit of a career. They might question the man's masculinity in his "allowing" his wife to work. The old question of "who wears the pants in the family" might become a point of discomfort for both husband and wife in their interactions with more traditional, married couples.

Family Issues

Despite the women's liberation movement, the ideology of romantic love and the "Prince Charming" legend still reinforce the myth that man was meant to be the provider and protector upon which the fair young woman was to depend. Ask the average teenage girl where she'll be in her late 20s and you're likely to hear something to the effect that it will be in a rose-covered cottage with a white picket fence shared with her two children and her husband who supports them. The result of such role socialization is that females accept the work in the domestic sphere and men enter the workplace. Society has long conceptualized the normal, mature man as the economic provider and the normal, mature woman as the housewife and mother. The division of labor by sex was declared biologically natural and universal and therefore functional. The dual-career marriage dramatically contradicts this notion.

If a mother commits herself to a career as does the father, who then should care for the children? In reviewing the literature on child development, it seems that much of the research was premised with the fact that care of the children is best met by the mother. With 60% of all mothers working, half of them with children under six years of age, it is no longer tenable to hold such a belief. More recently, it has been stressed that it is *what kind* of care, not *who* provides it, that determines healthy adjustment in children. Furthermore, Bernard (1974 p. 169) points out that professional mothers are anything but rejecting. "They find as much pleasure as other women in motherhood, if not more. They have read the books but are not cowed by them. They believe they are better mothers when they have relief from full-time attention to their children and they think it is better for their children as well. Without their work they could become resentful, sullen, angry, depressed." Nevertheless, as a mother, the career woman questions her ability to meet the demands of both job and family. She thinks she must be dependent and noncompetitive to be motherlike, and independent and competitive to be a professional. Some theorists worry that if the bright, achievement-oriented women are discouraged by rigid, institutional structures and sex-role stereotyping, they may opt to forgo childbearing, thus eventually depleting the gene pool. "We may be facing elective sterilization of the best and brightest among us" (Gurtin, 1980, p. 30).

Thus, if dual-career couples desire families, they are faced with the problem of finding good child care. Their employers have been remiss in acknowledging this dilemma, and have failed to develop plans to help couples with child care responsibilities. Couples have been forced to secure domestic help or find a suitable child care facility, both somewhat undesirable and costly solutions. Licensing of existing day care facilities has often been found inadequate, and systematic research as to whether licensing has preventive effects is lacking (Hoffman and Nye, 1975). Because the dual-career couple is a relatively new phenomenon, society has yet to accommodate them by assuring quality day care that is affordable and diverse enough to meet different couples' preferences. Instead, the couple is made to feel guilty for not devoting enough time to their children. Society frowns upon the increasing number of "latchkey" children (those who come home from school to an empty house), yet offers few, if any, alternatives to this solution. Also, it is not clear from the research whether this pehnomenon is positive or negative (Long and Long, 1983; Trumberger and MacLean, 1982; Ferguson and Cunnison, 1951).

In addition to the issue of child care responsibilities is the potential for unequal division of household tasks. If the mother in a dual-career marriage does not assume the role of housewife, who, then, does the housework? Goode (1982) states that while working wives devote fewer hours to housework than full-time homemakers (26–35 hours per week vs. 35–55 hours per week), husbands of *both* working wives and full-time homemakers only spend about 10–13 hours weekly doing housework.

Thus, while both spouses may pool their incomes and claim that each other's jobs are equally important, the woman still bears most of the responsibility for the traditional housework while the man is more likely to handle financial matters. Garza (1980) found further evidence that no differences exist in the domestic division of labor among conventional, quasiconventional (married, but lived together previously), and unconventional (living together) couples. He concludes that since all people acquire and retain traditional attitudes about the sexes, they will eventually surface in a close, continuous relationship regardless of the degree of conventionality. In other words, having intellectual insight into the problem of equality between the sexes does not guarantee egalitarian behavior. In fact, Kassner (1981) found that today's male university students still prefer traditional marriages while female students prefer egalitarian marriages becuase males feel the traditional marriage will help their careers and females feel it will hinder their careers. Indeed, females have been found to be more tolerant of working women than males (Powell and Steelman, 1982).

Still another source of conflict for the dual-career family is commuting. What happens if each spouse's job is in a different town? In attempting to accommodate the demands of both their employers and their families, dual-career couples have resorted to an alternative arrangement, the "commuter-marriage." Here the couple lives apart because they are unable to find desirable jobs in the same locale. The commuter-marriage is extremely trying, even for the most devoted of couples, with the greatest disadvantage being the children. Child care, phone bills, travel expenses and double residences, not to mention the lack of tax benefits for a long-distance arrangement, all combine to make commuter-marriages difficult to pull off. Inflexible personnel policies, particularly regarding leaves, vacations, work scheduling, and transfers make the commuting process extremely difficult and may breed resentment directed toward the company.

Traditionally, the wife agreed to share her husband's income, to care for the family and to follow him when his employer demanded a transfer, remaining flexible in her activities so as to meet her husband's needs. Commitment to a career on the wife's part would have jeopardized the husband's chances for success. It seemed that to keep peace within the family, each sex functioned in a separate role; the wife staying at home and the husband entering the marketplace. When both spouses commit themselves to a career, they can no longer adhere to separate roles, and issues such as child care, division of household tasks, commuting, and many more become sources of conflict and threaten personal health and well-being.

Organizational Issues

Only recently has it begun to dawn on corporations, and other organizations employing one or both members of a dual-career marriage, that the dual-career phenomenon is not just another passing fad. For instance, employees once obediently transferred for the sake of a promotion or to meet the employer's needs. The employee in a dual-career marriage may be more hesitant to uproot the family, especially because his or her spouse may hold a rewarding job in that same community. Cooper and Marshall (1977) speculate that the current trend of increasing childless marriages (or postponement of beginning families or adding to them) and an increasing number of men emphasizing the quality of family life over financial success will result in shifts in power both within the family and the company. Thus, dual-career couples are more frequently opting to forgo climbing the proverbial "corporate ladder" in order to balance job and family. For example, in a study of professional dual-career couples and their job-seeking experiences (they had recently obtained PH.D.'s in either psychology or the biological sciences), it was found that, contrary to previous research and prevailing myth, many sought egalitarian or nontraditional job-seeking patterns (Wallston et al., 1978). Are employers prepared to deal with these changes and the resistance they may encounter from dual-career couples when transfers are requested?

Two other areas of concern are *day care* for the children of dual-career couples, and *nepotism policies* affecting couples working within the same organization. For the most part, employers have been insensitive to the problem of providing quality child care when both parents work full-time. The strain inadvertently imposed upon these couples to provide such care is compounded by the ensuing guilt for spending long periods of time away from their children. Thus, failure by the employer to construct a program to help parents cope with child-care responsibilities is likely to contribute to job dissatisfaction among these employees.

When both spouses are employed by the same organization, they may have to contend with traditional nepotism policies (stipulations designed to prevent favoritism among relatives working together). For instance, it has generally been accepted that one spouse should not supervise another, nor enter into the evaluation process of that spouse. Even if the spouses do not have a supervisor-supervisee working relationship, employers may question the feasibility of promoting one and not the other because they expect resentment by the unpromoted spouse both toward the employer and the promoted spouse. Furthermore, other employees may not be sure how to relate to the dual-career couple. Can one spouse be trusted to hold a confidence from the other spouse? Maintaining good working relationships and social rapport with both employers and fellow employees can be a difficult task for the couple working together.

Hall and Hall (1979) claim that up until now, it has been the couple's responsibility to compromise and be flexible if both members of the family wanted to pursue careers; employers expected them to cope with problems, such as those mentioned above, by themselves. It is likely that employers' resistance to facing these problems will prove counterproductive. It is no longer practical to adhere to traditional company policies regarding transfers, promotions, child care assistance, and other issues facing the dual-career couple.

To summarize so far, we see that the dual-career marriage phenomenon has emerged in this country in response to changing economic and sociologic conditions, and with it has come new challenges to mental health. By definition, this form of marriage has important implications for not only the employee's interpersonal relationships but for occupational well-being as well. The historical precedents of employee assistance programs and occupational stress-management programs provide a suitable rationale for including other occupationally based interventions designed to further promote employee mental health and, in this case, to solve the dilemmas previously described and associated with pursuing a dual-career marriage. This is particularly relevant since the number of dual-career marriages is increasing every year and affecting increasing numbers of employees and employers. Because the dual-career marriage is essentially a new form of marital relationship, we simply do not have the role models to fall back upon for guidance. This lack of role models is even further reason for organizations to begin providing employees with some assistance in coping with the dual-career dilemma rather than letting them "fend for themselves."

GOALS OF A DUAL-CAREER MARRIAGE ASSISTANCE PROGRAM

Briefly stated, the goal of any dual-career marriage assistance program is to better enable the participants to successfully meet the challenges of pursuing a dual-career marriage. However, when one tries to operationalize just what that means from a behavioral perspective so that the effectiveness of any such program may be evaluated, problems may arise. Outcomes such as reduction of stress, improvement in occupational adjustment or satisfaction are all potential goals for a dual-career marriage assistance program. Unfortunately, there is a paucity of literature to draw upon for historical precedents. The majority of the literature is prescriptive and those programs that are described fail to mention evaluation. This is, no doubt, due to the fact that this phenomenon has only recently emerged.

One noteworthy survey investigation (Catalyst, 1981) utilized specific attitude, stress, and personality assessment questionnaires. While such a strategy seems viable, it is questionable whether the measures used in this evaluation were appropriate. For further guidance on this topic, the reader can refer to the final section of this text in which there is a discussion of general psychometric issues and viable assessment tools. In addition, the reader may wish to refer to two excellent compendia of potential research tools. *Measures of Social Psychological Attitudes* (J. P. Robinson and P. R. Shaver) and *Measures of Occupational Attitudes and Occupational Characteristics* (J. P. Robinson, R. Athanasiou, and K. Head). Both volumes were published in 1973 by the Institute for Social Research, University of Michigan.

GUIDELINES FOR PROGRAM DEVELOPMENT

In this section, we will describe the potential components of a dual-career marriage assistance program. The present model is based upon the work of Hall and Hall

Table 10.1 Components of a Dual-Career Couple Assistance Program

I. Personal strategies
   A. Developing androgynous attitudes
   B. Redefining "success"
   C. Exhibiting "situation-dependent behavior" patterns
   D. Stress management training
II. Interpersonal strategies
   A. Developing communication skills
   B. Conflict resolution through compromise
   C. Strategies for providing child care
III. Organizational strategies
   A. Flexitime
   B. Provision of child care services
   C. Facilitation of transfers
   D. Re-examination of nepotism policies

(1979), Catalyst (1981), and Newman and Everly (1983) and is summarized in Table 10.1. This model has applicability not only to the dual-career relationship, but to the "two-paycheck marriage" and the "two-person career" as well.

The program described in Table 10.1 is designed to be delivered through individual counseling, a small group format, or a larger, more formalized training program/seminar. The program's components are briefly described below.

Personal Strategies

Androgynous Attitudes. Attitudes and personal philosophies usually serve as the basis for one's actions, that is, human behavior is often times based upon an underlying attitude about the subject of the person's action. Therefore, when helping people adjust to the dual-career marriage, it is extremely useful to help them adopt an attitude toward dual careers that is adaptive and health-promoting. One concept that appears to be extremely useful in helping people adjust to the dilemmas of the dual-career marriage is that of androgyny.

Androgyny is a concept developed by Sandra Bem (Bem, 1974) in the early 70s as an alternative to sex-role stereotyping of people as solely masculine or solely feminine. Androgyny is the extent to which a person believes he or she possesses the desirable attributes of both sexes, as opposed to being exclusively masculine or feminine. Sex-typed masculine males and feminine females are thought to be restricted by rigid sex roles in how they express themselves. In American society, the traits considered appropriate for a woman have put her at a disadvantage in terms of achieving what society sees as being successful and important by inhibiting competitiveness, assertiveness, and competency in females. Likewise, the male role restricts behavior, especially in areas such as sensuality, tenderness, and child care. Bem's new standard of psychological health for the sexes allows for personal expression of the best traits of men and women. Bem argues that establishing an androgynous attitude is necessary if each partner is to benefit equally from the relationship.

The idea of applying the principle of androgyny to the business world is not new. Alice Sargent (1981), author and management consultant, feels that the big American problem is isolation and alienation in the male-dominated business world, and that androgyny is the solution to this problem. She has found that the male manager is best at expressing anger, frustration, competition, and aggression, rather than understanding, tolerance, and compassion, and that it is the former expressions that cause employee alienation. What she touts is a program designed to counteract sex-role stereotyping by raising managers' awareness of this issue and teaching them skills they did not learn due to their own sex-role socialization (e.g., expressing understanding).

Redefining "Success." One area in which attitude change may be extremely useful is in the area of a couple's concept of "success." It could be beneficial for the androgynous couple to redefine their notion of what success means to them. Peterson (1978) believes dual-career couples may have to choose their own "tangible goals which they can use as indicators for success, so that they avoid the trap where success is defined as something out there that can be attained, but is always yet to be attained" (p. 128). Thus, one or both spouses may decide to have their children before launching careers, delaying financial gratification. Or, conversely, they may forgo the personal satisfaction gained from having children until they are well established in their careers. It may also be possible to reach some midpoint between these two options. Whatever decision is made, the dual-career couple should try to determine what is best for them without having to rely on social convention or traditional goals defined by sex roles.

"Situation-Dependent Behavior." A behavior pattern that can grow out of an underlying androgynous attitude is that of "situation-dependent behavior." Eileen Newman and George S. Everly, Jr. (1983) describe "situation-dependent" behavior patterns and contrast them with "role-dependent" behavior patterns. According to these authors, situation-dependent behavior involves having partners behave in a flexible, situation-specific manner in order to preserve the well-being of the family unit. Each partner must consider three aspects of the situation:

1. The objective *demands* of the situation.
2. The *relative ability* of each partner to perform the task at hand.
3. The *preferences* of each partner, given the present task.

If, on the other hand, a couple exhibits role-dependent behavior, it means that the couple's behavior, both interpersonally and socially, is dictated by social convention and existing socially approved sex roles, thereby often disregarding what may be more situationally effective behavior.

From an operational perspective, the principle of situation-dependent behavior dictates that the woman may handle the family's finances, mow the lawn, or change the oil in the car if she can perform those activities better than her spouse, and/or she prefers to do such work, and she has the necessary time available. Likewise, the man may cook the meals, clean much of the house, or do the laundry if he can perform those activities better than his spouse, and/or he prefers to do such work, and he has the necessary time available. The principle of role-dependent behavior, on the other hand, would dictate that the woman perform the duties that are traditionally performed by the woman and the man perform the duties that are traditionally performed by the man.

Situation-dependent behavior appears to be a far more rational and effective method of interpersonal interaction for couples. It has the inherent advantages of allowing the demands of the situation to interact with the abilities and preferences of the individuals, reducing the difficulties encountered when behavior is simply role-dependent. For example, it eliminates the "need" for men to have aptitudes in traditionally masculine endeavors, and vice versa. It seems a far better schema to view behavior as neutral, neither masculine nor feminine. Behavior is behavior and should extract meaning from the situational context, not from the traditional sex-role stereotype which is now probably outdated.

Stress Management. Most authors agree that the dual-career marriage can be extremely stressful at times. For this reason, it is highly advisable to provide individuals with basic skills in stress management. Guidelines were provided in an earlier chapter, so they will not be detailed here.

Interpersonal Strategies

Two keys to interpersonal effectiveness in the dual-career marriage are communication skills and conflict resolution skills. Child care will also be discussed in this section as it has proven to be a key factor in effecting an arrangement where both spouses can enjoy job and family.

Effective Communication. The word communication comes from a form of the Latin word "communicare" meaning to share. In other words, communication is an action or situation involving mutual sharing between at least two people, a sender and a receiver. It is not talking *at* someone, but talking *with* him or her. This means the receiver actively listens to what the sender is communicating by paying close attention to the message and expressing interest in its content.

There are many obstructions to communication. Barriers are raised to defend against perceived threats. As the sender experiences a threat, he or she retaliates, thereby inducing feelings of threat in the receiver. The receiver likewise retaliates, so that the cycle spirals ever upward until communication eventually breaks down. To avoid this vicious cycle, be aware of feeling threatened and discuss these feelings openly before they interfere with communication.

Based on past experience, we also tend to anticipate what the sender is going to say and inaccurately perceive the sender's message. Our past experiences result in certain perceptual biases or prejudices that dictate how we interpret the sender's words, so that we hear only what fits our purposes. By evaluating or judging the meaning behind the sender's words, we miss part of the information being communicated. Avoid judgmental thoughts when attending to the sender's words, and become aware of your prejudices or biases.

Another difficulty is the difference in vocabularies between individuals. The same word may hold different meanings for two people. If you are unsure what the sender means, ask for an accurate definition. Similarly, if you sense the receiver has misunderstood your message, describe more fully what you mean.

Generally, most of the gaps in communication have psychological origins. We distort, forget, and exaggerate information because of its emotional impact upon us. Try to recognize emotional responses in the receiver. If they occur, stop and ask how that statement or question made the receiver feel. Then restate the information in a way that is more acceptable so as to reduce the discomfort and thus increase understanding.

Remember that information is transmitted nonverbally as well as verbally. Probably the most blatant cues are facial expressions. Try to detect facial changes and relate them to any verbal material the sender simultaneously communicates. Nonverbal, physical movements of expression include gestures and bodily postures. For example, restlessness indicates anxiety or irritation. Slow, listless movements suggest sadness or depression. Also, the physical distance the person maintains from another indicates how accepting or withdrawn he or she may be. The receiver must be aware of both his or her own nonverbal messages as well as those sent to the sender.

Conflict Resolution Through Compromise. The dual-career relationship can be fraught with conflicts. One way of resolving conflicts is compromise.

If each spouse accurately perceives the other as gaining as much as the other, the quality of the relationship will be enhanced (Peterson, 1981; Houseknecht and Macke, 1981; Spitze and Huber, 1982). If this can be achieved, the advantages are numerous: there is mutual supportiveness, the ability to share more experiences, a greater appreciation of each spouse's accomplishments, and a greater admiration for each spouse as both a professional and a family member. Runde (1980) relates the experience of Nobel Prize winner Rosalyn Yalow, a medical scientist, and her husband, a professor of physics. Both work in New York City. Rosalyn claims her marriage has not been adversely affected by the added responsibilities of her profession. Her husband says, "Our relationship has been collaborative . . . I've served as an extra pair of hands and eyes and another brain'' (p. 68). Unfortunately, most of us have not had parents like Dr. Yalow and her husband as models. Therefore dual-career couples must develop new approaches to solving problems. Provided below in Table 10.2 is a brief list of conflicts commonly encountered by dual-career couples and possible solutions to these problems.

Strategies for Providing Child Care. As noted previously, one of the interpersonal issues facing dual-career couples is providing child care. In general, there are four approaches to this problem: (1) the mother acts as the sole caretaker, (2) the husband acts as the sole caretaker, (3) the parents pay someone to provide child care, or (4) the husband and wife balance their schedules so that either parent is in the home at all times.

The first strategy, having the mother act as sole caretaker, has been the option historically chosen. While rich in historical precedent, this solution is an untenable

Table 10.2 Ways to Compromise

Potential Problem Areas	*Solutions*
Each spouse works in a different location	• Arrange a commuter marriage • Judge the distance between the two adjacent cities in which each spouse works and commute from a halfway point each day
One spouse is transferred	• If a spouse is considering transferring, list the pros and cons for each spouse in advance using a ''balance sheet'' approach; list the possible cities of relocation and compare opportunities
Career goals clash	• Combine careers, working full-time with one spouse or working in related fields • Become a ''company couple,'' possibly suggesting to the employer that he or she fill two positions (''spouse bargaining'') • Discuss alternative goals for one or both spouses (this is difficult if both are competitive achievers) • Consider alternating which spouse's career is more important at that time, with the idea that opportunities will arise at different times for each spouse • Reverse traditional roles so that, for example, the man becomes a ''house-husband''

one within the dual-career relationship. Because the mother is pursuing her career just as avidly as her husband, taking on the responsibilities of sole caretaker is equivalent to working two, full-time jobs. As a result she becomes physically and emotionally drained.

A new phenomenon has arisen, as women have gained earning power, in which the husband acts as the primary homemaker. These "house-husbands" ignore societal demands to enter the marketplace in favor of enjoying raising their children, and feel competent at completing household tasks. One example of this new type of father is David Townsend. His wife Kathleen (oldest daughter of the late Robert Kennedy) works as a poverty programs analyst for the state of Massachusetts. After four years of teaching classics at a college, David is now content to write novels at home while caring for their two children. Kathleen and David seem to share all aspects of their lives, from childbirth (David actually delivered one child at home when no doctor was available) to law school. When asked about the biggest difference between her family and the "old" Kennedy family, Kathleen noted that fathers spend more time with their children. Groups such as "Free Men" are springing up in recognition of these changes and are helping men to cope with the pressures of role reversal.

When both parents are committed to working outside the home, the last two strategies should be considered. Paying others to act as caretakers (grandmothers are not the answer to the day care dilemma) is probably the most common solution. Parents can either locate a child care facility or hire a domestic servant or babysitter to care for the child in the home, both cutting substantially into the family's income.

The final strategy, balancing parents' work schedules, is perhaps the best potential solution. The biggest barrier to the success of this arrangement has primarily been the father's misconception that he simply "doesn't belong in the nursery." Years of male sex-role socialization take their toll, and it is likely that the husband will feel emasculated by assuming the role of part-time caretaker. Society is unlikely to reward him for sharing in housework and child care, but his participation has a very positive effect upon his working wife and his children, despite society's disregard for his cooperative efforts (Parke, 1981). Furthermore, these parenting skills, previously reserved for women, can be learned, and it is only the father's negative perception of these tasks that prevents him from learning to be an effective caretaker.

The myth that the qualities distinguishing the male and female sex roles cannot coexist in the same person, is detrimental to both spouses. As fathers, men generally want to be close to their children, but their actual behavior is disappointing to the child. Men have feelings inwardly. When men are growing up, vulnerability is not an acceptable component of male friendship. If approached by a man professing a need for a kindred spirit, many, if not most, males would question whether this man was coming out of a closet! Women, on the other hand, are comfortable with sharing intimacies. Males usually end up sharing their emotions with a close female partner, if they share their emotions at all. This discomfort with expressing tenderness and nurturance makes the man's resistance to sharing the maternal role more understandable.

Organizational Strategies

Employers have traditionally and impractically dealt with the problems of the dual-career couple on a case-by-case basis. It is essential that they now build a conceptual framework to deal with these problems before policy is set. This means that both management and employee representatives must work together, carefully analyzing the problems and planning for the future. Let's explore some areas of concern.

Flexitime. Ways in which employers can attempt to meet the needs of an employee who is one half of a dual-career couple are providing flexible work hours (often called "flexitime") and responsible child care.

For years, universities have offered their faculty sabbaticals, and assembly lines have run on shift schedules. Is it necessary for administrative offices to rigidly adhere to the 9-to-5 routine? One possible solution, now considered in the experimental stage, is operating within a "flexitime" program (Bohen and Viveros-Long, 1981). For example, employers might set a general arrival time (say 7–9:30 A.M.), a midday lunch break (scheduled somewhere between 11:30 A.M. and 1:30 P.M.), and an afternoon leave time (3:30–6:00 P.M.). All workers would thus be at work from 9:30–11:30 A.M. and 1:30–3:30 P.M. A two-week notice to change a worker's arrival, lunch break, or departure time should be required. There are many possible alternatives to this example, but the advantage of such a program is the flexibility provided to employees for meeting family responsibilities in those early morning and late afternoon hours when children are not in school. Other benefits accrued are decreased stress on the mother and an incentive for fathers to share in domestic tasks and spend more time with their children.

Providing Child Care Facilities. Company officials also need to alter their views regarding parental responsibility. McCroskey (1982) recommends they thoughtfully develop a child care plan whether it be to provide benefits to cover offsite child-care expenses, to provide onsite child care themselves, or to join with other agents specializing in child care on a contractual basis. It is encouraging to note that the White House Group on Women is currently developing legislation to offer tax incentives to employers who offer day care programs for their employees.

Facilitation of Transfers. Comapnies can help couples faced with the possibility of one spouse being transferred. First, the employer should consider other ways of advancing the employee's position besides transferring him or her. If all else fails, however, companies can assist the couple in selling and/or buying a home, locating child care facilities and other support systems in the new community and helping the mate find employment in this new area.

Re-examination of Nepotism Policies. Another corporate strategy is reforming recruiting techniques by including both spouses in the preselection process or by seeking husband and wife teams. The latter strategy requires revision of nepotism policies that typically prevent supervision or evaluation by a relative. Also, travel policies could be revised for dual-career couples so that at least the financial burden of child care expenses is lifted.

Finally, companies could provide marital counseling, either through hiring a professional counselor or by joining with a professional in the community on a contractual basis. As noted earlier, seminars and workshops could also be arranged to educate couples in the problem-solving skills essential to dual-career management.

Summary

The dual-career couple is a growing phenomenon that has resulted from the influx of women into the workplace and changing attitudes about appropriate behavior for the sexes.

Currently, most employers and employees are inadequately prepared to deal effectively with the problems created by the dual-career marriage, lacking comprehensive guidelines or precedents for such problems. Left to work out these problems for themselves, and often chastised by traditional families, dual-career couples suffer extreme stress. Without existing role models, these couples must learn to deal not only with the conflicts inherent in balancing career and family responsibilities, but also with powerful societal norms working against adopting and maintaining role flexibility.

The key to success in these marriages is compromise. Problems of geographic mobility, child care and equal division of household tasks must be negotiated, and the solutions agreed upon must constantly be re-evaluated to insure against regression to sexually stereotyped behavior. Maintaining an androgynous philosophy, where family members are able to behave as the situation dictates, rather than as their sex role dictates, will result in more equal sharing of roles.

The dual-career couple is providing us with a family model that offers greater flexibility, variety, and effectiveness in the behaviors each spouse is able to exhibit. The advantages to these couples' children is that they may grow up less restricted by rigid sex roles than the children of traditional couples. Some feel this is an evolutionary process that may take generations, but it is a step forward in achieving equity between men and women at home and in the workplace.

RESOURCE GUIDE FOR DUAL-CAREER RELATIONSHIPS

Organizations

Association of Part-Time Professionals
P. O. Box 3419

Employee Relocation Council
1627 K Street NW

Alexandria, VA 22302
(703) 370-6206

Free Men
P. O. Box 920
Columbia, MD 21044

National Congress for Men
P. O. Box 147
Mendham, NJ 07945
(201) 543-6060

National Council for Alternative Work Patterns, Inc.
1925 K Street, NW
Suite 308
Washington, DC 20006
(202) 466-4467

National Council of Career Women
1725 K Street NW
Suite 607
Washington, DC 20006
(202) 347-1401

National Job Sharing Network
%New Ways to Work (see below)
Washington, DC 20006
(202) 857-0857

National Organization for Women
425 Thirteenth Street NW
Suite 1048
Washington, DC 20004
(202) 347-2279

New Ways to Work
149 Ninth Street
San Francisco, CA 94103
(415) 552-1000

Options for Women (also helps men)
8419 Germantown Avenue
Philadelphia, PA 19118
(215) 242-4955

Women for Racial and Economic Equality
130 E. Sixteenth Street
New York, NY 10003
(212) 473-6111

Journals

Family Relations

Journal of Marriage and the Family

Personnel Administrator

Personnel Journal

The Personnel and Guidance Journal

Books

Clarke-Stewart, A. *Daycare*. Cambridge, Ma.: Harvard University Press, 1982

Doyle, J. A. *The Male Experience*. Dubuque, Ia.: Brown, 1983.

Hall, F. S. and Hall, D. T. *The Two-Career Couple*. Reading, Ma.: Addison-Wesley, 1979.

Nieva, V. F. and Gutek, B. A. *Women and Work: A Psychological Perspective*. New York: Praeger, 1982.

Parke, R. D. *Fathering*. Cambridge, Ma.: Harvard University Press, 1981.

Pepitone-Rockwell, F. *Dual-Career Couples*. Beverly Hills, Ca.: Sage, 1980.

Renetzky, A., Jacobson, D. J., and Rudd, H. *Directory of Career Resources for Women*. Santa Monica, Ca.: Ready Reference Press, 1979.

Rice, D. G. *Dual-Career Marriage: Conflict and Treatment*. New York: The Free Press, 1979.

REFERENCES

Bem, S. L. "The measurement of psychological androgyny." *Journal of Personality and Social Psychology,* 1974, **42,** 155–162.

Bernard, J. *The Future of Motherhood.* New York: Penguin, 1974.

Bird, C. *The Two-Paycheck Marriage.* New York: Rawson-Wade, 1979.

Bohen, H. H. and Viveros-Long, A. *Balancing Jobs and Family Life: Do Flexible Work Schedules Help?* Phiadelphia, Pa.: Temple University Press, 1981.

Bryson, J. B. and Bryson, R., eds. *Dual-Career Couples.* New York: Human Sciences Press, 1978.

Cooper, C. L. and Marshall, J. *Understanding Executive Stress.* New York: Petrocelli Books, Inc., 1977.

Corporations and Two-Career Families: Directions for the Future. New York: The Staff of the Career and Family Center, *Catalyst,* 1981.

Ferguson, T. and Cunnison, J. *The Young Wage Earner.* London: Oxford University Press, 1951.

Garza, J. M. "Sex roles within cohabiting relationships." *American Journal of Psychoanalysis,* 1980, **40,** 159–163.

Goode, W. J. "Why men resist." In *Rethinking the Family: Some Feminist Questions,* B. Thorne and M. Yalom, eds. New York: Longman, 1982.

Gurtin, L. "The dual career family." *Journal of College Placement,* 1980, **40,** 28–31.

Hall, F. S. and Hall, D. T. *The Two-Career Couple.* Reading, Ma.: Addison-Wesley, 1979.

Hendrick, C. and Hendrick, S. *Liking, Loving and Relating.* Monterey, Ca.: Brooks/Cole, 1983.

Hiller, D. and Philliber, W. "Necessity, compatibility and status attainment as factors in the labor force participation of married women." *Journal Marriage and the Family,* 1980, **42,** 347–354.

Hoffman, L. W. and Nye, F. I. *Working Mothers.* San Francisco, Ca.: Jossey-Bass, 1975.

Houseknecht, S. K. and Macke, A. S. "Combining marriage and career: The marital adjustment of professional women." *Journal of Marriage and the Family,* 1981, **43,** 651–661.

Kassner, M. R. "will both spouses have careers? Predictors of preferred traditional or egalitarian marriages among university students." *Journal of Vocational Behavior,* 1981, **18,** 340–355.

Long, T. and Long, L. *The Handbook for Latchkey Chlildren and Their Parents.* Chicago, Ill.: Priam Press, Inc., 1983.

McCroskey, J. "Work and families: What is the employer's responsibility?" *Personnel Journal,* 1982, **61,** 30–38.

Newman, E. and Everly, G. "The Dual-Career Marriage: An Alternative Model of Family Life for the 80s." Unpublished manuscript, 1983.

Parke, R. D. *Fathering.* Cambridge, Ma.: Harvard University Press, 1981.

Parker, M. Peltier, S., and Wolleat, P. "Understanding dual-career couples." *The Personnel and Guidance Journal,* 1981, **60,** 14–18.

Perrucci, C. C. Income attainment of college graduates: A comparison of employed men and women. *Sociology and Social Research,* 1978, **62,** 361–386.

Peterson, C. "Equity, equality and marriage." *Journal of Social Psychology,* 1981, **113,** 283–284.

Peterson, S. ''Achievement and success as American Values.'' In *The Two-Career Family: Issues and Alternatives,* J. M. Richardson and G. V. Kreuter eds. Washington, D. C.: University Press of America, Inc., 1978.

Powell, B. and Steelman, L. ''Testing an undertested comparison: Maternal effects on sons' and daughters' attributes toward women in the labor force.'' *Journal of Marriage and the Family,* 1982, **44,** 349–355.

Rapaport, R. and Rapaport, R. N. *Dual-Career Families Re-examined.* New York: Harper Colophon Books, 1976.

Runde, R. ''Now, the Us generation.'' *Money,* November 1980, 62–70.

Sargent, A. G. *The Androgynous Manager.* New York: American Management Association, 1981.

Smith, R. E. *The Subtle Revolution: Women at work.* Washington, D.C.: The Urban Institute, 1979.

Spitze, G. and Huber, J. ''Accuracy of wife's perception of husband's attitude toward her employment.'' *Journal of Marriage and the Family,* 1982, **44,** 477–481.

Thomas, S., Albrecht, K., White, P., Faires, C., and Shoun, S. ''Determinants of marital quality in dual-career couples.'' Presented at the American Psychological Association, 90th Annual Convention, Washington, D.C., August 23, 1982.

Trumberger, R. and MacLean, M. ''Maternal employment: The child's perspective.'' *Journal of Marriage and the Family,* 1982, **44,** 469–475.

Wallston, B. S., Foster, M. A., and Berger, M. ''I will follow him: Myth, reality or forced choice job-seeking experiences of dual-career couples.'' *Psychology of Women Quarterly,* 1978, **3,** 9–21.

CHAPTER 11

Promoting Occupational Safety and Health

Robert H. L. Feldman

INTRODUCTION

In the meat department of a large supermarket workers began showing symptoms of asthma. An analysis of the workplace environment indicated that is was not the meat, nor any of the chemicals used to clean the workplace that was causing the health problem. As more workers became ill, including some that became temporarily disabled, further investigations of the workplace were carried out. Finally, it was found that workers who wrapped the meat in plastic wrap had the severest symptoms. The machine that cut the plastic wrap used heat that released polyvinyl chloride (PVC) fumes from the plastic. When a new, totally automated meat-wrapping machine was installed together with an improved exhaust and ventilation system, the PVC fumes were reduced to a safe level for most of the workers. Those workers who were still affected by the slight release of PVC fumes were transferred to other departments in the store (Smith, 1982).

In a shipyard employing over 20,000 workers an analysis of occupational injury records indicated that over 60% of the injuries were eye injuries. The workers experiencing the greatest number of injuries were shipfitters. To reduce these injuries a behavioral modification program was instituted. First-line supervisors were trained to give praise, that is, positive social reinforcement, for the wearing of safety glasses by shipfitters. The results of the behavioral modification program indicated a decrease in eye injuries after the program had been instituted. In addition, the group that received the program had a larger decrease in eye injuries than a control group that did not participate in the program (Smith, Anger, and Uslan, 1978).

These two examples illustrate different approaches to the field of occupational safety and health. In the first example, an examination of the workplace environment indicated that a chemical hazard, PVC, was cause of an occupational illness, and environmental controls (e.g., new equipment and exhaust system) were instituted to reduce the problem. For those workers still affected by the minimum release of PVC, administrative controls, that is, assignment to other work duties, were used.

In the second example, analysis of injury records indicated that eye injuries to shipfitters were the greatest number of injuries at a shipyard. Psychologists utilizing psychological techniques, that is, behavioral modification (social praise and reinforcement), were able to increase the wearing of safety glasses and reduce the number of eye injuries at the shipyard. Thus, various approaches such as environmental controls, administrative decisions, and psychological techniques are used in the implementation of occupational safety and health programs.

Since the passage of the Occupational Safety and Health Act in 1970, increased interest has focused on safety and health in the workplace. Government agencies, unions, management, the press, the courts and legal profession, and the general public have shown concern about worker safety and how to establish a safe and healthy workplace. In addition, safety and health committees involved with workplace hazards have been formed by unions, management, universities, and citizen activist groups. It is clear that occupational safety and health is a major public health concern.

Occupational safety and health problems are pervasive through all segments of industry. For example, the cancer-causing substance inorganic arsenic is found in such occupational settings as auto-repair garages, banking, electric and gas services, transportation equipment, stone and glass products, leather and leather products, amusement and recreation services, rubber and plastics products, petroleum and coal products, printing and publishing, apparel and other textile products, heavy construction contractors, forestry, agriculture services, and oil and gas extraction (Workplace Carcinogen Control, 1980). In terms of occupational injuries, the most common disabling injury, back injuries, is found in almost all types of work settings from construction workers to hospital workers (Stellman and Daum, 1973). Therefore, the purpose of this chapter is to describe occupational safety and health as a component of a health promotion program.

SCOPE OF THE PROBLEM

The field of occupational safety and health is concerned with the injuries[1] and illnesses received while individuals are engaged in work. Under the 1970 Occupational Safety and Health Act each employer is required to furnish to his or her employees "employment and a place of employment which are free from recognized hazards that are causing or are likely to cause death or serious physical harm to his employees" (Occupational Safety and Health Act, 1970, p. 4). Therefore, the issue is to determine workplace hazards and methods of reducing and eliminating these hazards.

[1]The term "injury" and "injury behavior" will be used throughout this chapter rather than the unscientific term "accident," which connotes chance, fate, and unexpectedness. The term "injury" refers to damage resulting from acute exposure to physical and chemical agents (Haddon and Baker, 1978).

The number of work-related injuries and illnesses that occur every year is startling. Statistics from the National Safety Council (1982) indicated that in 1981, 12,000 deaths from injuries were occupationally related and over 2 million disabling injuries occurred in the workplace. According to Gordon (1971), another 25 million serious injuries and deaths may go unreported every year. The Bureau of Labor Statistics (1978) reports that nearly 3 out of every 10 injuries involving lost workdays have a duration of 15 days or more away from work.

In terms of occupational disease, of the half-million workers who have been exposed to asbestos in their work, it is estimated that 100,000 will die of lung cancer, 35,000 will die of abdominal or chest cancers, and about 35,000 will die of scarring of the lungs (asbestosis) (Stellman and Daum, 1973). Over the next 20 years about 6,000 uranium workers are expected to develop cancers. Black lung disease (pneumoconiosis) afflicts about 100,000 miners of which about 4,000 die of it each year and about 17,000 cotton, flax, and hemp workers have brown lung disease (byssinosis) (Stellman and Daum, 1973).

Occupational hazards are not the exclusive problem of blue-collar workers but are found throughout white-collar occupations. Dentists, physicians, nurses, artists, beauticians, and office workers are all exposed to toxic substances. Hospital workers are exposed to toxic chemicals in the form of bacteriocides (fluorides, ozone, tin), disinfectants (acetaldehyde, benzyl chloride, ethylene oxide), health care materials (ethyl bromide, ethyl ether, isopropyl alcohol), lab materials (benzene, formaldehyde, toluene), laundry substances (hydrogen fluorides, oxalic acid, formic acid) and office chemicals (asbestos, fibrous glass, solvents) (Braver, 1978). In effect, the workplace is a major source for injuries and illness among American workers.

GOALS FOR AN OCCUPATIONAL SAFETY AND HEALTH PROGRAM

In order to determine suitable goals for an occupational safety and health program it is necessary to have data on the rates of work-related injuries and illnesses. Workers' compensation claims is one source of information on the rates of injuries and illnesses. Workers' compensation claims include such information as minor injuries requiring first aid or no medical treatment, major injuries requiring medical treatment and/or loss of work time, minor illnesses that involved little or no loss of work time, and major illnesses resulting in extended periods of absence. Employees report the minor injuries and illnesses in order to receive first aid, examinations to determine the seriousness of the injury or illness, reassurance and legal documentation of the occurrence in the event that unexpected medical complications develop. Such reporting will vary greatly among different groups of employees and different individual employees, depending upon such factors as awareness of coverage under workers' compensation, access to first aid and medical examinations at work, time-constraints at work, concern as to the seriousness of the entity and recognition of the occupational origin of the injury/illness. While the reporting of the major illnesses and injuries is less variable than the reporting of minoı entities, it is important to recognize that the occupational origin of many work-related illnesses is not suspected by employees or their physicians. Even when employees recognize that their injury/illness is related to work, they may seek private medical care without reporting to workers' compensation. Many employees have health insurance coverage from the first dollar so that they would not have a strong financial incentive to report occupational injuries and illness (Braver, 1978).

Additional sources of data on workplace injuries and illness include other government data sources (e.g., Bureau of Labor Statistics, Occupational Safety and Health Administration, National Center for Health Statistics), the National Safety Council, and industry sources. It should be noted that these sources of *recorded* occupational injuries and illnesses are conservative figures, since not all occupational injuries and illness are recorded.

Based upon government records and epidemiological studies it is possible to determine, at least, the relative rates of occupational injuries and illnesses, that is, which industries and work settings have higher than average rates and which have lower than average rates. Though the ultimate goal of occupational safety and health programs is the elimination of all occupationally related injuries and illnesses, it is useful at first to concentrate on those work settings which have higher than average rates of occupational hazards. Depending upon the particular work hazard, specialists from a variety of fields (industrial hygiene, safety, health education) could assist in the development of an effective occupational health and safety program.

SOLVING THE PROBLEMS OF WORKPLACE INJURIES AND ILLNESSES

The complex problems of occupational safety and health have been examined from a number of different perspectives. We will take a look at these different perspectives in helping to develop a comprehensive effective program to reduce and eliminate workplace injuries and illnesses.

Industrial Hygiene

The field that has played a major role in the area of occupational health and safety is industrial hygiene. Industrial hygiene is concerned with the protection of worker health through the control of the work environment (Clayton, 1973). Industrial hygienists are trained in engineering, chemistry, physics, and biological sciences. With this knowledge their functions are (1) the recognition of workplace conditions that may cause illness and injury, (2) the evaluation of these workplace conditions, and (3) the control of workplace conditions to reduce and eliminate illness and injury.

The recognition and evaluation of workplace hazards include examinations of work operations and processes detailing the nature of the work, materials, and equipment used, products and by-products, number and sex of employees, and hours of work (Industrial Hygiene, 1959). Control of workplace hazards include recommendations of suitable and effective means for reducing and eliminating workplace hazards and the preparation of rules, regulations, standards, and procedures for the healthful conduct of work.

The most frequent problems encountered by industrial hygienists are chemical hazards (liquids, dusts, fumes, mists, vapors, gases, smoke, and smog), physical hazards (electromagnetic and ionizing radiations, noise, vibration, and exceedingly great temperature and pressure extremes), biological hazards (insects, mites, molds, yeasts, fungi, bacteria, and viruses), ergonomic hazards (monotony, repetitive motion, anxiety, and fatigue) and occupationally related stress. Industrial hygienists attempt to reduce and eliminate these occupational hazards through environmental modifications and engineering techniques. Other professionals that have contributed to our understanding of these factors include health physicists, toxicologists, and safety engineers.

Medical Professionals

Traditionally, the role of the medical profession in occupational safety and health has been to treat medical emergencies. First aid and the operation of dispensaries have been the major responsibilities of nurses and physicians. With the more recent emphasis on preventive medicine the role of the medical professional has expanded. Dispensaries are not only used in the treatment of injuries, but are involved in the early recognition of work-related illnesses. Physicians and nurses are engaged in pre-entrance physical examinations, periodic medical examinations, health appraisals and surveillance, early disease detection, comprehensive risk-profiling, and the establishment of individual and group-health baselines (Hoskin, 1977; Healy, 1977). In addition, the medical professional is involved in the referral of workers for further treatment, clinical follow-ups, and rehabilitation. Thus, medical professionals play a role in the prevention and treatment of workplace injuries and illnesses.

Behavioral and Educational Approaches

The role of the occupational physician and occupational health nurse in the early recognition of occupational illness and the prevention of disease has been identified. Also, the importance of industrial hygienists and safety engineers in the protection of workers' health through the control of the work environment has been mentioned. However, it is becoming increasingly clear that human-factor considerations need to be addressed in the prevention of occupational injuries and illnesses. This is especially true in the following instances:

1. Personal protective equipment is sometimes needed as an interim means of reducing work hazards during the maintenance of equipment and until engineering controls are in place.
2. During work emergencies such as fires, spills, and machine malfunctions, special procedures are needed to avoid harm to workers and to contain the problem.
3. Increased worker awareness of occupational safety and health risks would be useful in increasing protective behavior against work hazards (Cohen, Smith and Anger, 1979).

That is, a number of workplace conditions exist where protective behavior can reduce the chance of injury. These situations require personal protective behavior, yet protective behavior does not always occur. To better understand the occurrence of protective behavior it is helpful to examine research in the area of risk perception.

Perceptions of risk in terms of the likelihood of the risk or hazard and the consequences of the risk are important factors in whether individuals take protective measures. Slovic and his associates (Slovic et al., 1977) conducted research that examined risk perception. Their measure of protective behavior was insurance-purchasing decisions. They found that individuals were more likely to protect themselves (insure) against higher-probability, low-loss hazards rather than lower-probability, high-loss hazards. That is, people prefer to protect themselves against high-probability events with lesser consequences. For example, people prefer to have low deductible automobile insurance for high-probability, low-consequences accidents. Less than 15% of the population uses seatbelts to protect themselves against fatal automobile accidents, which are viewed as low-probability, high-loss hazards (Slovic, 1978). Not only do high-probability, low-consequences hazards lead to greater protective behavior than low-probability, high-consequence hazards, but individuals often do not have accurate knowledge of the probabilities of hazards.

Research of Tversky and Kahneman (1974) and Fischhoff and co-workers (1981) indicate that we often base our judgment of the probability of frequency of hazardous event on the ease with which we can imagine that event happening or the ease with which we can recall past instances of that event. Factors such as recency, vividness, and emotional involvement may lead to overestimation of certain events. Lichtenstein et al. (1978) found that the likelihood of dramatic, sensational, hazardous events that receive heavy media coverage such as homicides are overestimated, while unspectacular events that take victims one at a time and are nonfatal, such as asthma and diabetes, are underestimated (Kahneman, Slovic and Tversky, 1982).

Workers' perceptions of the causes of occupational injuries may influence their protective behavior and are useful information in the design of occupational safety programs. Feldman (1979) conducted research on how employees perceive the causes of occupational injuries and how they perceive the various solutions to reducing injuries. As part of a program to develop an effective means of reducing occupational injuries in hospitals, 93 nurses were given descriptions of injuries in terms of the severity of the injury, personal factors of the injured party, and environmental workplace conditions. They were asked to determine responsiblility for the injury, the relative causes in terms of personal and environmental factors, and whether the injured nurse could have avoided the injury. The results of the study indicated that the personal factors of the injured nurse were the most important consideration in judging the occupational injury. In addition, the nurses were asked to evaluate the effectiveness of various personal and environmental solutions to reduce work-related injuries. In general, the nurses considered environmentally oriented solutions more effective in reducing occupational injuries than personally oriented solutions. Thus, the results of examining workers' perceptions may be useful in the development of occupational safety programs.

Another approach to improving employee safety and health is to focus on the communication process. How do we communciate to effect changes in knowledge and attitudes that would result in an increase in safety behavior? Examination of the communication process indicates that the communicator of a message is a key component in the acceptance of that message (McGuire, 1969).

In the workplace setting, the supervisor is most often the communicator of the message. Crisera, Martin, and Prather (1977) in a study of roofing supervisors found some evidence that the safety level of a work crew related to the job-safety attitudes of its supervisor. If the supervisor believed that occupational injuries were avoidable through proper safeguards then there was indication of a low injury rate among the work crew. On the other hand, if the supervisor considered occupational injuries to be inevitable, then there was indication of a poor safety record among the work crew. Since supervisors probably came up through the ranks, members of the work crew could readily identify with them. The perceived power of supervisors to reward or punish the workers is another factor and the work crew's respect for their knowledge is a third factor which probably supported the influence of supervisors.

Supervisors are an influential force in affecting the attitudes and behavior patterns of the workers who are supervised. They are respected for their knowledge of work and its hazards. Commonalities and similarities between supervisors and workers increase their influence. Workers are more likely to identify with a supervisor who has similar interests, background, and experience. Therefore, supervisors may be able to increase their influence by forming "linkages" with their workers, that is, by emphasizing their common interests and experiences.

Another component of the communication process is the content of the message. Communications using a high-fear-arousal appeal (which emphasizes the hazardous nature of the work) are better at influencing people to cease an unsafe or unhealthy act, while low-fear-arousal communications are better for encouraging people to submit to procedures for the early detection of harmful diseases (Leventhal and Watts, 1966).

Another component of the communication process is the medium of the communications. Informal, face-to-face meetings appear to be more persuasive than written communications (McGuire, 1969); therefore, a communication process utilizing (1) knowledgeable, respected supervisors (2) who have formed linkages with their workers, (3) who emphasize the hazardous nature of the work, and (4) who communicate face-to-face with their workers are likely to decrease occupational injuries and increase safe behavior-patterns.

Psychological approaches to occupational safety have emphasized behavioral techniques. As mentioned at the beginning of this chapter a study by Smith, Anger and Uslan (1978) of shipyard workers was able to reduce eye injuries among shipfitters by the psychological training of first-line supervisors. Specifically, the supervisors were trained in behavioral modification techniques. They were taught to accurately observe workers' use of safety glasses. Based upon their observations they recorded how often the glasses were used. Then the first-line supervisors gave praise and social reinforcement for the wearing of safety glasses.

Other types of reinforcements that have been used include money, goods, and coffee and donuts in a textile spinning plant where the wearing of ear protectors reduced the hazard of excessive noise exposure (Miller, 1978). In this situation records were kept and posted of the daily use of earplugs. Reinforcements were given to the wearers of earplugs and within two weeks all of the workers taking part in the program were wearing earplugs. However, the behavioral reinforcement program was terminated thereafter and no follow-up results were reported.

Role of Management

Psychologists at the National Institute for Occupational Safety and Health (NIOSH) have looked at the characteristics of successful safety programs. Smith et al. (1978) and Cohen (1977) examined safety-program practices of matched pairs of plants with low- and high-injury rates to determine factors that might account for the differences in safety performance. The most important factor was greater management commitment and involvement in the total safety program in plants with low-injury rates. This was reflected by the rank and stature of the company safety officer, regular inclusion of safety issues in plant-meeting agenda, and personal inspections of work areas by top management officials. In some plants, inspections were made almost every day. Interviews with top management in several plants with the best safety records indicated that they placed the same emphasis on safety as on production and sales (Cohen et al., 1979).

A second factor was a more humanistic approach in dealing with employees, stressing frequent, positive contact and interaction in the low-injury plants. More open, informal interactions between management and workers, and everyday contacts between workers and supervisors on both safety and other matters were found in the low-injury plants. These interactions provided increased opportunities for early recognition of hazards, freer exchanges of ideas in correcting such problems, and greater worker participation. By comparison, management contacts with workers in plants having poorer safety records were more formal and less frequent, being confined largely to safety committees or other worker-management meetings (Cohen et al., 1979). A third factor was better employee selection procedures in the low-injury plants.

Other characteristics of low-accident plants included more frequent use of senior-level workers to train employees versus supervisors, more efficient housekeeping practices, cleaner and better plant environmental qualities, lower employee turnover and absenteeism, and in general, a more stable workforce. It can be concluded from these studies that low-injury rates result when management is committed to safety, when management takes a more humanistic approach, and when management is willing to take time for thorough employee selection, training, and development.

GUIDELINES FOR PROGRAM DEVELOPMENT

Before a general discussion on how to implement an occupational safety and health program is presented, three issues need to be addressed: (1) workers' right to know about occupational hazards, (2) worker and union involvement in occupational safety and health programs, and (3) workplace size.

Right to Know

How useful is a government toxic-chemical standard limiting workers exposure to a chemical if the workers do not know which chemical they are exposed to? How are workers going to take action to protect themselves from occupational hazards if they do not know what the hazards are? In recent years state and city governments have considered this problem by passing "right-to-know" laws. In November 1983, the Occupational Safety and Health Administration (OSHA) promulgated a national "Hazard Communication" rule (29 CFR 1910.1200). The purpose of the rule was "to create a federal standard for communicating information about hazardous chemicals to employees of chemical manufacturers and importers" (OSHA, 1984, p. 1).

The rule does not apply to public employees and non-manufacturing workers, however, OSHA's reason for promulgating the rule was based in part upon illness and injury data complied by the Bureau of Labor Statistics. For example, almost 200,000 occupational illnesses reported during 1977 and 1978 were most likely due to chemical exposures, and roughly half of these illnesses occurred in the manufacturing sectors. In addition, studies indicate for every reported occupational illness 50 may go unreported (OSHA, 1984).

The extent of exposure to hazardous chemicals among Americans is startling. According to the National Occupational Hazards Survey conducted by the National Institute for Occupational Health and Safety (NIOSH) 40 to 50 million Americans, that is, 23% of the entire U.S. population, may have been exposed at some point in their life to one or more hazardous chemicals regulated by OSHA (Hazard Communication, 1983).

The Hazard Communication rule requires that a chemical must be labeled once it has been determined to be hazardous. Container labeling is an alert mechanism to inform workers of the hazardous material found in the container. In addition, the labels must direct the worker to appropriate material safety data sheets (MSDS) (OSHA, 1984). A MSDS is a technical bulletin usually two to four pages in length, which contains information about a hazardous chemical, its composition and characteristics, health and safety hazards, and precautions for safe handling and use (Hazard Communication, 1983). The MSDS is presently an important source of information on chemicals throughout industry, but until the promulgation of this rule workers were usually unable to obtain access and had no legal right to see a MSDS.

Another important component of the Hazard Communication rule is the training of workers (Denny, 1984). Employers must provide employees with information and training regarding hazardous chemicals in their work area at the time of their initial assignment and whenever a new hazard is introduced into their work area.

To inform workers about workplace hazards, programs need to be developed (1) that supply information about hazards and their impact on worker health, and (2) offer advice on what workers and management can do to reduce these hazards.

Worker and Union Involvement

As workers become more knowledgeable of workplace hazards they are playing a larger role in the prevention of occupational injuries and illnesses. Although technical expertise is needed to recognize and evaluate many workplace hazards, other-potential hazards are easily recognized by employees at their worksites. Workers have first-hand experiences of occupationally induced hazards such as gases, odors, chemical spills, noise levels, heat, and fatigue. Worker experience and knowledge of these day-to-day hazards are extremely valuable to the identification and determining of occupational injuries and illnesses (Grynbaum, 1981). Therefore, worker involvement in the various phases of an occupational safety and health program will ensure that those individuals closest to the potential problem and in many cases most familiar with the problem are utilized in the recognition, evaluation, and control of an occupational safety and health program.

As workers have become more involved in occupational safety and health issues, so have worker representatives. Unions are negotiating not only for wages, hours, and benefits, but for changes in unsafe and unhealthy working conditions (Goldsmith and Kerr, 1982). The United Auto Workers Union, the United Rubber Workers Union, the United Mine Workers of America and the Oil, Chemical, and Atomic Workers International Union have been active in the field of occupational safety and health. Therefore, in worksites where unions are present, union participation in occupational safety and health programs can assist in the implementation of these programs.

Worksite Size

The development, implementation, and evaluation of an occupational safety and health program requires the knowledge of a number of disciplines. Professionals in industrial hygiene, safety, medicine, health psychology, and health education are needed for a comprehensive program. Yet, small plants and worksites are not able to afford full-time professionals in all of these areas. Therefore, a number of alternative suggestions are recommended.

1. The Pooling of Similar Worksites. Factories, plants, and industries producing similar products and having similar processes could share industrial hygienists, occupational health nurses, and other specialists. For example, dry-cleaning establishments contain a number of health hazards including benzene, carbon tetrachloride perchloroethylene, and tricholorethylene (Stellman and Daum, 1973). Though most dry-cleaning establishments are too small to hire a full-time industrial hygienist to monitor the levels of these potential hazards, a consortium or association could pool their resources and share in the expense of an industrial hygienist. The industrial hygienist could develop a schedule to monitor the different worksites.

   Similarly, an occupational health educator could be utilized in educating and informing employers and employees about the various health and safety hazards associated with dry cleaning. Large meetings with employers and employees from many different establishments could be planned or small meetings with individual establishments could be arranged.
2. Consulting. If different worksites are not able to be pooled together on some common basis, then individual worksites could hire occupational safety and health professionals on a consultant basis. Numerous worksites have arrangements with consulting occupational physicians and safety engineers to provide their services on a needs basis. Since some problems occur only on occasion, occupational specialists could be brought in when the need arises.

Occupational Safety and Health Program

A thorough program to reduce and eliminate occupational injuries and illnesses involves a number of methods and techniques from a variety of specialities. The first step in any program is an extensive analysis of potential workplace hazards. The use of industrial hygienists and safety experts is central to this task. Industrial hygienists conduct "walk-through" inspections of actual workplace conditions. Specific problems due to toxic substances are examined by industrial toxicologists while problems due to human-machine interaction are investigated by specialists trained in ergonomics. When a plant or industry is concerned with the relationship between specific workplace conditions and the incidence of illness or injury then the expertise of the epidemiologist is needed. "Epidemiology is concerned with the patterns of disease occurrence in human populations and of the factors that influence these patterns. The epidemiologist is primarily interested in the occurrence of disease by time, place, and persons. He tries to determine whether there has been an increase or decrease of the disease over the years; whether one geographical area has a higher frequency of the disease than another; and whether the characteristics of persons with a particular disease or condition distinguish them from those without it" (Lilienfeld and Lilienfeld, 1980, p. 3). Therefore, in analyzing the physical aspects of the workplace environment we turn to the specialities of industrial hygiene, safety engineering, toxicology, ergonomics, and epidemiology.

Another component of occupational safety and health care is the medical profession. Occupational physicians and occupational health nurses are involved in a wide range of medical and health duties. Both medical specialities are responsible for (1) entrance physical examinations, (2) interviewing and medical-history taking, (3) emergency medical care, (4) health maintance and health appraisals, (5) rehabilitation, and (6) ordering of laboratory tests and x-rays (Hoskin, 1977; and Healy, 1977).

Health Education Program

A comprehensive occupation safety and health program includes examination of the workplace environment, medical surveillance, and the health and safety education of employers and employees. We have discussed the contributions of specialists such as an industrial hygienist to the recognition, evaluation and control of the workplace environment, and the role of occupational physicians and occupational health nurses. We will now describe how to implement a health education program. A health education program will draw upon behavioral and educational approaches to reducing and eliminating injury and illness. A number of behavioral and educational techniques to increase safety behavior has been described. In this section we will discuss how these techniques can be brought together and utilized into a health education program.

Health education as been defined, as "any combination of learning experiences designed to facilitate voluntary adaptations of behavior conducive to health" (Green, Kreuter, Deeds, and Partridge, 1980, p. 7). Therefore, the aim of an occupational safety and health education program is to facilitate the voluntary adaptions of behavior conducive to a safe and healthy workplace. The first step in this program is to determine the knowledge, attitudes, and perceptions of employers and employees. We have discussed the research concerning risk perception. People are more likely to take protective action in situations where they believe there is likelihood of injury or illness. Therefore, it is necessary to first examine employers and employees perceptions of the likelihood that certain job-related actions (or lack of action) will lead to injury or illness. Since research indicates that people have misperceptions of the likelihood of hazards, health education strategies would address these misperceptions.

The second issue is the perception of the causes and solutions of occupational problems. Additional material would be collected to determine the accuracy of employers' and employees' perceptions in these areas. Health and safety information would be also be disseminated addressing these issues. After an occupational health education program has been implemented, proper evaluation methods would be utilized to determine the effectiveness of the program. That is, has the program increased the knowledge of employers and employees? Do employers and employees have more accurate perceptions of occupational hazards, the causes of injuries and illnesses, and solutions to reduce these hazards? And finally, has the occupational health education program lead to changes in behavior and reduction in injuries and illness? Based upon the results of the evaluation, modification and changes in the program may be necessary.

An additional focus is on the communication process between supervisor and worker. An effective communication would utilize face-to-face meetings between supervisor and worker, rather than utilizing written communications and informal discussions rather than formal discussions. In addition, supervisors would be trained to be knowledgeable of their work, how to gain respect from their employees and how to form linkages with their workers.

Another contribution that the behavioral sciences and specifically psychology can make to the reduction of occupational injuries is the behavioral approach for increasing safety behavior. Psychological techniques exist that can "elicit, shape, and maintain specified target behavior patterns in concurrence with safe work practices or other actions affording worker protection against job hazards" (Cohen et al., 1979). For example, according to Goldstein (1975), the first step in a behavioral-training program is to inform workers about the need to learn safe and healthful work practices, emphasizing active worker involvement throughout the training. Secondly, the actual behavioral patterns to be learned must be clearly identified. Then, practice time should be spent on those behavior patterns that are weakest and need strengthening relative to safety. Behavioral techniques include making rewards explicit and contingent upon the performance of safe behavior patterns, the establishment of performance goals, and the presentation of the results or feedback. The feedback should include whether or not the employee has performed the proper behavior pattern, constructive criticism, and information that will increase the worker's understanding. Another behavior technique frequently used is over-learning. That is, repeated drills, beyond the point of competency, ensures that safe behavior is unaffected in times of stress. Therefore, in emergency situations these behavior patterns will become "reflexive."

The establishment of specified safety behavior should be paired with rewards or reinforcers, followed by partial reinforcement to sustain that behavior. To maximize the effect of a reward program, the rewards should be given immediately after the target behavior has been performed. To utilize these behavioral techniques, supervisors need to (1) stress the learning of safety behavior patterns as opposed to the unlearning of unsafe or harmful acts, (2) allow enough time for their employees to practice the safe behavior patterns, (3) set performance goals for safe procedures with frequent feedback to determine success, and (4) establish meaningful rewards (either material, e.g., money prizes; or social rewards, e.g., praise, recognition) with an adequate reinforcement schedule.

Management

Based upon the research of NIOSH's Behavioral and Motivational Factors Branch on low- and high-injury-rate plants, it is recommended that to decrease injuries management should make a greater commitment and involvement in safety programs and safety matters, and that management should exhibit a more humanistic approach in interacting with employees, stressing frequent, positive contact and interaction. A table summarizing the comprehensive approach to occupational safety and health follows.

In conclusion, occupational safety and health involves a team-work approach to be truly effective. Industrial hygienists, occupational physicians, occupational health psychologists, and occupational health educators all have a role to play to reduce and eliminate injuries and illnesses. Industrial hygienists and occupational physicians have traditionally played an active role in this field. It is only recently that the contributions of the behavioral and educational approaches to occupational safety and health are being recognized. Future advances in this field should aid the development of a safe and healthy workplace.

Summary

Occupational safety and health problems are pervasive throughout all segments of industry making it a major public health concern. Therefore, the goal of an occupa-

Table 11.1 A Comprehensive Occupational and Health Program

I. Physical environment
   A. Recognition, evaluation, and control (industrial hygiene)
   B. Safety problems (safety engineering)
   C. Toxic substances (toxicology)
   D. Physical problems (health physics)
   E. Human–machine interaction (ergonomics, human factors engineering)
   F. Incidence, prevalence, and distribution of injury and illness (epidemiology)

II. Medical problems (occupational medicine and occupational health nursing)
   A. Physical examinations
   B. Emergency care
   C. Rehabilitation
   D. Health maintenance

III. Increasing knowledge and changing behavior (occupational health education and occupational health psychology)
   A. Risk perception
   B. Communication
   C. Behavioral approaches

IV. Role of management
   A. Management commitment and involvement
   B. Management-worker interaction
   C. Management training

tional safety and health program is to reduce and eliminate occupationally related injuries and illnesses. To achieve this goal requires the effort of an interdisciplinary team of specialists. Industrial hygiene has played a major role in the rocognition, evaluation and control of workplace conditions. In addition, safety engineers, toxicologists, health physicists, ergonomists and epidemiologists have all made contributions to this area.

Medical and health problems such as entrance physical examinations, emergency medical care, health appraisals, and health maintance has been the responsibility of occupational physicians and occupational health nurses. Increasing knowledge and changing behavior within the workplace setting has involved occupational health educators and health psychologists. These specialities have been examining perceptions of risk, the communication process, and behavioral approaches to change. In addition, studies on the role of management indicate that a greater commitment and involvement in safety programs and safety matters, and a more humanistic approach in interacting with employees leads to industrial plants with few occupational injuries. Included in any occupational safety and health program is active employee involvement since workers have first-hand familiarity with their work situation and workplace environment. In conclusion, an occupational safety and health program requires a team-work approach. In this chapter we have discussed the contributions of the various disciplines in the development of an effective occupational safety and health program.

A RESOURCE GUIDE FOR OCCUPATIONAL SAFETY AND HEALTH

Organizations

American Academy of Occupational Medicine (AAOM)
150 N. Wacker Drive
Chicago, IL 60606

American Conference of Governmental Industrial Hygienists (ACGIH)
1014 Broadway
Cincinnati, OH 45202

American Federation of Labor-Congress of Industrial Organizations (AFL-CIO)
Department of Safety and Health
815 16th Street, NW
Washington, DC 20006

American Industrial Hygiene Association (AIHA)
475 Wolf Ledges Parkway
Akron, OH 44311

American Occupational Medical Association (AOMA)
150 N. Wacker Drive
Chicago, IL 60606

American Society of Safety Engineers (ASSE)
850 Busse Highway
Park Ridge, IL 60068

Human Factors Society
P.O. Box 1369
Santa Monica, CA 90406

Industrial Health Foundation, Inc.
5231 Centre Avenue
Pittsburg, PA 15232

National Institute of Environmental Health Sciences
P.O. Box 12233
Research Triangle Park, NC 27709

National Institute for Occupational Safety and Health (NIOSH)
5600 Fishers Lane
Rockville, MD 20857

NIOSH
4676 Columbia Parkway
Cincinnati, OH 45226

NIOSH
944 Chestnut Ridge Road
Morgantown, WV 26505

Occupational Safety and Health Administration
U.S. Department of Labor
200 Constitution Avenue, NW
Washington, DC 20210

Society for Occupational and Environmental Health (SOEH)
1341 G Street, NW
Washington, DC 20005

U.S. Environmental Protection Agency (EPA)
Washington, DC 20460

Women's Occupational Health Resource Center
Columbia University, School of Public Health
60 Haven Avenue, B-1
New York, NY 10032

Councils and Committees on Occupational Safety and Health

Bay Area Committee for Occupational Safety and Health (BACOSH)
Alameda County Labor Temple
2315 Valdez Street, Room 108
Oakland, CA 94611

Chicago Area Committee on Occupational Safety and Health (CACOSH)
542 S. Dearborn #502
Chicago, IL 60605

Electronics Committee on Occupational Safety and Health (ECOSH)
630 N. First #201
San Jose, CA 95112

Los Angeles Committee on Occupational Safety and Health (LACOSH)
1921 Glendon Avenue
Los Angeles, CA 90025

Maryland Committee for Occupational Safety and Health (MARYCOSH)
P.O. Box 3825
Baltimore, MD 21217

Massachusetts Coalition for Occupational Safety and Health (MASSCOSH)
Main Office
718 Huntington Avenue
Boston, MA 02115

New York Committee on Occupational Safety and Health (NYCOSH)
32 Union Square E. #404
New York, NY 10003

Philadelphia Area Project on Occupational Safety and Health (PHILAPOSH)
1321 Arch Street #607
Philadelphia, PA 19107

Rhode Island Committee on Occupational Safety and Health (RICOSH)
P.O. Box 95 Annex Station
Providence, RI 02901

Western New York Council on Occupational Safety and Health (WNYCOSH)
59 Niagara Square Station
Buffalo, NY 14201

Books

Crisis in the Workplace: Occupational Disease and Injury. Nicholas A. Ashford MIT Press, Cambridge, Ma, 1976.

Encyclopedia of Occupational Health and Safety. International Labor Organization, McGraw-Hill, New York, 1972.

The Human Side of Accident Prevention. Bruce L. Margolis and William H. Kroes, eds., Thomas, Springfield, Ill., 1975.

The Impact of OSHA. Herbert R. Northrup, Richard L. Bowan, and Charles R. Perry, Wharton School, Philadelphia, Pa., 1978.

The Industrial Environment: Its Evaluation and Control. NIOSH, U.S. Printing Office, Washington, D.C. 1973.

Industrial Health. Jack Peterson, Prentice-Hall, Englewood Cliffs, N.J., 1977.

NIOSH Publications Catalog. 4th ed., NIOSH, Cincinnati, Ohio, 1980.

Occupational Diseases: A Guide to their Recognition. NIOSH, U.S. Printing Office, Washington, D.C., 1977.

Occupational Safety and Health: The Prevention and Control of Work-Related Hazards. Frank Goldsmith and Lorin E. Kerr, Human Sciences Press, New York, 1982.

Women's Work, Women's Health: Myths and Realities. Jeanne Stellman, Pantheon, New York, 1977.

Work and Health. Robert L. Kahn, Wiley, New York, 1981.

Work Is Dangerous to Your Health. Jeanne Stellman and Susan Daum, Vintage, New York, 1973.

Journals

Accident Analysis and Prevention

American Industrial Hygiene Association Journal

Archives of Environmental Health

Ergonomics

Health Physics

Human Factors

Journal of Occupational Medicine

Journal of Safety Research

Occupational Health & Safety

Occupational Health Nursing

Occupational Psychology

Professional Safety: Journal of the American Society of Safety Engineers

Films

"More Than a Paycheck," U.S. Department of Labor, Occupational Safety and Health Administration, Washington, D.C., 16mm, color, 25 min., 1979.

"Song of the Canary," Josh Hanig and David Davis, Franklin Lakes, New Jersey, 16mm, color, 60 min., 1978.

REFERENCES

Braver, E. "Hospital occupational safety and health: A medical record study of housekeepers and a study of workers' compensation claims of hospital employees at a large university hospital." Unpublished manuscript, Johns Hopkins University, 1978.

Bureau of Labor Statistics reports occupational injuries and illnesses incidence rate remain virtually unchanged for 1976. (1978) *Job Safety & Health* **6.** 2–3.

Clayton, G. D. "Introduction." In *The Industrial Environmental—Its Evaluation and Control.* National Institute for Occupational Safety and Health, Washington, D.C., 1973.

Cohen, A. "Factors in successful occupational safety programs." *Journal of Safety Research,* 1977, **9,** 168–179.

Cohen, A., Smith, M. J., and Anger, W. K. "Self-protective measures against workplace hazards." In *National Conference in Health Promotion Programs in Occupational Settings,* Washington, D.C., 1979.

Crisera, R. A., Martin, J. P., and Prather, K. L. "Supervisory effects on workers safety in the roofing industry." *Management Sciences Company Contact No. CDC 99-74-35.* National Institute for Occupational Safety and Health, Division of Safety Research, Morgantown, West Virginia, 1977.

Denny, D. "Labeling standard may re-define health and safety responsibilities." *Occupational Health and Safety,* 1984, **53**(1), 30–32.

Feldman, R. H. L., Wang, V. L., Liskin, L. S. and Braver, E. R. "Reducing occupational injuries: A personal-environmental approach." Paper presented at the meeting of the American Psychological Association, New York City, September 1979.

Fischhoff, B., Lichtenstein, S., Slovic, P., Derby, S. L., and Keeney, R. L. *Acceptable Risk.* New York: Cambridge University Press, 1981.

Goldsmith, F. and Kerr, L. E. *Occupational Safety and Health: The Prevention and Control of Work-Related Hazards.* New York, Human Sciences Press, 1982.

Goldstein, I. L. "Training." In *The Human Side of Accident Prevention,* B. L. Margolis and W. H. Kroes, eds. Thomas: Springfield, Ill.

Gordon, J. B., Akman, A., Brooks, M. L. *Industrial Safety Statistics: A Re-examination.* A Critical Report Prepared for the U.S. Department of Labor, p. 144–151. New York: Praeger, 1971.

Green, L. W., Kreuter, M. W., Deeds, S. G., and Partridge, K. B. *Health Education Planning: A Diagnostic Approach.* Palo Alto, Ca: Mayfield, 1980.

Grynbaum, G. A. "Health Education in a center for municipal occupational safety and health." Paper presented at the meeting of the American Public Health Association, Los Angeles, 1981.

Haddon, W. Jr. and Baker, S. P. "Injury Control." In *Preventive Medicine, 2nd ed., D. Clark and B. McMahon, eds. Boston: Little, Brown, 1978.*

"Hazard Communications." Federal Register, 1983, **48**(228), 53280–53348.

Healy, B. "The role of the nurse." In *Occupational Safety and Health Symposia 1976,* Washington, D.C. National Institute for Occupational Safety and Health, No. 77-179, 1977.

Hoskin, W, D. "The role of the medical profession." In *Occupational Safety and Health Symposia 1976,* Washington, D.C. National Institute for Occupational Safety and Health, No. 77-179, 1977.

"Industrial Hygiene." *American Industrial Hygiene Association Journal,* 1959, **20,** 428.

Kahneman, D., Slovic, P. and Tversky, A. eds. *Judgement Under Uncertainty: Heuristics and Biases.* New York: Cambridge University Press, 1982.

Leventhal, H. and Watts, J. C. "Sources of resistance to fear-arousing communications." *Journal of Personality,* 1966, **34,** 155–175.

Lichtenstein, S., Slovic, P., Fischhoff, B., Layman, M. and Combs, S. "Judged frequency of lethal events." *Journal of Experimental Psychology: Human Learning and Memory* 1978, **4,** 551–578.

Lilienfeld, A. M. and Lilienfeld, D. E. *Foundations of Epidemiology* 2nd ed. New York: Oxford University Press, 1980.

McGuire, W. J. "The nature of attitudes and attitude change." In *The Handbook of Social Psychology,* G. Lindzey and E. Aronson, eds. Menlo Park, Ca.: Addison-Wesley, Vol. 3, Chap. 21, 1969.

Miller, L. M. *Behavior Management: The New Science of Managing People at Work.* New York: Wiley, 1978.

National Safety Council. *Accident Facts, 1982* Chicago, 1982.

Occupational Safety and Health Act, Public Law 91-596. S. 2193, 1970, p. 4.

OSHA. "Rule on Hazard Communication" *Details,* (American Lung Association, 1984, **7**(5), 1–7.

Smith, E. E. "For occupational asthma, the best treatment is prevention. *American Lung Association Bulletin,* July/August 1982, 12–15.

Smith, M. J., Anger, W. K., and Uslan, S. S. "Behavioral modification applied to occupational safety." *Journal of Safety Research,* 1978, **10,** 87–88.

Smith, M. J., Cohen, H. H., Cohen, A., and Cleveland, R. J. "Characteristics of successful safety programs." *Journal of Safety Research,* 1978, **10,** 5–15.

Slovic, P. ''The psychology of protective behavior.'' *Journal of Safety Research,* 1978, **10**(2), 58–68.

Slovic, P., Fischoff, B., Lichtenstein, S., Corrigan, B., and Combs, B. ''Preference for insuring against probable small loses: Insurance implications.'' *Journal of Risk and Insurance* 1977, **44,** 237–258.

Stellman, J. M. and Daum, S. M. *Work is Dangerous to Your Health.* New York: Random House, 1973.

Tversky, A. and Kahneman ''Judgment under uncertainty: Heuristics and Biases.'' *Science,* 1974, **185,** 1124–1131.

Workplace Carcinogen Control, Maryland Division of Labor and Industry, 1980.

CHAPTER 12

Back Injury Prevention Programs

Carole Lewis
Kay Schaefer

INTRODUCTION

Glenda Harrison is a nurse who learned the hard way about the most common industrial injury, lower back injury (Tabor, 1982). Until she suffered from low back pain herself, she did not fully understand the physical, emotional, and professional ramifications of a lower back injury.

Glenda hurt her back one night while caring for an obese, comatose person in the intensive care unit of a hospital. Instead of taking time off to allow her back to heal, she continued to work for three weeks until she experienced numbness in one leg and severe back pain. Her injury had developed into a full-blown herniated disc that was pressing on the nerve root into her leg. She was unable to return to work for seven months (Tabor, 1982).

Glenda returned to work better prepared to prevent injury of her back than she was that night in the intensive care unit. She was better prepared because she learned to ask for help with heavy jobs, she learned that back injuries often get worse when they are ignored and she learned how to protect her back while performing her work activities. It is too bad that Glenda did not acquire that knowledge before she spent seven months in pain and out of work (Tabor, 1982).

Glenda Harrison is not alone. Eighteen million working days are lost every year in industry due to lower back problems (Anderson, 1981). Back injury costs in the United States, including medical bills and compensation costs, exceed all other industrial injuries combined, running about 14 billion dollars annually (Tabor, 1982). Low back injuries are the most common type of cumulative trauma injury and are the injuries that are most frequently litigated (Tabor, 1982). In total, 75 million Americans suffer from back pain, with 7 million new victims each year, leaving 2 million unable to work and 5 million partly disabled (Toufexis, 1980).

Given these statistics, it would make sense for industry executives to pay attention to situations in the workplace that contribute to back injury and to offer educational programming that teaches workers and management alike how to prevent this most common industrial injury.

This chapter describes the scope of the problem of back injuries in the workplace, explains the necessity for back injury prevention programs, discusses several existing programs and describes a model back injury prevention program.

SCOPE OF THE PROBLEM

The statistics for low back pain and its impact on industry continue to grow (Anderson, 1981). With the growth of back injuries, employers are asking what the real effects of this problem are, and what can be done to improve the situation.

The effects of low back pain in industry are twofold. Socially and personally, the employee views the injury as a possible permanent disability, life-long discomfort and unending pain (Strasser, 1980). With this view in mind, the employee may look to the employer for assistance or alleviation of these fears. The employer, on the other hand, may see back pain in terms of absenteeism, worker's compensation payments, lawsuits, company apathy, and early retirement and disability pensions. None of these views of back pain, either by the employee or employer, are positive factors that work toward enhancing the quality of life of the employee and the productivity of the company (Strasser, 1980).

In recent years, companies have been alerted to the growing problem of back injuries (Larry, 1978). Periodicals for occupational safety in industry have been carrying more articles about back injury, and orthopedic journals have had more articles on industrial intervention. These publications have identified the adverse effects of back injury on the quality of life of the employee (Anderson, 1981; Garret, 1977).

Back injuries have recieved sufficient attention in recent years to be identified as an important health and social problem. Back injuries are viewed as the major cause of disability for persons age 20–45 (Chaffin, 1973).

In reviewing the causes for back pain in the worksite, two major categories were determined (Kelsey, 1979). Failing to lift objects properly and maintaining poor body mechanics in daily work routines are the two major categories of causes for back injury on the job (Kelsey, 1979). Both of these causes can be alleviated through education and skill development (Kelsey, 1980). Unfortunately, most employees do not receive adequate or appropriate education or skill development in back injury prevention. Failure to receive education in proper body mechanics and lifting procedures has been shown to result in fear, hesitancy, and avoidance in carrying out these activities by employees who have sustained back injury while lifting or moving. The potential consequences of insufficient or ineffective education and skill development to prevent back injuries results in costs to the employee, employer, and society.

Eighteen million working days lost per year due to back injuries affects the employee on a psychological, social, and financial level (Anderson, 1981). The employee who has a back injury may become less enthusiastic about work tasks and may feel that management is not concerned about employee safety. In addition, if an employee is injured with a resultant chronic disability, the employee's long-term income security is at risk. This work loss will affect the employee's income and may affect productivity of other workers by encouraging a negative and fearful attitude toward their worksite. Society will be faced with the possibility of providing social and psychological support to the employee as well as the employee's family and may also have to bear financial burdens if the back injury becomes chronic and the employee collects disability or worker's compensation.

Calliet (1975) states that the most important factor affecting the maintenance of a healthy back is "knowledge and practice of proper posture, body mechanics, and lifting techniques." A viable educational program for employees may mean the difference between a healthy productive employee and one who requires sick time and disability days along with expensive medical treatments.

The high incidence of low back pain and disability in industry causes increased absenteeism, decreased morale, increased disability, increased discomfort and dissatisfaction, and increased costs to the company. Practical attention given to the problem of back injury and back pain will ultimately reduce the unnecessary and costly dependence of injured employees in the future.

The following section outlines a back injury prevention program that has as its ultimate goal, a reduction in the incidence of low back pain and disability of the employees at a particular company. This goal is much too broad for effective action. However, the scope of the problem can be narrowed down by educating the employees that are especially vulnerable. Often, vulnerable employees are required to lift, move, and maneuver heavy pieces of equipment throughout the day. These particular employees have an increased incidence of back injuries and longer sick-leaves (Blair, 1981). "These particular employees also have a disproportionate amount of lawsuits for on-the-job injuries" (Troup, 1981).

The preceding discussion suggests that appropriate and timely intervention, that is, back injury prevention program, may prevent the physical, psychological, and financial problems at a particular company from developing or worsening.

RESEARCH REVIEW ON BACK INJURY PREVENTION

Industry has recently begun to recognize that human productivity is related to job satisfaction and optimum health (Larry, 1978). Ill health is a tremendous loss to a person's family, to his or her community and especially to the organization for which he or she works. Many illnesses, such as injury of the back and neck, coronary artery disease, strokes, and heart attacks, can be prevented through early detection, education, and fitness programs (Larry, 1978). Corporations are helping influence the prevention of these illnesses by sponsoring fitness programs and by encouraging participation in them (Larry, 1978).

As a matter of fact, industrial prevention-fitness programs are becoming so popular that symposiums are held on the development of fitness programs. One such symposium was held in Pittsburg, Pennsylvania in March of 1978. Participants came from a variety of corporations and institutions, such as United States Steel Corporation, ALCOA Corporation, Mobay Chemical Corporation, Harmarville Rehabilitation Center, and Blue Cross of Western Pennsylvania. Included in the symposium was a discussion of the need, the cost, the benefits, the program content, and the organization of prevention and fitness programs.

Physical therapists have joined with industrial trainers, physicians, human resource specialists, occupational nurses, corporation executives, and health insurance companies to develop programs that have an emphasis on prevention of injury and disease in the worker (Mattmiller, 1982). Bill Mattmiller, a physical therapist, and Cris Schulenberger, an industrial trainer, have combined forces and together consult with numerous industrial and health care companies in developing and operating successful injury prevention programs. Their consultations consist of a thorough review of the companies' injuries history, a survey of the plant and the offices for potential injury sites, recommendations for engineering and environmental changes, and training programs for employees (Mattmiller, 1982).

Mattmiller and Schulenberger walk through the companies' plant to inspect not only the physical conditions, but also the attitudes of employees and management. During the walk-through, they look at factors such as employee work posture and work patterns. The walk-through and a study of the companies' absentee and injury records help them design an appropriate prevention program.

The program recognizes that injuries cannot be eliminated, but they can be minimized for a reduction in frequency and severity of injuries. The program they develop includes instruction in proper body mechanics, exercise, equipment, modification recommendations, and exercise motivation. Through their experience, Mattmiller and Schulenberger have found that one program given per year will have an initial effect that quickly wears off. The prevention program, to be effective, must be ongoing and involve all employees, especially management and first-line supervisors (Mattmiller, 1982).

Mattmiller and Schulenberger have found that back injury prevention programs reduce accidents down-time, and improve morale, productivity, and profits (Mattmiller, 1982).

Their training program includes a pre- and posttest, anatomy, aging concepts, posture, proper use of the back, lifting skills, body mechanics, and discussion of specialized activities (Mattmiller, 1982).

Keith Klevin is a physical therapist who is a pioneer in developing back injury prevention programs in industry. He has developed programs for Trans World

Airlines and Hilton Hotels (Klevin, 1979). The emphasis of his programs is on employee motivation. He states that employees who feel that they are part of a team in which the social setting rewards safety and injury prevention are more motivated to stay well (Klevin, 1979). The responsibility for employee motivation lies both with management and with the employee. Klevin states that a work atmosphere that promotes teamwork, employee trust, and a belief in the employees' intention to prevent injuries reduces injuries. He has found that companies that have used his program have saved significant sums through fewer work days lost due to injury-related illnesses (Klevin, 1979).

Another concept in programming for managing the problem of back injuries in industry that is becoming popular throughout the United States, Canada, Australia, and Sweden is the Back School (Johnston, 1982). A Back School is a training program, usually sponsored by a physical therapist or physician specifically for those who have had a back injury, and may take place in a physical therapy clinic or at the workplace. Although the various programs differ, they each consist of a series of sessions that teach the individual how to improve his or her existing back problem and how to prevent a similar problem from recurring. The goal of the Back School is that each student is able to return to permanent employment or find a new opening in the labor market (Tabor, 1982).

Back Schools include both individual and group sessions and include the following components: assessment of individual strengths and weaknesses, instruction in general concepts of back pain, self-care, activities that encourage protection of the back, relaxation techniques, stress management, body control, stretching, strengthening, prevention, fitness, and nutrition. Individuals who attend a Back School are evaluated for reinjury follow-up at three-month, six-month, and one-year intervals. An annual refresher course helps maintain a high level of prevention awareness in the employee (Johnston, 1982).

George J. Moudry, a physical therapist in Mobile, Alabama, runs a Back Pain Prevention Program designed to prevent reinjury of the back. His program takes place in a physical therapist's office and consists of an evaluation of each employee's injury, specific treatment of that injury individually and in groups, education of prevention techniques, and follow-up clinical testing and treatment evaluation. This program emphasized the importance of exercise and posture in the prevention of reinjury (Moudry, 1980).

As employee-management theory changes from a dictatorial leadership style to a style of teamwork, managers are realizing that employees are their most productive when they are healthy and happy (Larry, 1978). Toward that end, many companies throughout the industrialized world are instituting back injury prevention and fitness programs. The benefits of these programs greatly exceed the costs in terms of worker productivity and dollars spent versus days lost due to illness (Wilson, 1978). The program presented in this chapter is one such industrial back injury prevention program.

GUIDELINES FOR PROGRAM DEVELOPMENT

The following is an example of a proposed back education program.

Many nonbehavioral factors contribute to the incidence of low back pain and injury. Some of these are

1. Weight and bulk of equipment.
2. Seat positions.
3. Counter and desk heights.
4. Daily work and leisure schedules.

Although most employers have little or no control over the above factors, there are various behaviors that can be controlled to aid in the prevention of back pain and injuries. These behaviors are stated in Table 12.1 in terms of Predisposing, Enabling, and Reinforcing Factors. Each of these factors is rated according to importance and changeability. The factors of highest importance and changeability will be incorporated into this proposal. The factors listed in Table 12.1 received a high marking in changeability because they have been changed in previous programs (Green, 1980; Shactman, 1979). These same factors recieved high rankings in importance because data is available linking these factors to back pain prevention (Calliet, 1975).

Once the possible factors that may contribute to low back injury have been identified and ranked in terms of importance and changeability, goals need to be set in terms of quantifiable action that will be taken.

Statement of Behavior in Terms of Action to be Taken

Knowledge: By the end of the program, 90% of the participants:

a) will identify body mechanics as a cause for back pain.
b) will think critically about posture and consequences of various movements on the back.
c) will be able to recognize the anatomy of the spine.
d) will be able to identify proper posture, body mechanics, and exercises for the lower back.

Beliefs: By the end of the program, 80% of the participants:

a) will believe that back problems are serious.
b) will believe that a program in proper body mechanics, posture and exercises is helpful.
c) will believe that they can take action to alleviate or prevent back pain.

Skills: By the end of the program, 70% of the participants:

a) will be able to evaluate and practice proper posture positions.
b) will be able to demonstrate proper body mechanics.
c) will be able to execute exercises properly.

Table 12.1 Factors Influencing Back-Care Behaviors Are Ranked According to Importance and Changeability

Factors	*Importance*	*Changeability*
Predisposing		
Knowledge		
1. Importance of back care	High	High
2. Proper body mechanics	High	High
3. Anatomy of the low back	Medium	High
4. Proper posture	High	High
Attitudes		
1. Self-confidence (sense of ability to care for onself)	High	High
2. Desire for a healthy body	High	High
Values		
1. Family practices in back care habits	High	Low
2. Cultural practices in back care habits	High	Low
Perceptions		
1. Belief that an educational program will help prevent back pain	High	High
2. Sense of control over back pain	High	Medium
Enabling		
1. Ability to learn proper posture, body mechanics, and exercise	High	Medium
2. Environment enabling the above to occur	High	High
Reinforcing		
1. Support of back care principles by peers, supervisor	High	Medium
2. Support of back care principles by supervisors	High	High
3. Support of back care principles by family	Medium	Low
4. Trust in back care information providers	High	Medium

d) design an individualized program that will help to prevent and relieve back pain.

Once the goals have been set in behavioral terms, then behavioral objectives specific to the goals can be constructed.

Statement of Behavioral Objectives

Who: employees at specific worksite.
What: a reduction in the incidence of low back injury.

How Much:	30% within the first year and an additional 20% within the next three years.
When:	by the time of the proposed follow-up evaluation—one year after completion of the course.
How:	Back injuries will be recorded for employees receiving and not receiving the educational program. These two groups will be compared. (See evaluation section.)

Predisposing factors are the motivating factors of attitude, values, information, and perception that an employee holds. Enabling factors are the skills and resources necessary to perform a behavior or complicated set of behaviors. Enabling factors also relate to availability and accessibility of resources. Reinforcing factors strengthen or weaken the motivations and actions of the employee. These factors have been presented previously in Table 12.1 in terms of changeability and importance. The following table is an analysis of these influencing factors of employees in general (not in terms of the specific desired behavioral outcomes). (See Table 12.2.)

Selection of Educational Components

Providing the most effective program for employees requires choosing educational methods that fulfill the needs outlined in the previous section. A model program should use at least three forms of educational interventions. In addition, the program should be aimed at providing information and intervention to fulfill the re-

Table 12.2 Factors Influencing Health Behaviors

Predisposing factors
- Demographic factors
  1. Age
  2. Gender
- Knowledge
  1. Lack of information about health, anatomy of the back, lifting procedures, body mechanics
- Attitudes
  1. Positive or negative attitudes about a sense of control over back pain

Enabling factors
1. Lack of time in daily schedule to take a course
2. No specialized back program available to employees

Reinforcing factors
1. Employers do or do not understand the needs of the employees
2. Lack of appropriate social support from family and peers

quirements of the Predisposing, Enabling, and Reinforcing Factors. For cost efficiency, the most inexpensive methods should be explored; however, since a back program does require a complex set of behaviors, several interventions are needed. The following strategies have been used in this course:

1. Lecture-discussion and audiovisuals—to provide knowledge about (Predisposing):
   (a) Back pain.
   (b) Back physiology.
   (c) Body mechanics.
2. Skill Development—to provide skills in (Enabling):
   (a) Body mechanics (i.e., lifting, etc.).
   (b) Posture.
   (c) Exercises.
3. Modeling and Behavior Modification—to encourage the use of the above (Reinforcing):
   (a) This phase will encourage long-term effects.

Selection of strategies depends on the audience, site, problem, financial considerations, and purpose of the program. For this particular program the above strategies were chosen to reach a large number of employees and yet provide some individualized instruction for a reasonable cost.

Administrative Diagnosis

This final phase of the program analysis *describes* how the back program will fit into the larger company system.

I. Within Program Analysis—the factors pertaining to the education program itself.

Class variables: The following variables are flexible and can be changed to meet the companies' needs.

1. The course will require 1 to 2 instructors—for class instruction.
2. 1 to 2 persons will be needed to conduct pre-instructional worksite evaluation.

Space and equipment: The following variables and flexible and can be changed to meet the companies' needs.

1. An 800 sq. ft. room (with mats, preferably).
2. An additional lecture room or 20 to 30 folding chairs for #1 above.
3. One screen.
4. One chalkboard.
5. Optional—(a) video tape recorder and monitor
   (b) slide projector

II. Within Organization Analysis—the following variables analyze the receptivity of the proposed organization to a back education program.

1. Newness of health education programs as part of the organization—if this is very new there may need to be some initial employee sensitization to deter any apprehensions on the part of the employees.
2. The degree to which health education programs are known and valued—if health education programs are seen to have little value by the employees, additional information (i.e., pamphlets and flyers) can be made available to gain support by the employees.
3. The effects of the program on other parts of the organization—if other parts of the company will be affected, additional preplanning will be done to alleviate any difficulties in accomplishing the goals of this program and to minimize any difficulties in the other areas.
4. Readiness of the organization for change—this can be the biggest selling point to the employee or employer. This will be assessed using the AVICTORY method. One list of readiness indicators has been labeled AVICTORY. It provides for a more detailed assessment of readiness for change.

   A = ability to carry out the change (capability, resources and social costs).
   V = values; compatibility with mission or goals.
   I = ideas or information about the qualities of the innovation (communicability, observability, susceptibility to successive modifications, divisibility, reversibility, scientific status).
   C = circumstances that prevail at the time (climate of trust, willingness to entertain a challenge).
   T = timing or readiness for consideration of the idea; early involvement of potential users.
   O = obligation or accepted need to deal with a particular problem; relevance, commitment; shared interest in solving recognized problems.
   R = resistance or inhibiting factors; skill in working through uncertainty and risks.
   Y = yield or perceived prospect of payoff for adoption; expected reward; belief in the efficiency of the innovation.

Any of the above variables may indicate the best method of convincing the employer or employee in the need for the program. This assessment also identifies potential areas of difficulty in program implementation (Green, 1980).

III. Interorganizational Analysis—the extent to which the outside world may influence the development of the program.

1. The proximity of other back-education programs will be analyzed.
2. The success of other back-education programs will be cited.

3. The reinforcement available in the community of the various objectives will be investigated.

Evaluation

The back pain prevention program is designed to decrease the incidence of back pain of the employees at the specific worksites. To show that there has been significant decrease in this problem as a result of the program, a controlled-experimental approach can be used to evaluate the results (see section on evaluation).

The procedure that will be used to conduct this evaluation is as follows:

1. The measure of Impact Assessment will be conducted by the use of a Pre-test, Post-test, Control Experimental Design R E_1 X E_2
   R C_1 C_2
2. The employees required to participate in the program will be identified.
3. The employees in #2 above will be randomly assigned to groups of 25.
4. All groups will be evaluated.
5. The experimental groups (groups 1 and 2) will receive the program.
6. All groups will be compared upon the experimental group at completion of the program.
7. All groups will be evaluated at 3, 6, and 12 months, after completion of the class to check for incidence of back injuries.
8. The two groups will be compared on item 7 above at 12 months.
9. The control group will receive the class after the 12-month evaluation.

The evaluation should be analyzed, compiled, and presented to the specific employer.

The program presented above is only one type of back injury prevention program that may be done. Designing other programs requires individualized assessment of the companies' specific problems. For example, a proposed program for a small legal office may simply include an onsite evaluation and lectures on body mechanics, posture, and back injuries. A program for a large chain of grocery stores, on the other hand, may require a more complex program than the model program cited above. The grocery stores may need an extensive evaluation of numerous types of job classifications ranging from truck loaders to line checkers. A helpful tip in conducting any onsite evaluation is to inconspicuously watch the daily activity and, if possible, use audiovisual to actually show the faulty body mechanics onsite. The most successful programs are those that deisgn the program to reflect the needs of the individuals in the company.

Summary

Low back injuries may at the present time be one of the greatest causes of disability in industry. Back injuries are costly to companies and to the individual. Programs specifically designed to meet the needs of the various companies are just beginning to be designed and implemented. Nevertheless, there is information available on the positive effects of such programs (Schactman, 1979) and as more time passes, more longitudinal data will be available on the long-range effects of these programs.

This chapter has presented some introductory information on the severity of the problem as well as the current state of the health educators' interventions to alleviate the back injury problem. A model program was described to provide an outline as a guide for planning this type of program in industry. The model program outlined the importance of identifying risk factor and behavior necessary for change. In addition, this section described methods of choosing behavioral objectives and educational components. Finally, the integration of a model program in the functioning of the entire company was outlined as well as the method of evaluating the back education program.

As mentioned earlier, the area of educational intervention for back injuries is in the infancy stage. There is a need for future programs and more quantifiable data on the effects of such programs.

This chapter has presented ideas and strategies relating the severity of the problem to the feasibility of a program in the industrial setting. It is up to health professionals to continue to create and pursue programs on an objective basis to enrich this area of improved employee satisfaction and safety.

RESOURCE GUIDE FOR PREVENTION OF BACK INJURY

Organizations

The American Back School
P.O. Box 1193
Ashland, KY 41105

APTA - American Physical Therapy Association
Sports Medicine
1111 N. Fairfax Street
Alexandria, VA 22314

Back Owner's Manual
Patient Information Library
PAS Publishing
Daly City, CA 94402

Body Theraputics
6317 Wilshire Blvd.
#503
Los Angeles, CA 90048

Orthopedic Physical Therapy Products
2305 Louisianna North
Minneapolis, MN 55427

OSHA—Occupational Safety and Health Administration
Publishing Distribution
Room North 4401
200 Constitution Avenue, NW
Washington, DC 20210

Journals

American Industrial Hygiene Association Journal

Physical Therapy

American Journal of Public Health

Health Education

REFERENCES

Anderson, Gunnar. ''Epidemiological aspects on low back pain in industry.'' *Spine,* 1981, **6,** 53–61.

Calliet, R. *Low Back Pain Syndrome,* Philadelphia, Pa.. Davis, 1975.

Chaffin, Don. ''A longitudinal study of low back pain as associated with occupational weight-lifting factors.'' *American Industrial Hygiene Association Journal,* December 1973, 513–524.

Garret, J. T. ''The industrial back problem.'' *American Industrial Hygiene Association,* October 1977, 560–571.

Goldberg, Henry M. ''Diagnosis and Management of Low Back Pain,'' *Occupational Health and Safety.''* June 1980.

Green, L., Marshall, K. Deeds, S., Partridge, K. *Health Education Planning—A Diagnostic Approach.* Palo Alto, CA., Mayfield, 1980.

Johnston, Ben. ''Industrial Health Program,'' Paper presented to a meeting of the American Physical Therapy Association, 1982.

Kelsey, J. ''The impact of musculoskeletal disorders on the population of the United States.'' *The Journal of Bone and Joint Surgery,* October 1979, 61A, **7,** 955–964.

Kelsey, J. ''Epidemiology and impact of low back pain,'' *Spine,* March 1980, **5,** 133–141.

Klevin, Keith. ''Industrial Physical Therapy Programs with TWA and Hilton Hotels.'' Paper presented to a meeting of the American Physical Therapy Association, 1979.

Larry, R. Heath. ''Foreword.'' *Fitness in Industry.* Monograph of the Proceedings of a Symposium, published by the Health Education Center of the Health and Welfare Planning Association, Pittsburg, 1978.

Mattmiller, William and Schulenberger, Chris. ''Back Injury Prevention in Industry.'' Paper presented to a meeting of the American Physical Therapy Association, 1982.

Moudry, George J. ''A Model Program for Back Pain Prevention.'' *Occupational Health and Safety,* June 1980.

Nordby, Eugene J. ''Epidemiology and Diagnosis in Low Back Injury.'' *Occupational Health and Safety,* January 1981.

Speigel, A. and Hyman, H. *Basic Health Planning Methods.* Germantown, Maryland, Aspen Systems Corporation, 1978.

Shactman, A. ''Back care in industry.'' *Physiotherapy.* August 1979, **65,** 249–252.

Tabor, Martha. ''Reconstructing the scene: Back Injury.'' *Occupational Health and Safety.* February 1982, 16–24.

Toufexis, Anastasia. ''That Aching Back.'' *Time,* July 1980, 30–38.

Troup, J. ''Back pain in industry.'' *Spine,* January 1981, **6,** 61–70.

Wilson, Phillip. ''Fitness in Industry: Costs, Benefits, Services Provided, Program Development.'' Monograph of the Proceedings of a Symposium, published by the Health Education Center of the Health and Welfare Planning Association, Pittsburg, 1978.

CHAPTER 13

Time Management Training: Overcoming a Major Barrier to Occupational Health Promotion[1]

George S. Everly, Jr.

INTRODUCTION

In an earlier chapter, we discussed the complexities involved in obtaining compliance in occupational health promotion programs. Clearly, compliance is a multifactorial phenomenon and will be improved most effectively through an individualized, multifaceted approach. And yet, if there is one prepotent barrier to achieving compliance/adherence for health-promoting behavior within the world of work, with its schedules, deadlines, ubiquitous urgency, and its "needed-it-yesterday" attitude, it would be the barrier of time—more correctly stated, the perceived lack of time to dedicate to health promotion. Successful health promotion at the worksite can only be achieved if the employee perceives enough time is available for the pursuit of health. In this chapter, we will address the issues of time and time management. Furthermore, we will use this chapter to conclude the present section on occupational health promotion programming, for it is our belief that time is the sine qua non of occupational health promotion. We believe that a lack of perceived time for participation is the most commonly encountered barrier to success in any and all of the occupational health promotion programs described earlier in this section. This chapter attempts to provide insight into this problem, thus enabling program coordinators to offer training in time management to health promotion participants, either as a separate health promotion program itself, or as a component of any other program, as to increase the participants' belief that they have enough time for the pursuit of health.

[1]Material in this chapter is adapted from: "Time management: A behavioral strategy for disease prevention and health enhancement," G. S. Everly, Jr. In *Behavioral Health: A Handbook of Health Enhancement and Disease Prevention,* J. Matarazzo, N. Miller, S. Weiss, et al. eds. New York: Wiley, 1984.

SCOPE OF THE PROBLEM

As we discussed earlier in this text, the Health Belief Model delineates a rather complex series of events that act as a foundation for an individual to actually make a conscious decision to pursue health-promoting behavior and to then act accordingly. A major criticism that has been levied against health professionals is that we fail to understand what motivates people to engage in health-promoting behavior. We naïvely ask the rhetorical question, "what is more important than health?" We then gaze in utter amazement as people continue to fail to practice any reasonable form of health promotion. This problem is especially relevant when applied to the working world.

The Health Belief Model and related models tell us that an individual will act in a health-promoting way if he or she perceives a personal need, perceives some reasonable likelihood of success, and perceives the availability of adequate resources—time is one of the most important of those resources. All of the requirements of the Health Belief Model may be met, but if an individual fails to perceive enough time as being available for the pursuit of health, no such pursuit will be undertaken. One of the more intriguing aspects of this whole dilemma of how to improve compliance/adherence is that the major factors that affect behavior are all perceptual factors. That is, reality, like beauty, is in the eye of the beholder. If a worker *perceives* a lack of time available for health promotion, he or she simply won't make the attempt. Similarly, if that worker perceives health promotion activities as undermining work-related activities or as creating a high degree of time pressure that is stressful, that worker is likely to quickly discontinue efforts at health promotion.

For instance, in a preliminary study of the relationship between time pressure and health-related behavior patterns, Everly (1982) found that subjective reports of time pressure were positively correlated with generally accepted behavioral risk factors such as caffeine consumption and smoking among 64 clerical personnel working in a high technology, multinational corporation. In this same study, the subjective feeling of time pressure was inversely correlated with the practice of health-enhancing behavior such as exercise and relaxation breaks. These data support earlier findings. Everly and Rosenfeld (1981) reported that their patients with psychophysiologic disorders were hesitant to practice relaxation techniques due to a lack of perceived time available. Furthermore, Carrington et al. (1980) found lack of time the major reason for lack of compliance in a worksite meditation program. Finally, Brunner (1969) found that adults who were unsuccessful in an exercise program responded that a perceived lack of time was the major reason for the lack of compliance.

Compounding this problem with occupational health promotion practices, we find that our society's virtual obsession with efficient time utilization seems to have its roots in the occupational setting. In 18th-century Europe, the Puritan work ethic seemed to combine with the industrial revolutions then being experienced in France and England. This combination of factors prompted the first known attempts at not only the encouragement of efficient time utilization, but its measurement as well (Minge-Klevana, 1980).

The process of analyzing work-related time appeared in the United States some years later. The best known of the time analysis projects were the time and motion studies of Frank Gilbreth. Gilbreth (1911) separated production-level tasks into component parts and analyzed each task while searching for ''wasted'' time and effort.

The 1950s and 1960s represented an era, both in Europe and the United States, when effective time-utilization was considered to be not only the key to economic success, but the ''sine qua non'' of business enterprise. Peter Drucker's influential text (1954) on management practices suggested that time was an individual's most perishable resource and effective time utilization was the key to effective management.

But this was an era of ''indictment,'' as well. In the 1960s, influential studies emerged (Sayles, 1963; Stewart, 1967) that indicated that most managers failed to use their time efficiently. Not coincidentally, the 1960s and 1970s were the temporal stage for a virtual plethora of time management training programs and textbooks that seem to be gaining popularity in the 1980s.

Thus, we see that the psychological stage was set for our society's quest to capture time. It is certainly a quest fueled by external forces, for example, our growing dependence upon high-technology data acquisition and effector systems as predicted in Alvin Toffler's *Future Shock* (1970). This concern for efficient time utilization has apparently carried over from work to home, leisure, and even eating activities, for example, witness the popularity of ''fast-food'' restaurants. This suggests that our apparent obsession for efficient time utilization may be as much self-imposed as environmentally dictated. Feelings of pressure and frustration arise when people attempt to ''beat the clock'' and time then escapes them. How can they cope with both the societally- and self-imposed pressure to be on time? The answer seems to lie not in a major conceptual shift in the organization's view of efficient time utilization, but rather in its recognition of the employees' inability to cope with a perceived lack of time. Then, all that remains is for a time management program to be devised and implemented. In order to overcome the barrier of lack of perceived time for health promotion, we can not only act to increase participants' motivation to promote health, but we can help participants find the time to promote their health as well.

TIME MANAGEMENT, SELF-RESPONSIBILITY, AND HEALTH PROMOTION

The term ''time management'' is a misnomer and may indeed lead one to reach an inappropriate conclusion regarding the nature of this health enhancement strategy.

The leading "technique-oriented" texts in the field agree that successful time management is based not upon *managing time,* per se, but is based upon *managing oneself* in relation to how one utilizes time. R. A. MacKenzie (1968, p. 3) in defining time management notes: "Time passes at a predetermined rate no matter what we do. It is a question not of managing the clock but of managing ourselves with respect to the clock." Douglass and Douglass (1980, p. 2) write: "Managing time really refers to managing ourselves in such a way as to optimize the time we have. It means conducting our affairs within the time available so that we achieve gratifying results." Finally, Dorothy Jongeward (1976, p. 77) concludes: " . . . managing time means investing time to get what you want out of life. . . ." These perscriptive definitions are congruent with the empirical observations of the present author (Everly, 1980a; 1980b) in the design and implementation of health promotion programs. Thus we see that the literal management of time is an exercise in futility, for time does not lend itself to control; therefore, time management really represents the management of self—the employment of self-responsibility and self-management skills for the utilization of time as a resource.

Rollo May has stated:

> "The more a person is able to direct his life consciously, the more he can use time for constructive benefits."

It seems reasonable to suggest that time management techniques can be employed to better enable a person to consciously direct his or her life and to better enable a person to avail him- or herself of the time to pursue health by way of a health-enhancing life-style. Matarazzo (1982) has delineated some of the "behavioral health challenges" awaiting us. He lists among them helping "currently healthy" citizens: exercise regularly, establish proper sleeping and rest habits, employ basic safety behaviors, use dental floss, and affect a more healthful diet. It seems a tenable notion that in the final analysis successfully meeting any and all of these "challenges" will be based, in part, upon the individual's perception that he or she has the *time* available to devote to such activities.

Thus, we see that the emphasis in time management is upon self-responsibility, as "individual responsibility" and "self-initiated" behavior lie at the foundation of behavioral health (Matarazzo, 1980). It may be argued that time management is a fundamental behavioral health-enhancing strategy; for even the most potentially effective behavioral risk reduction and health-enhancing strategies will lie impotent unless the individual perceives he or she has the *time* to employ them.

GUIDELINES FOR PROGRAM DEVELOPMENT

It has been suggested, above, that time management is a particularly important behavioral health strategy because it potentiates a general availability of time for the more specific practices of disease prevention as well as health enhancement. Unfortunately, no studies have been done (to the knowledge of the present author) that have isolated time management techniques and tested their effects upon risk factors, nor upon health-promoting behavior. On the other hand, Suinn (1978, 1980) and Gentry (1978) have reviewed behavioral interventions that contained varying degrees of time management strategies. These reviews are supportive of the notion that programs that contain some time management strategies may reduce coronary risk factors. Bhalla (1980) implemented a stress management program that contained a time management component. He found that such a program was able to reduce cardiovascular and electromyographic indices of stress arousal. In none of the reports cited, however, is it possible to determine to what degree the time management techniques alone were effective in affecting positive health outcome. Intuitively, it seems that time management interventions play an *interactional* role with other health-promoting strategies to account for the positive health outcomes reported in the aforementioned studies. Perhaps time management techniques do indeed provide the individual with the temporal opportunity to practice health-enhancing behaviors. This hypothesis is at least supported by the study conducted by Radar and Schabacker (1980) that found that generic time management techniques did increase the amount of time available for subjects practicing such techniques.

In order to fully understand the concept of time management, the question must be raised, "What does time management really do?" As noted in the previous section, time management is a generic term that applies to the employment of techniques that allow an individual to become more effective vis-à-vis his or her utilization of time. More specifically, however, it becomes of value to identify the active mechanisms that underlie the process of time management, they include:

1. Attending to the highest priority activities with a limited amount of time, thus potentially reducing demand while maximizing effort expenditure.
2. Increasing the amount of functional time available within which to attend to high priority activities.
3. Reducing the perception of time-urgent conditions.

In presenting the following model, it becomes necessary to provide a generic form of time management. This need arises due to the fact that time management techniques become most effective when they can be tailored specifically to the idiosyncratic needs and environment of the individual. Therefore, this generic model may be used as a basis from which more specific models can be developed. Finally, in preface, it should be recalled that this model is designed to allow an individual to manage his or her life vis-à-vis the resource of time. From a behavioral health perspective, this includes providing time for the practice of health-enhancing behaviors as well as reducing certain risk factors associated with time urgency and impatience, for example, stress and Type A behavior patterns.

Table 13.1 A Model for Time Management

I. Setting priorities
   A. Establish objectives
   B. List subgoals
   C. Conduct time analysis
   D. Prioritize

II. Increase functional time
   A. Identify time "wasters"
   B. Delegation
   C. Schedule

III. Reduce time urgency
   A. Minimize sources of time pressure
   B. Reduce procrastination

The generic model presented below represents a time management model integrating the work of Lakein (1973), Gentry (1978), Suinn (1978), Girdano and Everly (1979), Douglass and Douglass (1980), Porat (1980), and Rader and Schabacker (1980) and is summarized in Table 13.1.

Setting Priorities

Modern society is often characterized by its tendency to place a multitude of time-consuming demands upon an individual's limited supply of time and energy. In such conditions, where time demand exceeds supply, time management strategies may assist an individual by increasing his or her ability to determine and attend to the perceived highest priority demands, given a limited supply of time. This may be achieved as follows:

- Review/establish important personal (including health-related) objectives or goals. They could be short-term (less than one year) goals or long-term (one, five, or ten year) goals.
- Establish goal-path clarity by listing what subgoals or steps are necessary to achieve each major personal goal. In other words, determine what needs to be done to achieve the goals enumerated in the first step above.
- Analyze current time-consuming activities, that is, determine how current daily and weekly time is regularly spent. Ideally, current activities should support and lead to the ultimate achievement of personal goals, or objectives. If such is not the case, either the goals established in the first step or these time-consuming activities will most likely need to be altered. (Time

analysis may be achieved through the use of a daily "time log" used to chart daily time utilization.)

- Seeing that the first and third steps are congruent, it becomes necessary to prioritize current time-consuming activities. Prioritization is based upon primarily attending to only those current activities that yield the most significant "payoff" vis-à-vis the major personal goals, or objectives stated in the first step.

Increasing Functional Time

Once the determination of high-priority current activities has been made, it becomes desirable to allocate sufficient time to attend to such tasks. This is done by harnessing and increasing the functional time available as follows:

- Identify/reduce sources of "wasted" time, that is, current time-consuming activities that in no way or only minimally contribute to the achievement of important goals or objectives. By reducing time-wasting activities, the effect is to functionally increase the amount of time available for high-priority activities.
- Delegation of time-consuming activities to others will functionally increase the amount of time one has to spend on high-priority items. In order to delegate, support groups must be identified. One's spouse, children, friends, relatives, and co-workers (peers and subordinates) are all potential sources of support to which lower-priority activities may be delegated. If applicable, one might even hire or barter with others for work to be performed.
- Schedule high-priority activites. It seems to be commonly accepted in the time management literature that what doesn't get scheduled doesn't get done. The time analysis log mentioned earlier may assist in effective scheduling.

Reducing Perceptions of Time Urgency

Having determined high priority activities and allocated time for attending to them, the final step in this generic time-management model involves reducing the perception of time-pressure and time-urgency. This is accomplished by the following:

Analyzing and minimizing perceived sources of time pressure and time urgency. Not all perceptions of time urgency are warranted—some are arbitrary and needlessly self-imposed. Therefore:

1. Determine which sources of time urgency are environmentally dictated with valid reason.

2. Determine which sources of time urgency are environmentally dictated needlessly.
3. Determine which sources of time urgency are constructively self-imposed.
4. Determine which sources of time urgency are needlessly self-imposed.
5. Minimize needless time urgency.

Overcoming procrastination. Procrastination appears to be an "attitudinal" barrier to effective time management. It can frequently be traced to two major sources: (1) the reluctance to make a mistake, and (2) the inability to know how to start, or initiate, action. Success in overcoming procrastination may be increased once the individual accepts the notion that few decisions are, in reality, irrevocable and that making mistakes is actually nothing more than a part of "trial-and-error" learning. Similarly, learning to break a seemingly large or overwhelmingly complex task down into its smallest functional task-units may prove of extreme value in helping a person start a project he or she might have been procrastinating on because of its size or complexity. Lakein (1973) calls this strategy "swiss cheesing."

The generic time management model presented above has the potential for three obvious health-related applications:

1. As a general model for increasing the availability of time that could potentially be used for the practice of health-enhancing strategies.
2. As a model for reducing health-related risk factors that may be related to time urgency.
3. As a specific model for implementing any given health-enhancing or risk-reducing behavior.

Summary

In this chapter, we have briefly examined what we feel is the prepotent barrier to occupational health promotion—a lack of perceived time—and we see this barrier as affecting any and all of the health promotion programs discussed in this book. Therefore, in order to assist the health promotion coordinator better cope with this barrier, we have supplied a model for time management that can be used either as a health promotion program unto itself or as a component in any health promotion program as to assist participants in finding the time for the pursuit of health.

REFERENCES

Bhalla, V. "Neuroendocrine, Cardiovascular, and Musculoskeletal Analyses of a Holistic Approach to Stress Reduction." Doctoral dissertation, University of Maryland, 1980.

Brunner, B. C. "Personality and motivating factors influening adult participation in vigorous physical activity." *Research Quarterly,* 1969, **40,** 464–469.

Carrington, P., Collings, G., Benson, H., Robinson, H., Wood, L., Leher, P., Woolfolk, R., and Cole, J. "The use of meditation-relaxation techniques for the management of stress in a working population." *Journal of Occupational Medicine,* 1980, 221–231.

Douglass, M. and Douglass, D. *Manage Your Time, Manage Your Work, Manage Yourself.* New York: AMACOM, 1980.

Drucker, P. *The Practice of Management.* New York: Harper, 1954.

Everly, G. "The development of less stressful personality traits in adult learners: Preliminary findings." Proceedings of the Lifelong Learning Research Conference, College Park, Maryland: U.S. Maryland Education Association and the University of Maryland, 1980a.

Everly, G. "The development of less stressful personality traits in adults through educational intervention. *Maryland Adult Educator.* 1980b, 63–66.

Everly, G. "Time urgency and health-related coping behavior." Unpublished research report, Loyola College, Baltimore, Maryland, 1982.

Everly, G. and Rosenfeld, R. *The Nature and Treatment of the Stress Response: A Practical Guide for Clinicians.* New York: Plenum, 1981.

Gentry, W. D. "Behavior modification of the coronary-prone behavior pattern." In *Coronary-Prone Behavior,* T. Dembroski et al., eds. New York: Springer-Verlag, 1978, 225–229.

Gilbreth, R. *Motion Study.* New York: Van Nostrand, 1911.

Girdano, D. and Everly, G. *Controlling Stress and Tension.* Englewood Cliffs, N.J.: Prentice-Hall, 1979.

Jongeward, D. *Everybody Wins: Transactional Analysis Applied to Organizations.* Reading, Ma.: Addison-Wesley, 1976.

Lakein, A. *How to Get Control of Your Time and Your Life.* New York: Peter H. Wyden, 1973.

MacKenzie, R. A. *Time Trap.* New York: AMACOM, 1968.

Matarazzo, J. D. "Behavioral health and behavioral medicine: Frontiers for a new health psychology." *American Psychologist,* 1980, **35,** 807–817.

Matarazzo, J. D. Behavioral health's challenge to academic, scientific, and professional psychology. *American Psychologist,* 1982, **37,** 1–14.

Minge-Klevana, W. "Does labor time decrease with industrialization?" *Current Anthropology,* 1980, **21,** 279–298.

Porat, F. *Creative Procrastination.* San Francisco: Harper & Row, 1980.

Rader, M. and Schabacker, J. "The effectiveness of time management training." *Arizona Business,* 1980, **27,** 3–7.

Sayles, L. *Managerial Behavior Administration in Complex Organizations.* New York: McGraw-Hill, 1963.

Stewart, R. *Managers and Their Jobs: A Study of the Similarities and Differences in the Ways Managers Spend Their Time.* London: MacMillan, 1967.

Suinn, R. "The coronary-prone behavior pattern: A behavioral approach to intervention." In

Coronary-Prone Behavior, T. Dembroski et al., eds. New York: Springer-Verlag, 1978, 231–236.

Suinn, R. ''Pattern A behaviors and heart disease: Intervention approaches.'' In *The Comprehensive Handbook of Behavioral Medicine, Vol. I,* Ferguson and C. B. Taylor, eds. New York: Spectrum, 1980, 5–28.

Toffler, A. *Future Shock.* New York: Random House, 1970.

PART Three

THE PLANNING AND EVALUATION OF OCCUPATIONAL HEALTH PROMOTION PROGRAMS

In this third and final section of this text we shall address the issue of evaluation, and in doing so we must touch upon the issue of planning as it relates to evaluation. In Chapter 14 we introduce the reader to the generic concepts involved in planning and evaluation. In Chapter 15 we review behavioral principles for intervention and evaluation. Chapter 16 addresses the use of psychological tests in evaluation. Chapter 17 raises the issues of cost-effectiveness and cost-benefit analyses in the evaluation process. Finally, Chapter 18 concludes the text with a discussion of future directions in occupational health promotion.

CHAPTER 14

The Planning and Evaluation of Occupational Health Promotion Programs: An Introduction

Robert H. L. Feldman

INTRODUCTION

Health promotion programs in industrial settings have increased dramatically in the past few years. As has been shown in the previous chapters these programs have been involved in a wide range of health issues in a variety of work settings. Given this extensive activity, the question arises as to whether these programs work? That is, are they effective? Are workers benefiting from these programs? Is management obtaining the results that it desires? In order to answer these questions goals and objectives need to be stated and proper evaluations need to be conducted. Many companies have developed health promotion programs, but few of these programs have been evaluated and still fewer have been evaluated properly. A program evaluation measures the effects of a program against the goals it sets out to achieve. This evaluation process can contribute to an improvement in subsequent decision making about the program, including ways of enhancing the program (Weiss, 1972; Edwards, Guttentag, and Snapper, 1975). This chapter examines the process of planning and evaluating occupational health programs in workplace settings.

PLANNING OCCUPATIONAL HEALTH PROMOTION

Before a worksite health promotion program can be implemented and evaluated it needs to be planned. Thorough planning can enhance the likelihood of a program being successful. The steps in planning an occupational health promotion program are shown in Table 14.1. The first step in program planning is to conduct a needs assessment (Warheit, Bell, and Schwab, 1979). What is the prevalence of smoking among employees? Are enough employees interested in an exercise program to warrant establishing such a program? That is, the purpose of a needs assessment is

Table 14.1 Steps in Planning an Occupational Health Promotion Program

I. Needs assessment
   A. Does a health problem exist? (epidemiological data)
   B. Worker interest (surveys, questionnaires, interviews)

II. Goal setting
   A. Overall program goals
   B. Specific program objectives
   C. Specific behavioral objectives

III. Outcome measures
   A. Health knowledge
   B. Health attitudes
   C. Health behavioral intentions
   D. Health behavior
   E. Health status (physiological measures)
   F. Absenteeism
   G. Occupational injuries
   H. Health costs
   I. Worker attitudes about working conditions

IV. Setting priorities
   A. Importance (need)
   B. Changeability

V. Program resources (budgeting)
   A. People, staff
   B. Equipment, materials

to verify that a problem exists and that a program is desired. Green et al. (1980) propose a diagnostic approach to program planning. They recommend that an epidemiological diagnosis be conducted to determine whether a problem exists.

This diagnosis in a workplace setting would examine the incidence, prevalence, and distribution of health problems within a particular workplace or group of similar workplaces. For example, before instituting a smoking-cessation program a company would want to determine the prevalence and distribution of smoking among its employees. Other major health problems include hypertension, obesity, alcoholism, and stress (Cunningham, 1982). The main indicators of health problems are mortality, morbidity, fertility, and disability. To determine whether a health problem at a particular worksite is worse than at another workplace, comparisons can be made. Comparative data from such sources as the National Center for Health Statistics, other agencies of the Department of Health and Human Services, local and state health departments and planning agencies are useful in documenting a particular problem (Green et al., 1980; Health Resources Administration, 1977).

Once a health problem has been determined the second step is to find out whether employees are interested and willing to participate in a particular health promotion program. Therefore, a survey of employee attitudes, beliefs, and intentions is required as part of the needs assessment.

The third step in the process of program planning is the setting of goals and objectives. The planning process consists of the stating of overall program goals, specific program objectives, and even more specific behavioral objectives (Green et al., 1980). For example, a program goal in a hypertension program is to decrease consumption of salt. A specific program objective is to have 70% of diagnosed hypertensive employees reducing their salt consumption by the end of the first year of the program. A specific behavioral objective is to have 70% of diagnosed hypertensive employees at Company X reducing their salt consumption by 50% at the end of the first year of the program. That is, the specific behavioral objective asks, "*Who* is expected to achieve *how much* of *what* behavior by *when?*" (Green et al., 1980, p. 65). An important aspect in setting behavioral objectives is the choice of quantifiable outcome measures (Sechrest and Cohen, 1980). In the above example, the quantifiable outcome measure is the reduction in salt consumption by 50%. In an alcoholism program the quantifiable outcome measure may be abstention from alcohol or having less than three drinks a day. In a stress reduction program a quantifiable outcome measure may be an increase in knowledge about stress (e.g., 50% increase), a more favorable attitude toward reducing stress (e.g., 35% more favorable), and greater practice of stress reduction techniques (e.g., 20% increase).

As the above example illustrates, outcome measures may be measures of knowledge, attitudes, or behavior. The measurement of knowledge change is useful to find out what people have learned; however, an increase in knowledge is not always followed by an increase in the practice of a certain behavior. Although knowledge is necessary for behavior change, it seldom is sufficient (Fishbein and Ajzen, 1975). Therefore, in addition, it is useful to examine health attitudes. Health attitudes measure how an individual feels toward health. Is the person in favor of smoking? Does the individual feel positively about exercise? How does a person feel about his or her alcohol consumption? Though changes in attitudes sometime predict changes in behavior, behavioral intention is a better predictor of behavior (Ajzen and Fishbein, 1980; Fishbein and Ajzen, 1975). An employee may *know* that exercise is beneficial, *feel* favorable toward exercising, but have *no intention* of exercising. Therefore, is is useful to obtain not only measures of employee knowledge, but also measures of behavioral intention.

Do you intend to stop smoking? Or being more specific, "Do you intend to abstain from smoking for the next six months?" Statements of behavioral intent have been found to be one of the best predictors of health behavior (Jaccard and Davidson, 1972; Sejwacz, Ajzen, and Fishbein, 1980).

Since a major goal of an occupational health promotion program is to make changes in employee health behavior, careful consideration should be given to the measurement of health behavior. For the many health behaviors that are not performed on the job, self-report measures may be used. Self-report questions on smoking, drinking, eating, and exercising should be phrased in a nonthreatening matter, so that, employees do not feel the social pressure to distort or exaggerate these actual behavior patterns. For particularly sensitive topics, careful wording of the questionnaire and good interviewing techniques can reduce response bias (Kidder, 1981).

When health behaviors are performed on the job, then observations and utilization rates may be determined. Measures can be taken of how often a company gymnasium is utilized, how often a jogging path is used, which snack foods (nutritious versus less nutritious) are bought in vending machines, which foods are bought in the company cafeteria, how many employees smoke on the job, and so forth.

Since health behavior patterns are an important influence on health status, the next step is to measure the health effects of any program. In addition to the health status measures of mortality, morbidity, fertility, and disability, measures of acute and chronic conditions and measures of injuries may be used.

Health measures also include specific physiological assessment such as blood lipids, blood pressure, body fat, weight, and estimated maximum oxygen uptake (Wilbur, 1982). The particular health status measures should be based on the specific goals of the program, the health and medical literature, and workplace conditions.

In addition to health status measures, occupational health promotion programs frequently examine absenteeism, sick days, occupational injuries, hospitalization and medical costs, and employee attitudes toward work. Employee attitudes such as satisfaction with working conditions, job involvement, and commitment to the organization have been shown to improve as a result of health promotion programs (Wilbur, 1982). Thus, when stating the goals and objectives of an occupational health promotion program, it is recommended that consideration be given to each of the following: health knowledge, health attitudes, health behavioral intentions, health behavior, health status, health costs, and employee attitudes about working conditions. Each of these factors need not be included in the planning and evaluation of every program. Different programs may have different aims, and therefore, different measures would be appropriate.

The next step in the program-planning process is the setting of priorities. Decisions have to be made concerning which programs are to be developed and which programs are not. Assuming that obesity among employees is an important health problem, what is the likelihood that a weight control program will produce any reduction in weight? That is, another factor to be considered in planning a program is the ''changeability'' of the recommended health behavior associated with the program. Highest priority should be given to those programs that have both the greatest need and the greatest likelihood of making changes in health behavior and health status.

The final step in program planning is a determination of program resources. Decisions have to be made regarding how the program will be structured, how many people are needed to staff program, and the type of equipment and materials that are required. Budgeting is a method of determining whether available resources are sufficient to meet program goals. Since a budget is an estimate, revisions are usually made as a program is developed. Preparing a budget is a valuable exercise because it requires the components of a program to be clearly determined and specified.

In summary, to properly plan health promotion programs in occupational settings, a number of steps should be taken. First, a needs assessment is necessary to document that a problem exists and that employees are interested in participating in the program. Second, program goals and objectives, specific behavioral objectives and quantifiable outcome measures should be clearly stated. Third, consideration should be given to choice of outcome measures—which measures will be examined: health knowledge, health attitudes, health behavioral intent, health behavior, health status, health costs, or employee attitudes toward working conditions? Fourth, once the goals and objectives have been stated, program priorities can be developed based upon need and the likelihood that the program will be successful. Fifth, determination of allocation of resources and budgeting ensures a realistic assessment of what can be accomplished within a given situation. Thus, a complete program-planning process increases the probability that a program will be implemented as specified, and offers the opportunity for a thorough evaluation of program effectiveness.

EVALUATING OCCUPATIONAL HEALTH PROMOTION

Program Monitoring

Evaluations are conducted for many purposes. (see Table 14.2) One purpose of a program evaluation is to monitor an ongoing program. Is the program reaching the appropriate target population? For example, are employees with alcohol problems signing up for the alcohol abuse program? Are smokers participating in the smoking cessation program? Also, program monitoring is needed to determine whether the program is being implemented in the way it was designed. Sometimes a program may fail to have an impact because the wrong type of program is presented, a weak version of the intended program is presented or an unstandardized version of the

Table 14.2 Purposes of Evaluation

- I. Program monitoring
  - A. Reaching the appropriate target population
  - B. Implementing properly
  - C. Fiscal accountability
  - D. Legal accountability
  - E. Degree of frequency
    1. Continuous
    2. Periodic
    3. One-shot
- II. Program feedback
  - A. Do workers like the program
  - B. Does the program meet expectations
  - C. Recommend to co-worker
  - D. Worthwhile, useful
  - E. Learn anything new
  - F. Able to implement recommendations
- III. Program effectiveness
  - A. Achieving its goals and objectives
  - B. Plausible alternative explanations for results
  - C. Effects
    1. Primary
    2. Secondary, spillover effects

program is presented to the different participants. Also, program monitoring may include fiscal accountability. Are the funds designed for the program being used properly? An additional aspect of program monitoring is legal accountability. In the implementation of the program are relevant laws being observed, such as the privacy of employee records (Rossi and Freeman, 1982)?

Program monitoring can occur with different degrees of frequency. Continous monitoring usually involves an onsite individual who is continuously checking the program to determine whether the target population is being reached, whether the program is being implemented properly, and whether the program is fiscally and legally sound. One-shot monitoring is a one-time monitoring of the program, usually due to a crisis situation in the program. Periodic monitoring which falls between the extremes of one-shot and continuous monitoring, is a process of collecting information about a program on a systematic basis over the life of the program. These three methods have different advantages and disadvantages. Continuous monitoring supplies the most information about a program, but at a high cost in resources. In contrast, one-shot monitoring is least costly, but supplies the least amount of program information. Therefore, periodic monitoring is seen as a compromise. It is not as costly as continuous monitoring, yet produces more information than one-shot monitoring. The frequency of program monitoring should depend upon the particular program and the state of the program. In addition, it should be pointed out that the distinction regarding one-shot, periodic, and continuous monitoring also applies to one-shot, periodic, and continuous evaluation of the effectiveness of health promotion programs discussed below.

Program Feedback

A second purpose of a program evaluation is to obtain feedback about the program from the program participants (Posavac and Carey, 1980). How do employees like the program? Did it meet their expectations and would they recommend the program to a co-worker? Was the program worthwhile and did the employees learn anything new from the program? Was the material and information presented useful and will they be able to implement what they have learned? Information on participants' perceptions and assessments of the program are necessary data and can be obtained by questionnaires and interviews.

Program Effectiveness

A third purpose of an evaluation is to assess the effectiveness or impact of a program. Is the program achieving its goals? Is there a plausible alternative explanation for the obtained result of the program? Have unintended results occurred due to the program?

One distinction that should be made is between primary and secondary effects of a program. An example of a goal of occupational health-promotion program is to increase exercise among employees. A positive primary effect of an exercise program is an increase in exercise behavior among employees. An additional positive effect of the occupational-based health promotion program is an increase in exercise among employees' families. That is, not only are employees engaging in more exercise activities, but their spouses and children are also participating in more exercise. This spread of effect or spillover effect is referred to as a secondary effect. However, secondary effects can also have negative consequences. If 90% of employees exercising at a company facility had previously exercised at a local ''Y,'' then the ''Y'' has lost membership and the company has engendered ill will in the community. Or if employees who used to exercise with their families at home are now exercising at work without their families, then a negative secondary effect of an occupational health promotion program is less family interaction. Though most secondary effects of programs are positive, negative consequences can occur. Therefore, it is recommended that secondary as well as primary effects of a program be examined.

To assess the effectiveness of an occupational health promotion program, a number of evaluation designs have been developed. The following section examines three evaluation approaches to measuring program effectiveness: true experimental designs, quasi-experimental designs, and approximate methods.

True Experimental Designs

The best approach for assessing program effectiveness is the use of true experimental designs. True experimental designs are designs in which individuals are randomly assigned to program (or experimental) groups and control groups. Random assignment gives each individual the same chance as any other individual of being assigned to any given group. The chance assignment of individuals to different groups precludes the possibility of initial systematic differences between the groups. However, this does not mean that the groups will be alike, only that whatever differences do exist are a result of chance alone (Selltiz, Wrightsman and Cook, 1976).

True experimental designs allow the testing of causal hypotheses to reduce the plausibility of rival explanations (Selltiz, Wrightsman and Cook, 1976). For example, a causal hypothesis may state that if employees enroll in a weight reduction program run by company A, then they will reduce their weight by 25% at the end of six months. If at the end of six months the employees completing the program have lost 25% of their weight, has the program proved effective? Are there plausible rival explanations to explain the loss of weight? If a true experimental design was *not* carried out, plausible rival explanations include (1) the possibility of another event or situation influencing the employees to lose weight, or (2) the initial weighing-in (or initial testing) influencing and motivating the employees to lose weight. However, if a true experimental design is used, then the above plausible rival explanations as well as other explanations are ruled out.

Plausible rival explanations are factors jeopardizing the internal validity of a particular program. Internal validity refers to whether a program made a difference, for example, in changing the health behavior or health status of employees participating in the program. Factors jeopardizing internal validity include (1) history, (2) maturation, (3) testing, (4) instrumentation, (5) statistical regression, (6) selection, and (7) mortality. (Campbell and Stanley, 1966; see Table 14.3).

History refers to any event outside the program occurring at the same time as the program, which may be a plausible rival explanation for the success or effect of the program. For example, if a corporate executive had a heart attack at the same time a high blood pressure control program was in progress, then that unfortunate event could influence the outcome of the program. Therefore, the success of the high blood pressure control program could be contributed to the fear gendered by the executive's heart attack, that is, by an event outside of the program. If a true experimental design was used with random assignment to program and control

Table 14.3 Factors Jeopardizing the Internal Validity of Programs

I.	History
II.	Maturation
III.	Testing
IV.	Instrumentation
V.	Statistical regression
VI.	Selection
VII.	Mortality

groups, then both groups would have had exposure to this ''historical'' outside event, but only the program group would have recieved the program. Using this design one can eliminate history as a plausible rival explanation.

A second factor that can jeopardize internal validity is maturation. Maturation refers to a process that occurs within an individual due to the passage of time, such as growing older, growing hungrier, and growing more tired (Campbell and Stanley, 1966). For example, a trucking company may develop a driver safety program for its new new employees, since injury rates are higher among young inexperienced drivers. If atfer several years the injury rate declines, one plausible alternative explanation is that the younger inexperienced drivers have matured (and gained experience), and therefore maturation and not necessarily the safety program may be reason for the lower injury rates.

A third factor that should be considered is testing. If pre- and posttests are given, then the initial pretests scores may affect the posttest scores regardless of the program. For instance, a pretest questionnaire asking workers about their smoking knowledge, attitudes, and behavior may act as a cue to action and cause some workers to reduce or stop smoking.

A fourth factor is instrumentation. Changes in the calibration of measuring instruments or changes in observers or scorers may produce changes in the measurements. For example, instruments used to measure psychophysiological responses in a worksite stress reduction program may show changes in responses if careful calibrations are not maintained. In addition, observers may tire and lose interest observing over long periods of time, and therefore health programs requiring extensive observation (e.g., programs observing safety behavior, programs monitoring food intake) are prone to changes in instrumentation.

A fifth factor is statistical regression. If individuals chosen for a health promotion program have been selected on the basis of their extreme scores (e.g., extremely obese), then postprogram scores would tend to be less extreme. The tendency for high scores to decline (and low scores to increase) is called regression to the mean. This statistical artifact produces apparent rather than real change. Therefore, a group of very obese workers are likely to show reductions in weight over a short period of time regardless of the weight control program. If a health promotion program does not use extreme groups then regression can be ruled out as a plausible rival explanation.

A sixth factor is selection. If two nonrandomized groups are compared, one group self-selecting a health program and one group not receiving the program, then differences between the two groups may be due to a selection factor. For instance, a comparison between workers participating in exercise programs with nonparticipating workers may indicate greater exercising behavior among the program group. However, since the participating exercisers self-selected the program, they may have increased their exercise behavior regardless of the program. Therefore, their increase in exercise may be due to their own initiative rather than the program.

A seventh factor is program mortality or the differential loss of respondents from comparison groups. For example, an alcohol rehabilitation program may show short-term success because those able to control their alcohol use remained in the program, while those unable to control their alcohol use dropped out of the program. Therefore, differential dropout rather than a program effect may be the explanation for the "success" of the program. Thus, seven factors have been shown to jeopardize the internal validity of programs by offering plausible rival explanations. By using true experimental designs (described below) all seven threats to validity can be eliminated.

Since true experimental designs require random assignment of individuals to program and control groups it has sometimes been mentioned that it is not possible to utilize a true experimental design in a corporate, industrial, or other workplace setting. Yet, this is not the case (Boruch, 1976). The use of "waiting-list control groups" derived from clinical psychology research is an ethical and practical way of treating this problem. For example, a company announces a smoking-cessation program for its employees and 80 employees sign up for the program. However, the program can only accommodate 40 participants. A fair and practical way of solving this problem and at the same time providing the opportunity to conduct a rigorous evaluation is to randomize the 80 volunteers into two groups: 40 as program participants and 40 as waiting-list individuals (the control group).

For example, if a program lasts ten weeks and a second program is scheduled shortly after the conclusion of the first program, then the waiting-list control group can be a control group for the time period prior to the start of the second program. The waiting-list group then has the opportunity and first priority to join the second program.

All true experimental designs are characterized by the randomization of individuals to program (or experimental) and control groups. In this chapter we will consider the two most common true experimental designs: (1) the pretest-posttest control group design and (2) the posttest-only control group design. A third true experimental design, called the "Solomon Four-Group" combines the two designs mentioned above; however, it is infrequently used in the evaluation of programs (see Table 14.4).

The pretest-posttest control group design (Campbell and Stanley, 1966) is described below:

R O_1 X O_2 Program Group

R O_3 O_4 Control Group

The Rs refer to the randomization of participants, Os represent the observations or measurements of the participants, and X refers to the program. For example, the smoking-cessation program discussed above with a waiting-list control group could be evaluated with this design. Volunteers would be randomly assigned to either the program group or the waiting-list control group. Measures of number of cigarettes smoked and other relevant information would be taken before the beginning of the program and at the end of the program. Determination would be made of whether indnviduals in the program group reduced their smoking behavior more than individuals in the control group.

Table 14.4 Assessing Program Effectiveness

I. True experimental designs
   A. Pretest-postest control group designs
   B. Posttest-only control group design
   C. Solomon four-group design

II. Quasi-experimental designs
   A. Nonequivalent control group design
   B. Time series design
   C. Multiple time series design

III. Approximate methods
   A. Before-after design without a control group
   B. Postprogram follow-ups
   C. Record keeping
   D. Comparisons with other programs, national results, and established norms
   E. Use of outside experts, program participants and program administrators

IV. Economic assessment
   A. Cost-benefit analysis
   B. Cost-effectiveness analysis

A second true experimental design, the posttest-only control group design, can be instituted without pretest measures as follows:

R X O_1 Program Group

R O_2 Control Group

There are situations when a pretest measurement may influence the attitudes or the behavior of both the program and the control group. For example, a pretest questionnaire on stress may influence the stress beliefs and attitudes of a control group or program group.

Both true experimental designs eliminate the seven factors that threaten internal validity. First, history is not a threat since within a work setting both the program and control groups would experience the same historical events. Second, both program and control groups would mature (grow older, grow hungrier, more tired) at the same time, therefore ruling out maturity as a threat to validity. Third, since both the program and control groups will be tested, testing will be common to both groups, and can be eliminated as a threat to validity. Fourth, since instrument calibration or human observers would be the same for both program and control groups, instrumentation can be eliminated as a threat to validity. Fifth, if individuals with extreme scores (e.g., extremely obese, very heavy smokers) are targeted for a program, randomization of participants to program and control groups will make it very unlikely that statistical regression will occur only in one group. That is, regression effects if they were to occur would probably occur in both groups, thus regression would not be a threat to validity.

Sixth, since true experimental designs require randomization of participants to program and control groups, self-selection is not permitted, therefore selection as a threat to validity can be discounted. Seventh, mortality, lost cases, or dropouts may be dealt with in a number of ways. One of the more common ways of treating dropouts is to only consider those individuals who completed the program. Then comparisons are made between those completing the program and the control group. However, careful consideration should be given to the characteristics of those individuals who drop out of the program. That is, do the people who drop out of the program differ in terms of demographic, psychological, or health status from those who remain in the program? Also, considerations should be given to whether an individual drops out of the control group in order to join a similar group. Information about dropouts is important in order to determine whether a bias is present. For example, if a substantial number of motivated employees in the control group leave the group to join another smoking-cessation program, then differences between the program group and the control group may be due to this differential dropout. That is, the remaining, less motivated individuals in the control group would be compared to the participants in the program group, and motivation rather than program itself may be the explanation for the difference between the groups. Therefore, dropout rates should be examined as part of the evaluation process.

Although true experimental designs are advantageous to implement, they can only be used when the target population is much larger than the actual program participants, that is, when the demand exceeds the supply (Cook and Campbell, 1976). A health promotion program designed to reach *all* employees cannot be evaluated by these designs because there will be no one available to be assigned to the control group. Also, for health promotion programs that have been conducted at a work setting for some time it may be difficult to find a sufficient number of volunteers who have not participated in the program to form a control group, thus making it unfeasible to use a true experimental design. In addition, in a small work setting where experimental and control group participants are likely to interact, contamination of effects may result, therefore making this setting unsuitable for a true experimental design.

Quasi-Experimental Designs

When the use of true experimental designs is not possible, then quasi-experimental designs (Campbell and Stanley, 1966; Cook and Campbell, 1979) are the next best approach. One type of quasi-experimental design is the nonequivalent control group design. When randomizing volunteers to program and control groups is not possible, then it is useful to find a nonrandomized comparison group to compare to the program group. Curtis Wilbur (1982) reports that when Johnson & Johnson Company initiated their "Live for Life" health promotion program they found that is was not possible to institute a true experimental design evaluation. In this situation it was not possible to institute the program over an entire factory and at the same time keep a randomly assigned control group from being exposed to the program. Random assignment of sites within a company was also not feasible. Therefore, from among the Johnson & Johnson family of companies, four companies were chosen as the program group and five additional companies were chosen as the control group. All program and control groups were located in the same geographical region.

The nonequivalent control group design is diagramed as follows:

O_1 X O_2 Program Group

O_3 O_4 Control Group

As with the previous design, the Os refer to observations or measurements and X is the program. The dotted line indicates that the groups are not randomized. Because the groups are not randomized it is important that the control group closely resembles the program group in all essential aspects. In the evaluation of the Johnson & Johnson health promotion program it was important that the companies assigned to the program were as similar as possible to the companies assigned to the control group. Matching for particular characteristics at either the individual or aggregate level reduces bias; however, it cannot insure that a difference between the program and control group is due to a program effect rather than the selection of program and control group participants (selection effect). Therefore, though matching is useful, randomization, if at all possible, is a superior method since it is not possible to match individual or groups on every relevant factor.

A second quasi-experimental design is the time series design that is diagramed below:

$$O_1\ O_2\ O_3\ O_4\ X\ O_5\ O_6\ O_7\ O_8$$

When longitudinal measures are available, program participants can be used as their own controls. For example, if good records of employee absenteeism or work-related injuries are available, a time series design can be used to determine the effect of a new program on subsequent absenteeism or injuries. A word of caution about this design is in order. If another event occurs at the same time as the program (for example, installation of new managers or a community or public-health campaign), it may not be possible to determine whether the program or the other event is causing the changes in the employees.

One way of improving both the time series design and the nonequivalent control group design is to combine the two quasi-experimental designs to form the multiple time series design (Campbell and Stanley, 1966) diagramed below.

$$O_1\ O_2\ O_3\ X\ O_4\ O_5\ O_6$$

$$O_7\ O_8\ O_9\quad O_{10}\ O_{11}\ O_{12}$$

If data from another factory, company, or institution is available and has been collected in a similar manner, then the multiple time series design is a very powerful evaluation tool. This design increases one's confidence in the results, because the effectiveness of a program is, in a sense, demonstrated twice—once against the control group (the nonequivalent control group design) and once against the pre-program measures (the time series design). The multiple time series is an excellent design and has clear advantages over the two previous designs. However, it is only feasible when (1) preprogram measures are available, (2) a second comparison group is available, and (3) all the measurements have been collected in a similar manner throughout the evaluation process.

Approximate Methods

Although true experimental and quasi-experimental designs are the most appropriate ways of evaluating program effectiveness, there are occasions when less rigorous methods are suitable. Since rigorous methods demand time, resources, and technical knowledge, it may be more appropriate to utilize approximate methods for demonstration programs, small projects, or programs of modest budgets. One approximate method is the before-after design without a control group. It can be used to document whether a program is having an impact; however, it does not offer definitive evidence that the program and no other factor is the real cause of any change. A second method is postprogram follow-ups that supply useful information about how individuals have been affected after participating in a health promotion program.

A third method is a record-keeping procedure that monitors the progress of participants. It can yield graphs and charts that demonstrate how a program is doing (Green, 1982). A record-keeping procedure using periodic data collection over a set time-interval provides additional benefits. A fourth method is to make comparisons between the results of a particular program and the results of other programs, national results, and well-established norms. Additional methods of evaluating programs are by obtaining assessments of the program from a variety of sources. The use of outside experts, the use of program participants and the use of the program administrators all contribute to the understanding of how well the program is meeting its goals. It should be noted that these approximate methods of evaluating programs should not be used where the program is costly, where there is a high likelihood that the program will be expanded to the regional or national level, or where sensitive issues concerning company policy are involved.

Economic Assessment

After a health promotion program has been shown to have an impact on the health and well-being of employees, cost-benefit or cost-effective analyses may be conducted. With the spiraling rise of health care costs, more attention is being paid to the costs, benefits, and efficiency of industrial health promotion programs. If the tangible and intangible benefits and costs of a program can be clearly specified, then a cost-benefit analysis can be made. A cost-benefit analysis measures (usually in monetary terms) the economic efficiency of program as a relationship of costs and benefits (Rossi and Freeman, 1982).

For programs whose benefits or goals cannot be easily converted into a monetary unit or some other common measure, cost-effectiveness analysis would be more appropriate than cost-benefit analysis. Cost-effectiveness analysis measures the efficacy of a program in achieving its goals in relationship to program costs (Rossi and Freeman, 1982). Benefits derived from a health promotion program may include job satisfaction, job involvement, improved morale, commitment to the organization, ability to handle job stress, better sense of well-being, and a higher level of wellness. These benefits are not easily converted into monetary terms, and therefore, cost-benefit analysis will not be appropriate. If more than one program is being evaluated, then cost-effectiveness analysis allows a comparison and ranking of the magnitude of the effectiveness of each program relative to its costs. For example, a group of programs may have as their goals to reduce sick days and increase a sense of well-being among workers. Three programs are offered: a comprehensive health promotion program, lecture and discussion groups on health, and health education pamphlets. A cost-effectiveness approach could determine which of the three programs is most effective in terms of program goals relative to program costs.

Summary

In summary, the evaluation of occupational health promotion programs requires thorough program planning, careful monitoring of the implementation of programs, accurate feedback from participants, and valid assessments of program effectiveness. In determining the effectiveness of programs, consideration should be given to attrition or dropout rates as well as to the primary and secondary effects of a program. Program effectiveness can be best measured by true experimental designs. Quasi-experimental designs would be used when randomization of participants is not feasible. For newly introduced modest programs, approximate evaluation methods would be suitable. Once health promotion programs have been shown to have an impact on employee health, cost-benefit or cost-effectiveness analyses would be used. Thus the evaluation process offers the best approach for rational decision making concerning the future of occupational health promotion programs.

The remainder of the chapters in this section will examine: planning the behavioral intervention, psychological assessment, and economic assessment.

RESOURCE GUIDE FOR PLANNING AND EVALUATION

Organization

Evaluation Research Society

Books and Articles

Campbell, D. T. "Reforms as experiments." *American Psychologist,* 1969, **24,** 409–429.

Conner, R. F., ed. *Methodological advances in evaluation research.* Beverly Hills, Ca.: Sage, 1981.

Cooper, A. M. and Sobell, A. M. "Does alcohol education prevent alcohol problems? Need for evaluation." *Journal of Alcohol and Drug Education,* 1979, **25,** 54–63.

Rossi, R. H., Wright, J. D., and Anderson eds. *Handbook of survey research* New York: Academic Press, 1982.

Webb, E. J., Campbell, D. T., Schwartz, R. D., Sechrest, L., and Grove, J. B. *Nonreactive measures in the social science,* 2nd ed. Boston: Houghton Mifflin, 1981.

Wortman, P. M., ed. *Methods for evaluating health services.* Beverly Hills, Ca.: Sage, 1981.

Journals and Reviews

Evaluation and the Health Professional
Evaluation News
Evaluation Quarterly
Evaluation Review
Evaluation Studies and Program Planning
New Directions for Program Evaluation

REFERENCES

Ajzen, I. and Fishbein, M. *Understanding attitudes and predicting social behavior.* Englewood Cliffs, New Jersey: Prentice-Hall, 1980.

Boruch, R. F. ''On common contentions about randomized field experiments.'' In *Evaluation studies review annual,* G. V. Glass, ed. Beverly Hills, Ca.: Sage, 1976.

Campbell, D. T. and Stanley, J. C. *Experimental and quasi-experimental designs for research.* Chicago: Rand McNally, 1966.

Cook, T. D. and Campbell, D. T. ''The design and conduct of quasi-experiments and true experiments in field settings.'' In *Handbook of industrial and organizational research,* M. D. Dunnette, ed. Chicago: Rand McNally, 1976.

Cook, T. D. and Campbell, D. T. *Quasi-experimentation: Design and analysis issues for field settings.* Chicago: Rand McNally, 1979.

Cunningham, R. M. *Wellness at work: A report on health and fitness programs for employees of business and industry.* Chicago: Inquiry Books, 1982.

Edwards, W., Guttentag, M., and Snapper, J. A. ''A decision-theoretic approach to evaluation research.'' In *Handbook of evaluation research,* E. L. Struening and M. Guttentag, eds. Vol. 1. Beverly Hills, Ca.: Sage, 1975.

Fishbein, M. D. and Ajzen, I. *Belief attitude, intention and behavior: An introduction to theory and research.* Reading, Mass.: Addison-Wesley, 1975.

Green, L. W. ''Evaluation. In *Managing health promotion in the workplace,* R. S. Parkinson and Associates. Palo Alto, Ca.: Mayfield, 1982.

Green, L. W., Kreuter, M. W., Deeds S. G., and Partridge, K. B. *Health education planning: A diagnostic approach.* Palo Alto, Ca.: Mayfield, 1980.

Health Resources Administration. *Baseline for setting health goals and standards.* DHEW (HRA) 77-640. Washington, D.C., 1977.

Jaccard, J. J. and Davidson, A. R. ''Toward an understanding of family planning behavior: An initial investigation.'' *Journal of Applied Social Psychology,* 1972, **2,** 228–235.

Kidder, L. H. *Selltiz, Wrightsman, and Cook's Research Methods in Social Relations* 4th ed. New York: Holt Rinehart and Winston, 1981.

Posavac, E. J. and Carey, R. G. *Program evaluation: Methods and case studies.* Englewood Cliffs, N.J., 1980.

Rossi, P. H. and Freeman, H. E. *Evaluation: A systematic approach,* 2nd ed. Beverly Hills, Ca.: Sage, 1982.

Sechrest, L. and Cohen, R. Y. ''Evaluating outcomes in health care.'' In *Health psychology—A handbook: Theories applications, and challenges to a psychological approach to health care system,* G. C. Stone, F. Cohen, and N. E. Adler. San Francisco: Jossey-Bass, 1980.

Selltiz, C., Wrightsman, L. S., and Cook, S. W. *Research Methods in Social Relations,* 3rd ed. New York: Holt, Rinehart and Winston, 1975.

Sejwacz, D., Ajzen, I. and Fishbein, M. ''Predicting and understanding weight loss: Intentions, behaviors, and outcomes.'' In *Understanding attitudes and predicting social behavior,* I. Ajzen and M. Fishbein. Englewood Cliffs, N.J.: Prentice-Hall, 1980.

Warheit, G. J., Bell, R. A., and Schwab, J. J. *Needs assessment approaches: Concepts and methods.* Rockville, Maryland: National Institute of Mental Health, DHEW Publication No. (ADM) 79-472, 1979.

Weiss, C. H. *Evaluation research: Methods of assessing program effectiveness.* Englewood Cliffs, N.J.: Prentice-Hall, 1972.

Wilbur, C. S. *Live for Life: An epidemiological evaluation of a comprehensive health promotion program.* Workshop presented at the meeting of the American Psychological Association, Washington, D.C., August 1982.

CHAPTER 15

Planning Health Promotion Intervention and Evaluation: A Behavioral Perspective

Samuel Berkowitz

INTRODUCTION

In order for occupational health promotion programs to be deemed effective, they will probably have to demonstrate their ablility to elicit increases in health-promoting behavior on the part of their participants. Stainbrook and Green (1982) argue that there is a stronger mandate than ever before for health promotion efforts to "improve health practices and reduce unnecessary morbidity and mortality" (p. 14). Similarly, King (1982) points out that the ultimate aim of health promotion efforts is to support the achievement of positive health, and this aim is unlikely to occur on the basis of educative practices alone! That brings us to the issue of how behavior can be changed, that is, the techniques for behavior change. Having provided an overview of planning and evaluation in the previous chapter, we will address the specific issue of planning and implementing behavioral interventions and, at the same time, evaluation.

Since before the turn of the century, behavioral scientists, especially psychologists, have been developing and subsequently refining technologies for changing human behavior. These technologies have drawn extensively from the experimental laboratory findings of learning psychology (Berkowitz, 1982; Kazdin, 1978). The early works of Pavlov and Thorndike laid the original groundwork for the development of behavioral technologies, principles, and terminology. Later work by Watson, Guthrie, Hull, and Miller in the period between 1910 and the 1960s demonstrated how basic scientific research can lead to behavioral technologies that, in turn, can favorably impact health-related problems. The more recent operant conditioning research of B. F. Skinner (1953) has had the greatest impact on the development of behavioral programs for health promotion so far. Such behavioral technologies applied to the alteration of human behavior have clearly demonstrated their usefulness in affecting health-promoting behavior (American Psychological Association, 1982; Stainbrook and Green, 1982). Yet these technologies and the principles underlying them have yet to realize their full potential in health promotion efforts despite the urgings of public health scholars such as Mayhew Derryberry as early as 1954, and more recent writers as Everly and Girdano (1980), Groden and Cautela (1981), King (1982) and Stainbrook and Green (1982).

It is therefore the purpose of this chapter to provide the reader with some general principles and specific procedures for behavioral program development. These may serve as a framework within which to utilize the specific content guidelines for the occupational health promotion programs which appeared in the previous section. These general principles are relevant to any program where change in health-related behavior is a criterion for success.

STEPS IN IMPLEMENTING BEHAVIORAL INTERVENTION AND EVALUATION

Step 1—Define the Target Behavior When developing a program designed to change behavior, the behavior which is targeted for change should be explictly described. This step is called defining, or more specifically, simply describing the target behavior (i.e., the behavior to be the focus of the program). Therefore, if the focus of the program is to increase some health-promoting behavior, it should be pinpointed and described explicitly. For example, *what behavior,* under *what conditions,* and *when* or *how often* (if applicable). If the focus of the program is to decrease some health-eroding behavior or some risk factor, again, it should be described explicitly rather than in broad, general or vague terms (see Figure 15-1). Target behavior descriptions must be defined in readily observable *and* measurable terms (see Figure 15.2). This will assist not only the program coordinator, but participants as well in (1) defining goals and expectations before the program, (2) assessing progress during the program, and (3) evaluating results after the program.

Step 2—Select the Measurement Procedure/Technique Once the target behavior has been explicitly described in objective terms that can be observed, measured, and recorded (see Figure 15.3), some form of measurement procedure or technique must be selected in order to count and assess pre-intervention baseline (see *Step 3*), monitor progress, and evaluate success. Based on these measurements, modifications and revisions in the program can also be made. It is critical that the measurement tool actually reflects the goal of the program, that is, the target behavior. The measurement can show the frequency, rate, or intensity of the target behavior. Each of the previous program chapters offered recommendations for measurement. Subsequent chapters in this section will expand on psychological measurement and economic assessment (cost-effectiveness and cost-benefit analyses.)

Step 3—Obtain a Baseline Having decided on a measurement technique to count the target behavior, it is advisable to generate a baseline (see Figure 15.4). A

What Is Behavior? Behavior is:

1. what people *DO*.
2. *observable.*
3. *physical movement.*
4. something specific that can be *seen, heard,* or *felt.*
5. *countable.*
6. *measurable.*
7. *recordable.*

Good Examples	**Poor Examples**
Speaking	Trying
Writing	Involved
Throwing	Communicating
Touching	Fantasizing
Smiling	Hostile
Typing	Disrespectful
Kissing	Insecure
Chewing	Crazy
Running	Lazy
Lifting	

Figure 15.1. Defining and describing behavior.

Describing behavior helps us to. . .

1. Observe and objectively *count* and *quantify* how often the targeted behavior occurs.
2. Obtain a *baseline* of behavior.
3. Observe *antecedents* and *consequences* of the behavior.
4. Determine to what *extent* the pinpointed behavior is a problem or not.
5. Eliminate *subjective impressions* and *labels.*
6. Communicate the targeted behavior's frequency of occurrence clearly to others.
7. Have *others* observe and accurately count the same behavior.
8. Develop *behavioral objectives.*
9. Develop an *intervention* program for the targeted behavior.

Figure 15.2. Advantages of describing behavior.

Counting Behavior

Objective observation and counting of the targeted behavior gives us a clear, unbiased measure of the behavior in question.

1. Helps us focus on *observable* behavior.
2. Indicates at what level to *start* working with the target behavior.
3. Encourage *quantification* of observations.
4. Gives constant *feedback* (evaluation) on the effect that procedures are having.
5. Helps to *evaluate, revise,* and *improve* procedures when necessary.

Figure 15.3. Observing and counting behavior.

BASELINE

Baselines consist of:

- the measure (frequency, rate) of the target behavior prior to intervention.
- the period of time *before* any modification procedures are introduced during which the target behavior is *counted.*

THE BASELINE:

1. Tells us the extent (frequency, rate, level) of the behavior *before* we make any behavioral changes.
2. Assesses where the individual is *right now* in relation to the target behavior.
3. Tells us where to *begin* our program.
4. Can aid in selecting *techniques* to use depending upon level of behavior.
5. Helps us evaluate the changes in the behavior after we implement our intervention program.
6. Encourages *objective observation,* quantification, and measurement.

EXAMPLES:

- Weight *before* going on a diet.
- Blood pressure level *before* initiating a control program.
- Distance run to exhaustion *before* beginning a conditioning program.

Figure 15.4. Baseline and goals.

baseline recording may be simply defined as (1) the beginning or existing level of the behavior or (2) the number of times a specific target behavior is engaged in *before* any behavior change intervention is applied. The baseline gives the program coordinator, as well as participants, an objective criterion that can be used later as a point of comparison to evaluate progress and ultimately to evaluate success. It also indicates at what level the program should begin.

Behavior does not occur in a vacuum, however. There are events that precede the behavior and events that follow the behavior. These events can be physical stimuli or cues in the environment, or most often, the behavior of other people.

Antecedents are those events that precede behavior. A traffic light turning green is an antecedent for stepping on the accelerator. A commercial on TV might be an antecedent for getting a snack to eat. Someone asking us a question is usually an antecedent for an answer. Those events that follow the behaviors are called consequent events. These can be in the form of social consequences—smiles, praise, "thank you," and so on, or more concrete consequences—money, prizes, food, and so on.

The ABCs of behavior, then, are

*A*ntecedent (preceding) events

*B*ehavior

*C*onsequent (following) event (see Figure 15.5)

When recording baseline rates of the target behavior, it is desirable to record the antecedents and the consequences for the specific behavior each time it is emitted. These steps are particularly important in dealing with behaviors such as eating and smoking (for example, see Figure 5.1 in the chapter on weight reduction). A typical baseline period lasts about one week and should represent the typical frequency, rate, or intensity of the target behavior.

Step 4—Establish a Terminal Goal (Goal of the Program) Once a baseline has been established, it is helpful to establish a terminal behavior. A terminal behavior is defined as the desired measured *goal,* or objective, of the terminal behavior at the end of the change program. The goal should be clearly stated so that

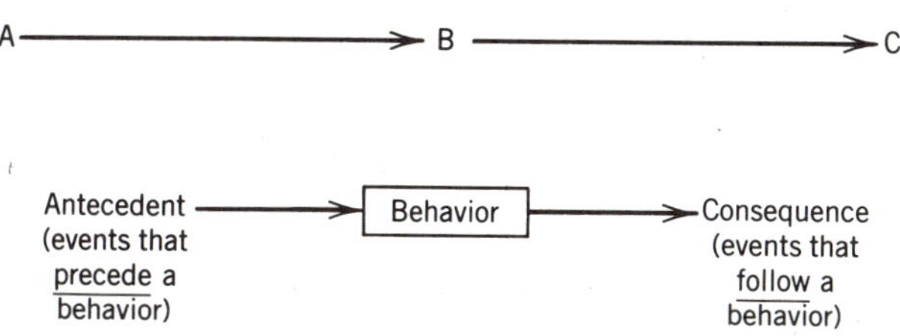

Figure 15.5. ABCs of behavior change.

the program participants as well as the program coordinator are very aware of it. Once achieved, this goal will define how successful the program has been. Systematic and continuous evaluation of progress throughout the course of the program also becomes possible and is highly recommended. Revisions and changes can then be made to improve the program. It is certainly appropriate to have multiple goals or to break a large goal into more readily defined, smaller component parts.

Step 5—Intervention After selecting and describing the behavior, deciding how to measure it, recording the baseline, and setting goals, behavior change can be facilitated through the use of operant behavioral technologies. Specific procedures

1. **Define the target behavior to be modified:**
   Describe the desired target behavior. What should the person be doing or not doing?

   Describe the *ontask behavior,* that is, the desirable health-promoting behavior to be *increased.*
   Describe the *off-task behavior,* that is, the undesirable health-eroding behavior to be *decreased.*

2. **Select measurement technique:**
   How will the behavior be counted or measured (frequency, duration, sampling, interval, products)?

   Select or construct a recording instrument to measure the behavior.

3. **Baseline behavior:**
   a. Measure or count the behavior under the present conditions prior to making any changes.
   b. Measure or count the behavior plus the antecedents and consequences of the behavior, prior to making any changes.

4. **Terminal behavior:**
   Establish *goals* for the target behavior (i.e., what is the measurable *goal* of the health promotion program?)

5. **Intervention:**
   Use of behavioral technologies (refer to the specific behavioral procedures in the next section).

6. **Continuous and Follow-up assessment and evaluation.**

Figure 15.6. An outline in the steps in conducting a behavioral intervention.

utilized during intervention programs are discussed in the following section. Ideally, we would like to believe that by merely informing, that is, "educating" individuals about the importance of practicing health-promoting behaviors, an increase in those health-promoting behaviors would actually result. Unfortunately, this is seldom the case. It then becomes important for the health promotion program coordinator to be familiar with behavioral technologies, other than simply education, that may be used to facilitate behavior change. Specific behavioral intervention procedures are discussed in the section following Step 6.

Step 6—Follow-up Assessment and Evaluation The final step in conducting a behavioral health intervention is follow-up assessment and evaluation. The follow-up assessment may be thought of as a longitudinal assessment that may occur at numerous points in time. Here the selected measurement tools are used to assess progress. Terminal evaluation occurs at the end of the entire intervention. The previous chapter discussed various evaluation methodologies.

These, then, are the six major steps in the planning and implementation of a behavioral health intervention. These steps are summarized in Figure 15.6.

SPECIFIC BEHAVIORAL PROCEDURES

Having examined the overall process of implementing a behavioral intervention, let's briefly review several of the behavioral technologies most likely to be of value in an occupational health promotion intervention program.

Positive reinforcement refers to the presentation of a desirable, rewarding stimulus or consequence after the emission of a behavior. Positively reinforcing a behavior is usually the most effective way of increasing the frequency of that behavior (See Figure 15.7). Many individuals believe that merely providing people with information that a given behavior is "good" for them will serve as sufficient

Learned Reinforcers

Primary Reinforcers	*Social Reinforcers*	*Token Reinforcers*	*Activity Reinforcers*
Food Touch Physical contact	Attention Praise Smile "Thank you" Approval Affection	Money Coupons Stars Check marks Points Poker chips Green stamps (These reinforcers are exchanged for other backup reinforcers.)	Listening to music Visiting a friend "Break time" Watching TV Reading a book Playing a video game Vacation days

Figure 15.7. Reinforcers.

motivation to engage in the behavior. This may be presumptuous in the case of smoking cessation, weight reduction, and physical fitness programs. Behaviorally speaking, the reason that these programs have such high dropout rates is twofold: (1) initially these programs "punish" the participants by removing something that is presently positively reinforcing and pleasurable for them (nicotine, food, oral gratification, etc.) or by inflicting something that is usually initially discomforting (muscular soreness, fatigue, etc.), and (2) the traditional positive reinforcement for these programs (feeling better, being more fit, looking healthier, and increased chances of living longer) are long-range reinforcers and usually seem, at first, to be simply abstractions or, at times, unobtainable. So, the long-range payoff seems minor in comparison to the discomfort resulting from the removal of the more immediate reinforcer—say, a cigarette. Therefore, the health promotion program coordinator should consider using short-term (immediate) and intermediate-term positive reinforcement techniques to improve compliance in the target behavior and to maintain successful attainment of the goal behavior once reached. So, for example, an individual's successful participation in a health promotion program can be immediately rewarded at various points on the way to the goal behavior by the presentation of positive reinforcers. Initial attempts and smaller steps can be reinforced. Shaping strategies can be extremely useful in weight reduction, smoking cessation, and physical fitness programs (see the section below on shaping). Refer to Figures 15.8 and 15.9.

Negative reinforcement also increases behavior. In this case, the increase in behavior results from the removal of an unpleasant stimulus when the desirable behavior is engaged in. So, for example, an individual's successful participation in an occupational health promotion program can be negatively reinforced through the removal of some undesirable aspect of his or her job as long as compliance is maintained. Negative reinforcement is commonly confused with punishment. Punishment results in a *decrease* in behavior through the presentation of some undesirable stimulus, for example, a monetary fine or penalty. It is important to remember that punishment reduces behavior, but negative reinforcement increases behavior. Punishment is generally not to be considered in a health promotion program due to its utilization of undesirable stimuli.

Stimulus control refers to the direct manipulation of antecedent events so as to increase or decrease target behaviors. Stimulus cues, or "reminders," can be used to help people remember to practice health-promoting behavior. For example, a colorful, self-adhesive dot applied to a wristwatch can be a reminder to a person to meditate once a day. Stimulus narrowing can be used to reduce undesired behavior. It is a stimulus control strategy often found in weight reduction and smoking cessation programs. For example, eating or smoking behavior may be gradually restricted to certain times or certain places.

Behavioral contracting is a way of obtaining a written commitment regarding behavior. In the contract, two or more parties (although a contract can be with

Consequences

Behavior is determined by its *consequences.*

Reinforcement

1. A *reinforcer* is any consequence following the behavior that *increases, strengthens,* or *maintains* the behavior.

Antecedent	*Behavior*	*Consequence*	*Result*
Request →	Ontask behavior →	Reinforcer →	Behavior increases

2. Reinforced behavior is *maintained* or *strengthened.*

Punishment

1. A *punisher* is any consequence following the behavior that *decreases* or *weakens* the behavior.

Antecedent	*Behavior*	*Consequence*	*Result*
Request →	Offtask behavior →	Punisher →	Behavior decreases

2. Punished behavior is *reduced* or *weakened.*

Note: The *behavior* is reinforced, not the individual.

Figure 15.8. Behavioral Consequences

Rules of Reinforcement

1. Reinforce *immediately.*
2. Reinforce *frequently.*
3. Reinforce only *desired behavior.* Do not inadvertently reinforce offtask behavior.
4. Start *where individual is at*—reinforce behavior at, or just below, the *baseline* level.
5. In the beginning, reinforce behavior continuously *each and every time* it occurs.
6. Later, reinforce only *intermittently.*
7. Start small; reinforce *improvements* (shaping).
8. Pair *social reinforcer* (descriptive praise) with other reinforcement.
9. *Ignore* all undesirable behaviors.
10. Don't "impose" reinforcers; determine what is important to the individual.

Figure 15.9. Reinforcement—strengthening behaviors.

oneself or even a group) agree on the terms surrounding the emission of a given behavior. The contract clearly states the desired behavior, the conditions for the emission of the behavior, and the consequences. Behavioral contracting is often used in weight reduction, smoking cessation, physical fitness, and counseling programs. It is also a useful tool in working out difficulties in relationships. Figure 15.10 provides an example of a blank contract. Such a contract could be used to increase compliance in any health promotion program. For example, in a physical fitness program, the contract would stipulate the level of desired behavior (jogging two miles, three times a week), where the behavior is to be engaged in (at the high school track), at what times (after work on Monday, Wednesday, and Friday), and what the reinforcers will be (earning so many points towards some finite point total that will result in some positive reinforcer, e.g., being treated to a steak dinner). The behavioral contract seems to increase compliance because it clearly defines the target behavior, provides a structure for those who need a more highly structured

---

I, ______________________________, agree to perform the following:

1.

2.

These behaviors will be performed at ______________________________, will be observed by ______________________________ and recorded in this manner: ______________________________.

In return for engaging in the above behaviors, I will receive the following from ______________________________:

1.

2.

If either party does not adhere to the terms of this contract, then the entire contract is considered null and void until the next renegotiation.

The duration of this contract is from ______________________________ to ______________________________. The contract will be renegotiated on ______________________________.

(Signed) ______________________________ (Date) ______________

(Signed) ______________________________ (Date) ______________

(Signed) ______________________________ (Date) ______________

Figure 15.10. Behavioral contract.

program, provides for positive reinforcement, and allows the parties to make an explicit "public" commitment. (See Figure 15.11.)

Shaping (see Figure 15.12) can be an extremely useful behavioral technology designed to facilitate increases in health-promoting behavior. Perhaps the most common reason people fail at personal health promotion is because they overcommit themselves, that is, they "bite off more than they can chew." Shaping involves breaking the goal behavior down into a series of steps, or smaller, successive approximations, each moving closer and closer to the desired goal behavior. Reinforcement is applied after each successive approximation. For example, the goal behavior for a physical fitness program may be training for four hours per week.

What Is a Contract?

A behavioral contract is a written agreement specifying the means of arranging the exchange of *behaviors* and *reinforcers* between two or more persons. The contract specifies:

A. Where and under what *conditions*
B. Who is to *do* what
C. What *reinforcers* will follow the behavior

What Goes into a Contract?

1. Where and when the behavior will take place.
2. Clearly stated description of target behavior(s) to be emitted.
3. Specific consequences to follow each specific behavior.
4. Who will record and count behavior.
5. Amount and kind of reinforcer(s) and bonuses to be used.
6. Who will give out rewards.
7. Delivery schedule of reinforcer(s).
8. Penalty for noncompliance with contract terms.
9. Date contract is to begin, end, and, if necessary, be renegotiated.
10. Date to review progress.
11. Name of outsider who will monitor contract periodically.
12. Signatures of all parties.

Why a Written Contract?

1. To make explicit the agreements and expectations of all the parties involved in the contract.
2. To provide the opportunity for all parties to negotiate and have input into the contract.
3. To act as a reminder (antecedent) or cue to engage in the agreed-to behaviors.

Figure 15.11. Behavioral contracting.

What is shaping?

Shaping: A gradual change method of reinforcing a behavior presently in the individual's repertoire (baseline behavior) that resembles or approximates a specific terminal behavior not presently in the repertoire. Successive reinforcements are made contingent upon a hierarchy of behaviors that more successively approximate the terminal behavior than the previous reinforced behavior, until the terminal behavior is reached.

1. The goal behavior is broken down into smaller component steps or approximations.
2. Each component behavior is reinforced one step at a time.
3. After the individual succeeds at one step, the behavior of the next step is reinforced until the goal behavior is achieved.
4. Each step, in turn, successively approximates and gradually resembles the goal behavior.

When is shaping used?

- Shaping is used when a specific target behavior is not presently in the individual's repertoire.
- Shaping changes behavior—the form, frequency, or duration of the behavior.
- Shaping changes the behavior from baseline behavior to terminal behavior.
- Shaping is also referred to as successive approximations.

Example:

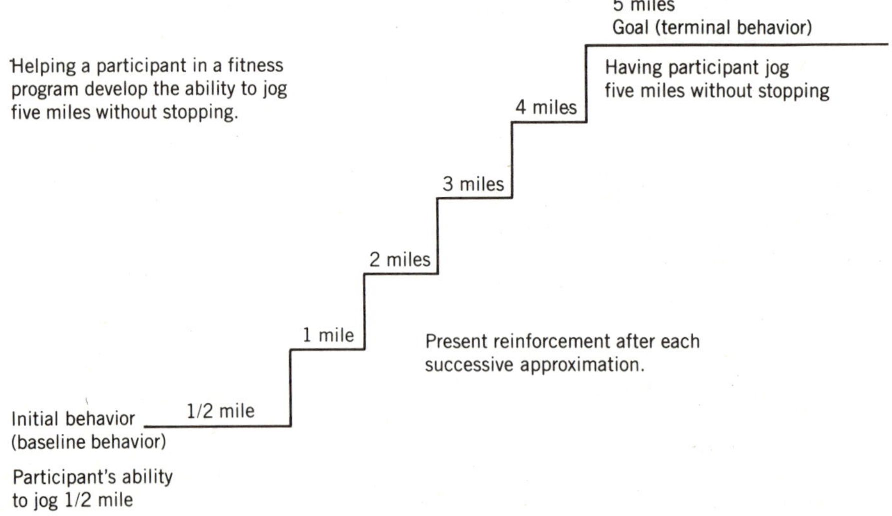

Figure 15.12. Shaping.

Using the individual's baseline, the present level, rather than starting with four hours a week, the individual is reinforced for one hour a week for the first month. Once the person is used to training at that level, performance is increased say to one and one-half hours per week for a month; then perhaps two hours or more a week, and finally up to four hours a week. This approach reduces the chances of frustration due to the inability to reach the goal behavior from the initiation of the program and increases reinforcement opportunities, as well as feelings of success and achievement. Shaping is often used in stress management, weight reduction, smoking cessation, and physical fitness programs.

The interested reader should refer to two excellent introductory sources for further elaboration on designing behavioral programs: O'Leary and Wilson, *Behavior Therapy: Application and Outcome,* Englewood Cliffs, N.J.: Prentice-Hall, 1975; and Whaley and Malott, *Elementary Principles of Behavior,* Englewood Cliffs, N.J.: Prentice-Hall, 1971.

Intermittent reinforcement pertains to *how often* and *when* a certain behavior is reinforced. When a behavior is first being learned, it is weak and unstable. Reinforcing each and every occurrence establishes the behavior; later, reinforcing it intermittently will maintain the behavior.

Once the behavior is established, reinforcing it periodically will *maintain* the behavior over a longer period of time. In order to maintain the behavior once it has been established, it is also best to gradually change from concrete reinforcers (tokens, money, food) to social reinforcers (praise, acknowledgment, attention).

Token economies can be extremely effective as tools for increasing compliance in occupational health promotion programs.

Tokens are symbols. They are usually a physical object (poker chip, check mark, stars, etc.) that can be accumulated and exchanged at some later time for other reinforcers. The reinforcers recieved for tokens are called back-up reinforcers.

A token economy is simply an "economic" system that utilizes tokens as reinforcers. A token economy can be established as a means of increasing compliance in any health promotion program. For example: tokens can be awarded to participants for jogging prescribed distances. Or, tokens can be awarded for attending various classes. They can even be awarded for losing weight. Contracting can be used to specify the particular behaviors and the rewards. Once the tokens have been earned, they must have some redeemable value, that is, back-up reinforcers are exchanged for tokens. Back-up reinforcers would include something selected by the individual, that is meals, certain privileges, or even money (which is a token itself).

Summary

In this chapter, we have specifically outlined the steps involved in planning the health promotion interventions. These steps transcend specific health-related problems and organizations. Clearly, in the final analysis, it remains for the health promotion coordinator to select the appropriate implementation guidelines and content components for utilization. These decisions will be based upon the nature, problems, and resources at the occupational environment itself.

REFERENCES

American Psychological Association. *Journal of Consulting and Clinical Psychology,* 1982, **50,** entire issue, No. 6.

Berkowitz, S. "Behavior therapy." In *The newer therapies: A sourcebook.* Abt, L. E. and Stuart, I. R. eds. New York: Van Nostrand Reinhold, 1982.

Everly, G. and Girdano, D. "Implementing behavior modification in a weight control program." *Health Education,* 1980, **11,** 15–17.

Groden, G. and Cautela, J. "Behavior therapy: A survey of procedures for counselors." *Personnel and Guidance Journal,* 1981, **60,** 175–181.

Kazdin, A. *History of behavior modification: Experimental foundations of contemporary research.* Baltimore University Press, 1978.

King, K. "Selected behavioral strategies for the health educator."*Health Education,* 1982, **13,** 35–36.

Skinner, B. F. *Science and human behavior.* New York: Macmillan, 1953.

Stainbrook, G. and Green, L. "Behavior and behaviorism in health education." *Health Education,* 1982, **13,** 14–19.

CHAPTER 16

Psychological Assessment in Occupational Health Promotion

Catherine J. Green

INTRODUCTION

Initially, psychologists developed psychological assessment instruments to answer questions about mental disease or dysfunction. The assessment tools were not constructed to provide answers concerning the nonpsychiatric problems of the individual, nor were the majority of these tools of any value in the assessment of healthy people or healthy functioning. Fortunately, psychological assessment techniques have now expanded into newer, nonpsychopathologically related areas. These developments have coincided with the growing need of occupational health promotion programs to offer some evidence as to their efficacy, or at least to be able to gauge their impact, if any, upon their target populations. In order to be able to gauge effect, however, one needs tools, that is, measurement instruments sensitive enough to measure the expected outcome. Psychological tests have begun to be used for just such purposes. Psychological assessment procedures can now give us insight into physical, as well as mental, disease and dysfunction. Similarly, such tools can give us further insight into the condition we call health. For example, by using psychometric instruments, we can gain insight into a person's risk of coronary heart disease as a function of the Type A behavior pattern. We can indirectly assess stress, coping styles, and we can even gain insight into how to improve an individual's compliance within an occupational health promotion program. Therefore, the rationales for including psychological assessment procedures in occupational health promotion programs include:

1. Psychological assessment may be used to help assess and increase compliance within occupational health promotion programs.
2. Assessment procedures can be used to screen individuals who are currently at high risk for emotional and physical disability so that effective preventive actions can keep such individuals as productive members of the work force.

3. Assessment procedures may be used to help assess participant needs and suitability for various health promotion interventions.
4. Assessment procedures may be used as part of any evaluation process used to assess the effectiveness of health promotion interventions.
5. Finally, assessment tools can be used to ensure the standardization of assessment and evaluation processes.

There exist two potential problems with assessment within the psychological/behavioral domain, however. First, assessment techniques of *any* kind are potentially subject to misinterpretation or misuse. These potential problems will be eliminated by simply following the guidelines spelled out in *Standards for Educational and Psychological Testing* published by the American Psychological Association (Washington, D.C.). The second potential problem exists in that not all psychological assessment instruments are of equal value or equally appropriate for various populations. We shall address this issue within the context of this chapter.

In the following sections of this chapter, we shall (1) provide a general discussion of how to evaluate psychological assessment instruments, and (2) provide a brief list of selected assessment instruments that may be of value within occupational health promotion settings.

GENERAL CRITERIA USED TO EVALUATE ASSESSMENT INSTRUMENTS

The two most commonly used criteria for evaluating psychometric instruments are *reliability* and *validity*. Reliability may be generally defined as the test's ability to yield stable and consistent results from one administration to another, as well as the extent to which the test instrument is free from such internal defects that would produce errors of measurement. Common forms of reliability are test-retest reliability, reliability across alternate forms of the same test, and internal test item consistency. Reliability is generally deemed to be the sine qua non of test construction, because if a test is not reliable, it cannot be valid. Furthermore, statistically, the highest validity that a test can have will be limited to the square root of the reliability coefficient.

Validity may be simply defined as the degree to which a test measures what it purports to measure, when compared to accepted criteria. Common forms of validity are construct validity (the degree to which a test measures some hypothetical construct), concurrent criterion-referenced validity (the degree to which the test can correctly identify existing status on some variable), predictive criterion-referenced validity (the degree to which the test can predict the future status on some variable), and content validity (the degree to which the test is a representative sample of the behavior domain to be assessed).

Other things to consider when evaluating a psychometric tool would be the completeness of the manual and instructions for administration of the test and interpretation of the test results. A final question that one should ask before selecting an assessment tool is "are the norms appropriate for use with the intended population?" In other words, has the test shown an ability to be used with the population currently being considered for assessment?

SELECTED INSTRUMENTS

State-Trait Anxiety Inventory (STAI)

This inventory was originally developed as a research instrument for investigating anxiety in normal adults (Spielberger, Gorsuch, and Lushene, 1970). The instrument is composed of two scales of similar format. The A-State scale asks the subject to indicate how he or she feels "right now." This scale is said to tap state-dependent, transitory anxiety. The A-Trait scale asks the subject to indicate how he or she "generally" feels. This scale purports to measure trait anxiety. Thus the two scales were designed to tap feelings of tension, nervousness, worry, and apprehension. Four-point Likert response sets are used on both scales.

The STAI was designed to be self-administered and may be given individually or in groups; complete instructions are printed on the test form.

Subjects respond to each STAI item by rating themselves on the four-point scales; most subjects require 20 minutes or more to complete the form. Templates and machine scoring are available.

The normative sample, i.e., those individuals to whom the test subjects are to be compared, is composed of college students or varied groups of psychiatric and medical patients. Although a variety of student groups were utilized in various stages of construction, the normative samples are narrow in scope, including very few noncollege or clinical populations. Given the difference between male and female responses on this inventory with high school and college samples, it would be unwise to view the medical sample as an adequate comparison group.

Interpretation is straightforward; high scores on A-Trait indicate higher levels of anxiety proneness. High A-State scores are conceptualized as transitory or characterized by consciously perceived feelings of apprehension. The instrument is not intended to be used for individual evaluation, but for assessment of group differences.

As expected, test-retest reliability (reliability of the instrument administered twice to the same group of individuals) for the A-State was low: .20 to .40, regardless of time elapsed. A-Trait anxiety test-retest was higher at about .80. Content validity (estimated by evaluating the relevance of the test items, individually and as a whole) and empirical validity (obtained by comparing to some external criterion measure of the trait assessed) of the instrument are two critical issues.

Scores obtained on these inventories have not been compared fully to nontest evaluations of anxiety, although hypothetical-conditions data were employed in item selections. The STAI was built on the assumption that the initial item pool covered the full domain of anxiety, and that further selection on the basis of internal consistency would not distort this circumstance.

While the STAI's specific focus on one aspect of the individual (state and trait anxiety) would cause it to be of limited value in assessing the complex interrelationships of the individual, his or her interaction with the job and his or her colleagues in the work setting, the STAI might be of some value in assessing the state and trait anxiety domains for the purposes of screening or for pre- and postintervention evaluation.

Life Experiences Survey (LES)

The Life Experiences Survey, developed by Sarason, Johnson, and Siegel (1978), was formulated to assess an individual's perception of the life stresses experienced during the preceding 12 months. The notion that life stresses lead to an increase in the frequency of illness is well established in the medical literature. Holmes and Rahe (1967) were the first to address this issue in their early Schedule of Recent Experience (SRE), an objective self-report measure of significant events in the respondent's life. They posited that each life event, regardless of its positive or negative impact, requires adaptive coping on the part of the individual. Events were chosen for the SRE because their advent required adaptation on the part of the individual. Standardized weights were then determined for each event reflecting the amount of social readjustment the event appeared to require. The absolutist approach to events contrasts with that of Lazarus (1977) who states that it is the individual's perception of the meaning of the event that is of critical importance.

Sarason et al. developed the LES as a means of gauging both the occurrence of an event and its perceived significance, thus attempting to compensate for the perceived failure of the SRE to consider the latter. The 57-item LES self-report inventory has two sections. The first, completed by all respondents, contains 47 specific events plus three blank spaces where individuals can add unlisted events. These events cover a wide range of experiences. The second section with 10 items is designed primarily for use with students. In both sections, the respondent is asked to indicate which events were experienced during the past year, and in addition (a) whether they veiwed the event as positive or negative on a 7-point scale, and (b) the perceived impact of the event at its time of occurrence on a 7-point scale.

The inventory is self-administered. Respondents are asked to check which events have occurred, when in the last year they took place, and then to indicate the type and extent of impact the event may have had. The LES takes 20–25 minutes to complete. A positive-change score is obtained by summing the impact ratings of events designated as positive; negative-change scores are derived by summing the impact of negative events. The total-change score is the sum of both scores.

LES items were chosen to represent life changes frequently experienced by individuals in the general population. Thirty-four of the 57 items are similar in content to those of the SRE. Specific items were reworded so as to be applicable to both male and female respondents. No further details are given as to the development of the item pool.

According to the authors, a stressful event calls for adaptation, and adaptation rests in part on the individual's perception of the event. It is further assumed that this level of adaptive behavior will be related to illness events. Although efforts were made to be representative, it was impossible to tap the entire domain of potentially significant life events. Leaving a section of blanks to be filled in addresses this problem, but recall is likely to be a less adequate process than recognition as a means of identifying events.

There is no indication that external criteria were utilized in constructing the instrument, although numerous correlational studies were employed to evaluate the relationship between LES scores and relevant personality indices. (One hopes that a formal manual, when published, will address a number of these issues.) Test-retest data with two samples show that positive-change scores correlate between .29 and .53 for five to six weeks, whereas negative change-scores correlated .56 and .88, and total-change scores .63 and .64.

LES scores were correlated with a number of other instruments. Negative scores correlated .19 with trait anxiety on the STAI and .46 with state anxiety; an almost zero correlation was found between positive scores and state or trait anxiety. Crown-Marlowe Social Desirability correlations were also in the zero range, suggesting that the LES is free of social desirability bias. Employing the Psychological Screening Inventory, correlations of .20 between neuroticism and negative LES scores, and a correlation of .28 between extraversion and positive LES scores. From these data it would appear that personal maladjustment is marginally associated with negative change scores. No studies were reported on the relationship between these scores and illness onset, the posited sequel to life stress.

This instrument proposes to address the issue of life events only. It is posited that life events have an impact on the incidence of illness and psychological difficulties. Although no construction validity studies have been reported on the efficacy of this instrument in making such predictions, research (Yunik, i980) indicates that negatively weighted events tend to have a high correlation with future illness; negative-impact scores correlated .42 with the total number of illness problems reported. On the basis of this preliminary study, it appears that the LES may prove to be useful in predicting illness behavior. Looking directly at the individual's perceptions of life stressors, it may highlight an "at-risk" population that may serve as an intervention focus in the work setting.

Internal-External Scale (I-E Scale)

The effects of reward or reinforcement on behavior depends to some measure on whether or not the individual perceives a causal relationship between his or her own behavior and the reward. The I-E Scale was developed to measure the degree to which an individual feels that reinforcement is or is not contingent upon his or her own action.

Brought to its final form by Rotter, Liverant, and Crowne (1961), the I-E Scale is composed of 29 items in which the respondent selects which of two statements is more strongly believed. Twenty-three of the items refer directly to the subject's "externality"; six serve as filler in an effort to disguise the intent of the instrument somewhat. The authors see this as an instrument suitable to measure group differences; it is not meant to be used for individual or clinical prediction. No specific population was targeted and items are applicable to many life settings. No manual has been written for the I-E Scale and users must rely on the 1966 monograph for information on construction and utilization of the instrument (Rotter, 1966).

Self-report instructions for the I-E Scale are printed at the top of the questionnaire and are geared to an upper high school reading level. Subjects are told to select which of a pair of statements is more strongly believed. The testee is admonished to select the one that seems to be more true rather than the one he or she thinks should be chosen or wishes to be true. The 29 items are usually completed in 15–20 minutes.

Scoring is accomplished by totaling the number of "external" responses. One appendix provides a frequency distribution for male and female samples of university undergraduates; no statement is made regarding the generalizability of this distribution nor its applicability to other samples or populations.

Means and standard deviations are reported for a number of samples; full distributions are not described and no attempt is made to present a set of norms. Users are encouraged to create their own local norms. Regarding interpretation, the higher the score, the more reinforcement is seen to be the product of luck, chance, or under the control of others.

The I-E Scale evolved through a number of developmental stages. The first instrument developed by Phares (1957) was composed of 26 items selected *a priori* with 13 stated as external attitudes and 13 as internal. Subsequently, researchers sought to broaden the test by developing a series of subscales. A 100-item scale was analyzed and reduced back to 60 items. Control for social desirability bias was considered and a forced-choice format was instituted. Item analyses showed that the subscales failed to generate separate predictions. A series of further construction steps sought to reduce the intrument's social desirability and to improve its internal consistency. The final 29-item scale is composed of 23 real and six filler items.

At no point does it appear that considerations of empirical vaidity entered into the construction of the instrument. Subsequent to the construction stage, it was found that correlation between interviewer ratings of externality and the I-E Scale score were in the .60 range. Having gone through a sequence of construction steps, the I-E is internally consistent and reliable. There is some question, however, as to what the instrument is really measuring and whether, in fact, it adequately covers the domain specified. The absence of normative data is a serious problem, forcing users of the instrument to make their own judgments as to the meaning of given scores. These concerns appear more serious in light of the research literature, since many studies find that the instrument fails to predict expected behaviors consistent with the measured construct (Lambley and Silbowitz, 1973; Marston, 1969).

Internal consistency (KR20) ranges from .69 to .73, while test-retest reliability at 1–2 months ranges from .49 to .83. Significant efforts had been made to reduce the relationship between the I-E Scale and social desirability. This appears to have been accomplished. In one study the correlation between the I-E Scale and Taylor Manifest Anxiety Scale was .24; in another, the correlation was .00 (Efran, 1963). Nonquestionnaire "projectives" correlated modestly with the I-E Scale scores (Adams-Webber, 1963), and Cardi (1962) found a significant correlation between a semistructured interview measure of locus of control and the I-E Scale.

The I-E Scale evalutes a single concept—externality. Although this construct is an appealing one, the instrument makes no effort or claim to tap any of the vast number of other variables that relate to health behaviors. In spite of this, it might serve well in treatment planning for an individual if it truly measured what it intends to measure. However, the general weakness of the instrument and its value as a group measure only, effectively eliminate it as a tool for individual prediction.

Other researchers, in an attempt to overcome the weakness of the I-E Scale, have developed scales specific to the assessment of locus of control among medical patients. The most widely used of these is the Multidimensional Health Locus of Control (Wallston and Wallston, 1981; 1982). This brief, face-valid instrument assesses whether one believes personal health is a function of internal or external factors, i.e. the belief that health is self-determined or a function of powerful external factors or fate, chance or luck.

Symptom Check List-90 (SCL-90)

This 90-item self-report symptom inventory was based on the Hopkins Checklist and was designed to reflect psychological symptom patterns of psychiatric patients. Derogatis (1977), one of the test developers, suggests that it also has utility for assessment of medical patients. Each item is rated on a five-point scale of distress from "not-at-all" to "extremely." These responses are then interpreted along nine primary psychological symptom dimensions: somatization, obsessive-compulsive, interpersonal sensitivity, depression, anxiety, hostility, phobic anxiety, paranoid ideation, psychoticism. In addition, three global indices of distress are calcualted:

global severity, positive-symptom distress index, positive symptom total. The SCL-90 is intended as a measure of current psychiatric symptoms and is designed to be interpreted on three levels. The first level is the global, and the "global severity index" (GSI) is employed as the gauge. The primary symptom dimensions address the patient's level of psychopathology. Individual items are used to relate the presence or absence of specific symptoms. Instructions are simple and written on the Test form. There is a note that asks the patient to record how much discomfort a particular problem has caused during the last *X* number of days. Most patients take 20–30 minutes to complete the task.

Scoring requires either templates or transferring the 90-item scores from the test paper to profile sheet, where scores are summed to arrive at distress scores for each of the nine symptom dimensions. The global severity index, positive symptom total and symptom distress index are also calculated. Raw scores are transformed into T-scores utilizing nonpsychiatric patient norms.

The nonpsychiatric normative group (females, N=480; males, N=493) was a stratified random sample that was balanced to be broadly representative of the general population and drawn from one county in a mid-Atlantic state. No determination was made as to their status as medical patients. Brief descriptions of the clinical significance of each scale and global indices are provided. The GSI is considered an indicator of overall distress while the nine primary symptom dimensions provide a profile of the patient's status in psychological terms. Discrete symptoms may also be noted. Small-sample profiles are provided as guides.

The SCL-90 was constructed to serve as a checklist of psychiatric symptomatology. It was developed through a combination of clinical/rational and empirical/analytic procedures. The clinical/rational procedure involved the development of items that are intended to reflect specific psychological symptom patterns of psychiatric patients, and are based on clinical experience and observation. The empirical/analytic procedures involve a variety of structural assessments as well as validation approaches utilized in the development of this instrument. The original item pool was drawn from the Hopkins Symptom Checklist (Derogatis et al., 1974a; 1974b), which in turn can be traced back to the CMI item pool (Wider, 1948). The method of developing new items is not detailed in the manual, nor is the method for the initial reduction of the Hopkins checklist. Internal consistency (referring to the extent to which the results obtained throughout the test are consistent when administered once) ranged from .77 to .90. These high measures may indicate the homogeneity of scale dimensions, but closer examination suggests that they may be the product of several items restating the same emotion or behavior in slightly different words. One-week test-retest reliability ranges from .78 to .90.

The SCL-90 includes both medical and psychiatric symptoms, but is geared primarily to the level and nature of psychopathology present. Its major utility in occupational health promotion settings would be as a psychiatric screening instrument, but clearly the SCL-90 is more a clinical scale than anything else. The problems of utilizing a test developed for clinical psychiatric patients in a non-psychiatric setting is a serious one. Even a well-designed psychiatric instrument is likely to provide information that is distorted when applied to a general health-oriented population. Problems arise because of the unsuitability of norms, the questionable relevance of clinical signs, and the consequent inapplicability of interpretations.

Beck Depression Inventory

The Beck Depression Inventory (Beck, 1972) is an instrument that seeks to approximate clinical judgments of depression intensity. Efforts were made also to clearly differentiate depressed from non-depressed psychiatric patients. It should be noted that the inventory was designed for research purposes and was conceived as appropriate for discriminating levels of depression only in psychiatric populations. The inventory is composed of 21 multiple-choice items reflecting specific behavioral signs of depression that were weighted in severity from 0 to 3.

Originally designed to be administered by a trained interviewer who read each statement and asked the patient to select the statement that fit best, the instrument is now often presented as a self-administered inventory. No information is given regarding the possible impact upon norms and scores of this modification to a self-administered format.

The total score is obtained by adding the weighted values for each response endorsed by the patient. No attempt is made to transform raw scores. Scores of 0–9 are considered normal, the 10–15 range is seen as representing mild depression, 16–19 represents mild to moderate severity, 20–29 judged as moderate to severe, and 30–63 as severe depression.

Two patient samples were utilized in the development of the inventory. Both groups of patients were drawn from routine admissions to the psychiatric outpatient department of a university hospital and from the psychiatric outpatient department and psychiatric inpatient service of a metropolitan hospital.

The test developer sought to develop explicit rather than inferred behavioral criteria for evaluating depression. To accomplish this, items were selected from the literature and clinical experience. Each subcategory describes a specific behavioral manifestation of depression and consists of a graded series of self-evaluative statements. The items were chosen on the basis of their relationship to overt behavioral manifestations of depression and do not reflect a particular theory regarding etiology or viewpoint concerning the psychological processes underlying depression.

Test-retest reliability was not employed in the traditional manner due to the assumption that change would be occurring due to treatment and that this factor would significantly alter the interpretation of results. Indirect methods were utilized to assess change scores over time; in general, total depression scores tended to parallel clinical changes. Such a pattern is deemed desirable.

Concurrent validity (indicating the extent of the test's agreement with other criteria measuring similar psychological operations or traits) was addressed by comparing scores with clinical assessment of levels of depression (correlations between test scores and clinical judgments averaged .66). It is of some note that prediction of clinical change was accurate in 85% of the cases.

The total depression score seeks only to address level of severity among psychiatric patients, having been developed expressly for this purpose. It can serve a function withother individuals in this regard if clinical levels of depression are suspected. It is unlikely to prove sensitive to the more moderate levels of dysphoria that, although affecting an individual's work effectiveness and satisfaction, would not be categorized as clinical depression. Thus the major utility of the Beck Depression Inventory in health promotion would be as a screening instrument or as a posttreatment clinical evaluation.

Jenkins Activity Survey (JAS)

The Jenkins Activity Survey (Form C), developed by Jenkins, Zyzanski, and Roseman (1979), is the latest version of a 52-item self-report questionnaire designed to measure the Type A behavior pattern. This pattern is characterized by extreme competitiveness, striving for achievement, aggressiveness, self-imposed responsiblility, impatience, haste, restlessness, and feelings of being challenged and under the press of time. The behavior pattern is not conceived of as a personality trait or a standard reaction to a challenging situation, but rather the reaction of a predisposed person to a situation that challenges them. A large body of research has been built around the significance of this pattern in the development and persistence of coronary heart disease. Although the instrument is actively marketed to be utilized for individual diagnostic assessment, the test developers recommend that the instrument be used primarily for research into group differences and, given the multifactorial pathogenesis of CHD, they state that the test should not be used by itself to predict individual risk. The instrument proposes to tap three factors within the Type A individual: speed and impatience, job involvement, and hard-driving competition.

The JAS is easily administered and the majority of subjects complete the instrument in 15–20 minutes. Each response is assigned numerical points based on the product of the item regression weight and the optimal scaling weight for that response. The sum of the points for all items constitute the raw score. Hand-scoring templates are avalilable, but machine scoring is encouraged.

Normative data are based on a 2588 male sample drawn exclusively from individuals in middle- and upper-echelon jobs. One serious flaw in this area of evaluation is the total absence of female subjects in both the development and normative stages of test construction. This further compounds the rather narrow socioeconomic group that was employed in the normative population.

Prior to the development of this instrument, Rosenman et al. (1964) designed a structured interview protocol to assess the Type A behavior pattern. Items for the JAS were derived from this interview as well as Jenkins' observations of interview behavior and the theory of Type A behavior. A major flaw in the development sequence of the JAS was the use of discriminant function as the prime tool of scale construction. Each successive form of the test was changed substantially insofar as specific item content. Without adequate data on cross-validation, items were dropped, while others, retained on the basis of discriminant utility, had to be dropped later when new samples were employed. Discriminant function can always be made to separate construction sample groups, but these discriminations may not hold up on cross-validation.

Test-retest reliabilities at 4- to 6-month intervals ran from .65 to .82 on the four factors, with internal consistency ranging from .73 to .85.

Turning to validity, the construct underlying the test has some intuitive logic as well as empirical support. Initially developed to maximize its correlation with the structured interview, the JAS attempts to adhere closely to its major features. As development progressed, however, particularly in later stages, it became uncertain as to whether the original Type A domain was still being fully addressed, that is, the items selected and the factors produced may have become a product of statistical manipulation and sample idiosyncracies. Each step in the construction phase appeared to move the instrument further away from the original base in the structured interview.

Although the developers recommend that the instrument be used primarily for research into group differences, it has been utilized for assigning individuals to preventive intervention groups in occupational as well as health care settings. It has been employed as a predictor of CHD, with some success, although recent literature shows that it does not improve on the predictive accuracy of the structured interview itself (Brand, Rosenman, Jenkins, Stoltz, Zyzanski, in press.) The instrument does not propose to provide information on the progress and management of ongoing disease processes.

16 Personality Factors Inventory (16PF)

The 16PF (Cattell, Eber and Tatsuoka, 1970), one of the oldest personality tests currently in use, was first published in 1949. It is composed of a multidimensional set of sixteen scales arranged in omnibus form and designed to assess 16 personality traits. If the supplement is used, it seeks to tap 23 personality dimensions. The most commonly used form, comprising 187 trichotomous items, is designed to measure what Cattell terms "source traits," rather than syndromes. The traits evaluated are said to have withstood critical examination of over thirty years of factor-analytic research. Since the scales are seen as factorially pure, there is no item overlap; consequently, each scale is gauged by responses to between 10 and 13 items.

A paper and pencil self-report inventory, the 16PF takes generally less than 45 minutes for the subject to complete and requires no assistance from the individual administering the test. Hand-scoring templates, as well as machine-scoring and computer-generated narratives are available. Raw scores are transformed into sten or standard ten scores for the 16 traits. A variety of normative samples have been developed over the years. Standardization of the instrument involved a sample of 15,000 normal adults; however, it is unclear from the manual as to the nature and size of the original construction samples.

The traits are viewed as bipolar and scores are evaluated in terms of their location on the trait continuum. A variety of publications are available to help in interpreting scores (Karson and O'Dell, 1976), including a book specifically addressing the issue of the medical patient (Krug, 1977).

A significant body of literature has been developed utilizing the 16PF; nevertheless, the meaning of these results is often unclear. Regarding initial construction, Rorer (Buros, 1972) states that the scales are of indeterminate origin and unknown significance. One of the continuing difficulties in evaluating this instrument is the atypical manner in which data are presented in the handbook (Cattell, Eber, Tatsuoka, 1970). Although large quantities of data are presented, they are not comparable to standard psychometric evaluation techniques.

Test-retest reliabilities are reported to be in the .70–.90 range short term and drop to .50–.80 at two months. Split-half reliabilities used to assess internal consistency by dividing the whole test into two halves that should be equivalent, or nearly so, have indicated considerable within-factor heterogeneity. There is a high likelihood that several of the supposedly pure factors subsume a number of different traits. This finding is especially troublesome since the instrument rests on the factorial purity of the 16 traits. No effort is made to provide correlation data with instruments purported to measure similar traits. Furthermore, the factoring method that Cattell has used is but one of many available, each of which might have produced different results. Correlations between identical scales on different forms of the test are sometimes very low, ranging from .15–.82; in fact, some scales correlate more highly with other scales than with their matched scales. This puts into question the meaning and reliability of these scales; at the very least, it makes comparison across forms impossible.

The 16PF has been proposed as suitable for assessing the personality of normal individuals. It may actually be more suitable with so-called normals than with general psychiatric populations for whom it has also been claimed to be useful.

One is left in reviewing this instrument with a sense of considerable methodological sophistication. At the same time, one is struck by the idiosyncratic conception of personality and the reluctance to compare this instrument and its results with anything other than itself. These deficits and idiosyncrasies may reduce the confidence one may have in the 16PF's construct validity and consequent usefulness.

Minnesota Multiphasic Personality Inventory (MMPI)

The MMPI is an empirically derived instrument that was constructed by Hathaway and McKinley in 1939 to serve as an objective aid in a psychiatric case work-up and as a tool for determining the severity of specific psychiatric conditions. In its original development the MMPI had little to do with personality traits. As stated by Dahlstrom, Welsh, and Dahlstrom (1972), the MMPI was developed and validated as a psychiatric nosologic categorizing device leading to dichotomous discrimination between psychiatric patients and normals. They wrote

> Although the content covered in the MMPI item pool included by far a larger array of personological topics than in any other instrument available, subsequent studies have indicated that—while some areas of emotional maladjustment may be overrepresented (Block, 1965)—items referring to values, to primary group relationships, and to mood, temperament, and various special attitudes are probably too scarce to provide a well-balanced coverage of the domain of personality (see Schofield, 1966) (p. 6).

The 566-item true-false questionnaire has 10 basic clinical scales: hypochondriasis, depression, hysteria, psychopathic deviance, male sexual introversion, paranoia, psychasthenia, schizophrenia, hypomania, and social introversion, along with the following validity scales: cannot say, lie scale, confusion scale or straight validity, K or suppressor factor.

The test was developed so that those 16 years old or above with six years of schooling would be able to complete the inventory. The instrument takes an average of one and one-half hours to complete, although there are no limits to the time alloted to complete the inventory.

Hand-scoring templates are available, as is machine scoring: scores are often transformed into profiles. Raw scores are converted to T-scores with separate norms for males and females. Unfortunately, T-score scaling assumes a normal distribution of the trait, an unlikely assumption given the highly variable prevalence rates of the syndromes involved.

The original normative group was drawn from samples of Minnesota adults with separate male and female groups. Most lived in rural or semirural areas, worked in skilled or semiskilled trades, and had an eighth-grade education. The test itself has not been renormed since this 1940 sample, although such efforts are currently under way.

Originally, only single-scale elevations were interpreted. Over the years extensive clinical data have led to its increased utility. A number of codebooks have been written to aid in profile interpretations. (Gilberstadt and Duker, 1965; Good and Brantner, 1974; Marks and Seeman, 1963).

Construction criterion groups were selected from adult psychiatric clinics and wards of the University of Minnesota Hospitals; ''normal'' subjects were families and visitors of the patients. Items were selected for inclusion in a scale based on their capacity to discriminate between normals and each of the criterion groups. No effort was made at that time to examine scale overlap or to develop a rationale for item selection. Cross-validation samples were employed to evaluate the stability of the obtained separation and the generality of initial scale findings. Three hundred and fifty-one of the 550 items (16 are repeated) are utilized in scoring the initial 10 clinical scales.

The MMPI is without doubt the most thoroughly researched personality instrument available. One problem in evaluating the inventory is that its manual has not been updated since 1967, and vast quantities of information on the test's empirical utility and validation reside in hundreds of journal articles often beyond the reach of all but the most diligent students. However, a significant amount of data is published concisely in the two-volume *Handbook* (Dahlstrom, Welsh, and Dahlstrom, 1972).

Test-retest reliability in one sample at three to four days ranged from .56 to .88 with the majority in the low 80s. Psychiatric patients show one-week test-retest ranging from .59 to .86 with the majority in high 70s. At one year, test-retest correlations drop to the range of .36–.72. Test-retest can be difficult to interpret because it is not possible to discriminate real change from reliability error, particularly in the case of psychiatric patients undergoing treatment over extended periods of time. Kuder-Richardson internal consistency estimates are reported on a sample employing the KR 21 formula. These ranged from .36 to .93 with a median of .70. Validity data are best summarized in the *MMPI Handbook* (Dahlstrom, Welsh, and Dahlstrom, 1972).

Undoubtedly, the best instrument of its generation, the MMPI has assumed an almost mystical impregnability as a function of its age. For its time and purpose, its excellence of construction was unmatched. Yet the MMPI has not gone unchallenged (see Benton and Probst, 1945; Schmidt, 1945; Ellis, 1959; Rotter, 1945; Adcock, 1965; and Lingoes, 1965).

The MMPI has been employed in a variety of health applications with some equivocal results. Most frequently it has been used to differentiate psychosomatic from organic disease, to delineate psychological factors associated with psychosomatic disorders, and to predict the outcome of surgery or recovery from illness (Dahlstrom, Welsh and Dahlstrom, 1972). As Butcher and Owens (1978) note however, the MMPI does not appear to successfully differentiate psychosomatic types, and the commonly noted 1-3/3-1 elevation may indicate a neurotic overlay that does not preclude the diagnosis of organic disease. (Schwartz, Osborn, and Krupp, 1972). Although utilized in a number of studies with medical patients, the MMPI appears to serve as an aid only if psychiatric issues are prominent.

Millon Behavioral Health Inventory (MBHI)

The MBHI was developed specifically with medical-behavioral decision-making issues in mind (Millon, Green, and Meagher, 1982). A major goal in constructing the MBHI was to keep the total number of items comprising the inventory small enough to encourage use in all types of diagnostic and treatment settings, yet large enough to permit the assessment of a wide range of clinically relevant behaviors. Geared to an eighth-grade reading level, it contains 150 items.

Diagnostic instruments such as the MBHI have increased usefulness if they are linked systematically to a comprehensive clinical theory or are anchored to empirical validation data gathered in their construction (Loevinger, 1957). The eight basic ''coping styles'' comprising the first eight scales of the MBHI are derived from a theory of personality (Millon, 1969). The six ''psychogenic attitude'' scales were developed to reflect psychosocial stressors found in the research literature to be significant precipitators or exacerbators of physical illness. The final six scales comprising the present form were empirically derived for the MBHI either to appraise the extent to which emotional factors complicate particular psychosomatic ailments or to predict psychological complications associated with a number of diseases.

Self-administered, with instructions printed on the questionnaire, the great majority of individuals can complete the inventory in 20–25 minutes. Multiple-keyed, the test is best machine-scored, computer-synthesized and interpreted.

Norms for the MBHI are based on several groups of nonclinical and numerous samples of medical patients involved in diagnosis, treatment, or follow-up. The nonclinical groups involved in the construction phases of test development consisted of subjects drawn from several settings (e.g., colleges, health maintenance organizations, nursing schools, medical schools, factories, etc.) and was composed of 212 males and 240 females; the test construction patient group, drawn from diverse clinical populations (e.g., surgical clinics, pain center, dialysis units, cancer program, etc.) consisted of 1194 of which 130 males and 170 females were selected as a representative cross section for purposes of developing construction norms. An additional series of patient populations comprised of 437 males and 482 females were involved in development and cross-validation of the six empirically derived scales. Hence there was a total of 2113 patients and 452 nonclinical individuals.

Interpretation is based on both profile configurations and single-scale elevations. The first eight scales, the basic coping styles, characterize individuals regarding interpersonal and personality traits. For most individuals, these characteristics blend with other features in a configural, or cluster, pattern of several scales.

The ''psychogenic attitude'' scales represent the personal feelings and perceptions of the person regarding different aspects of psychological stress presumed to increase psychosomatic susceptibility or aggravate the course of a disease already present. Some details concerning the six scales may be useful at this point.

The ''chronic tension'' scale gauges level of stress, a factor that has repeatedly been found to relate to the incidence of a variety of diseases. More specifically, qualitative studies of chronic stress, such as persistent job tensions or marital problems have been carried out with particular reference to their impact on heart disease, often addressed as Type A–Type B behavior (Friedman and Rosenman,

1974; Gersten, Frii, and Lengner, 1976; Jenkins, 1976; Rahe, 1977). Constantly on the go, they live under considerable self-imposed pressure and have trouble relaxing. Frequently endorsed for certain aspects of this behavior, such individuals are often found in positions of key responsibility and are well targeted for preventative intervention.

The ''recent stress'' scale addresses the patients' perception of events in the recent past that were experienced as stressful. This is a phenomenological assessment similar to the Social Readjustment Rating Scale (Holmes and Rahe, 1967) and Sarason, Johnson, and Siegel's Life Experience Survey (1978). High scorers on this scale are assumed to have an increased susceptibility to serious illness for the year following test administration. Recent marked changes in their life predicts a significantly higher incidence of poor physical and psychological health than in the population-at-large (Andrew, 1970; Yunik, 1980; Head, 1979). Such issues are of particular importance in work settings, where preventive interventions may prove both beneficial and cost-effective.

The ''premorbid pessimism'' scale represents a dispositional attitude of helplessness-hopelessness that has been implicated in the appearance or exacerbation of a variety of diseases such as multiple sclerosis, ulcerative colitis and cancer (Mei-Tal, Meyerowitz and Engel, 1970; Paull and Hislop, 1974; Schmale, 1972; Stavraky, Buch, Lott, and Wanklin, 1968). It differs from other ''depression'' indices by noting characteriologic tendencies toward viewing the world in a negative manner. High scorers on this scale are disposed to interpret life as a series of troubles and misfortunes and are likely to intensify the discomforts they experience with real physical and psychological difficulties (Yunik, 1980; Levine, 1980). Again, this scale is of particular salience in identifying individuals whose personal and work performance may suffer as a consequence of these feelings.

The ''future despair'' scale focuses on the individuals' willingness to plan and look forward to the future (Engel, 1968; Wright, 1960). This is more likely than the previous scale to tap the person's response to current difficulties and circumstances rather than a general or lifelong tendency to view things negatively. High scorers do not look forward to a productive future life and view medical difficulties as seriously distressing and potentially life-threatening.

''Social alienation'' looks at level of familial and friendship support, both real and perceived, which appears to relate to the impact of various life stressors (Cobb, 1977; Rabkin and Struening, 1976). This sense of aloneness had been detailed in sociological literature (Berkman, 1967; Comstock and Partridge, 1972; Moss, 1977; Parkes and Fitzgerald, 1969). High scorers are prone to physical and psychological ailments. A poor adjustment to hospitalization is also common. These individuals perceive low levels of family and social support and may not seek medical assistance until illness is extremely discomforting. Being able to target these individuals would allow the development of alternative support systems for such individuals.

All of the above stressors seem to be significantly modulated upward or downward by the preoccupations and fear that individuals may express about their physical state, a characteristic addressed in the "somatic anxiety" scale. Studies of what may be called somatic anxiety reflect the general concerns that people have about their bodies (Lipsitt, 1970; Lowy, 1977; Lucente and Fleck, 1972; Mechanic and Volkart, 1960). High scorers on this scale tend to be hypochondriacal and susceptible to various minor illnesses. They experience an abnormal amount of fear concerning bodily functioning and are likely to overreact to the discomforts of surgery and hospitalization (Green, Meagher, and Millon, 1980). Such preoccupations often lead personnel to excessive absences from work.

The next set of three scales was derived empirically. They have been labeled the "psychosomatic correlates" scales and are designed for use only with individuals who have previously been medically diagnosed as exhibiting one of the following specific disease syndromes: allergy, gastrointestional problems, cardiovascular difficulties. The scores of each scale gauge the extent to which the person's responses are similar to comparable diagnosed patients whose illness has been judged substantially psychosomatic or whose course has been complicated by emotional or social factors.

The last three of the empirically derived scales, these labeled "prognostic indices," seek to identify future treatment problems or difficulties that may arise in the course of the individual's illness. The scores of each scale—"pain treatment responsivity," "life threat reactivity," and "emotional vulnerability"—gauge the extent to which the person's responses are similar to patients whose course of illness or treatment has been more complicated and unsatisfactory than is typical.

The MBHI was developed following procedures recommended by Loevinger (1957) and Jackson (1970). A detailed explanation of test development may be found in the MBHI manual.

At four and one-half months the coping style scales showed reasonably high test-retest reliabilities, with most in the range of .77–.88 and a mean of .82. The psychogenic attitude scales also show high reliabilities, averaging around .85, as do the empirically derived scales at about .80 with the single exception of emotional vulnerability. KR20s were calculated as the optimal method for addressing internal consistency. The KR20 coefficients for all scales ranged from .66 to .90 with a median of .83.

Correlational data have been obtained employing a variety of different and often homogeneous patient and nonpatient samples. Among the inventories used were the MMPI, the SCL-90, I-E Scale, Beck's Depression Inventory, the Personal Orientation Inventory, the Life Events Survey, the Webber-Johansson Temperament Survey, and the California Personality Inventory; the results are reported at length in the MBHI manual (Millon, Green and Meagher, 1982).

The MBHI addresses interpersonal style, attitudes shown to be significant to the management of health concerns, likelihood of psychological components of medical problems, and specific prognostic issues. It uses this information to make probable statements about the individual's behavior in relation to illness, its management, and health care personnel. Directly addressing specified disease processes and their management it provides the basis for making recommendations across a variety of medical and health problems regarding the likelihood of illness occurring and probable progress as well as optimal management of the disease process. These facts make the MBHI an instrument for prime consideration in occupational health promotion efforts.

Summary

The role of psychological evaluation of individuals within the work setting has a long, well-established history as regards worker satisfaction and productivity. Utilizing psychological assessment tools for predicting and optimizing the physical health of workers is in its earliest stages.

This review of self-report inventories provides the practitioner with an overview of the state-of-the-art. Ranging from simple checklists to single-trait assessments and multidimensional personality inventories, these instruments provide a variety of diagnostic options for the clinician. Unfortunately, even within this selected group, construction and postconstruction validation results have often proved disappointing. It is critical that the clinician utilizing a given instrument demonstrate caution in both application and interpretation. Employing tests developed and normed with psychiatric populations may fit old habits and may be expedient, but only rarely have such instruments proven useful in nonpsychiatric health studies. Psychological assessment utilized in this fashion is new and instruments must be carefully evaluated regarding not only their construction, but their suitability to provide answers to diagnostic and decision-making requirements that are relevant to the populations that are to be tested.

REFERENCES

Adams-Webber, J. "Perceived locus of control of moral sanctions." Master's thesis, Ohio State University, 1963.

Adcock, C. J. "Review of the MMCI." In the *6th Mental Measurements Yearbook,* Oscar Buros, ed. Highland Park, N.J.: The Gryphon Press, 1965.

Andrew, J. M. "Recovery from surgery, with and without preparatory instruction, for three coping styles." *Journal of Personality and Social Psychology,* 1970, **15**(3), 223–226.

Beck, A. T. *Depression: Causes and Treatment.* Philadelphia: University of Pennsylvania Press, 1972, 186–207.

Benton, A. L. and Probst, K. A. "A comparison of psychiatric ratings with MMPI scores." *Journal of Abnormal and Social Psychology,* 1945, **41,** 75–78.

Berkman, P. L. "Spouseless motherhood, psychological stress, and physical morbidity." *Journal of Health and Social Behavior,* 1967, **10,** 323–334.

Brand, R. J., Rosenman, R., Jenkins, C., Stoltz, R., and Zyzanski, S. "Comparison of coronary heart disease prediction in the VCGS using the structures interview and the

JAS assessments of the coronary-prone Type A behavior pattern.'' *Journal of Chronic Diseases,* in press.

Brook, A. ''Mental health of people at work.'' In *Occupational Health Practice,* R. Schilling, ed. London: Butterworths, 1972.

Butcher, J. N. and Owens, P. L. ''Objective personality inventories: Recent research and some contemporary issues.'' In *Clinical Diagnosis of Mental Disorders: A Handbook,* B. Wolman, ed. New York: Plenum Press, 1978.

Cardi, M. ''An examination of internal versus external control in relation to academic failures.'' Master's thesis. Ohio State University, 1962.

Cattell, R. B., Eber, H. W., and Tatsuoka, M. M. *Handbook for the Sixteen Personality Factor Questionnaire (16PF).* Champaign, Ill.: Institute for Personality and Ability Testing, Inc., 1970.

Cobb, S. ''Epilogue: Meditation on Psychosomatic Medicine.'' In *Psychosomatic Medicine: Current Trends and Clinical Applications,* Lipowski, Z. J., Lipsitt, D. R., and Whybrow, P. C., eds. New York: Oxford University Press, 1977.

Comstock, G. W. and Partridge, K. B. ''Church attendance and health.'' *Journal of Chronic Disease,* 1972, **25,** 665–672.

Dahlstrom, W. G., Welsh, G. S., and Dahlstrom, L. E. *An MMPI Handbook,* Vols. I and II. Minneapolis: University of Minnesota Press, 1972.

Derogatis, L. R. *SCL-90 R (Revised) Version Manual-1.* Baltimore: 1977.

Derogatis, L. R., Abeloff, M., and McBeth, C. ''Cancer patients with their physicians in the perception of psychological symptoms.'' *Psychosomatics,* 1977, **17,** 197–201.

Derogatis, L. R., Lipman, R. S., Rickels, K., Uhlenhuth, E. H., and Covi, L. ''The Hopkins Symptom Checklist (HSCL): A self-report symptom inventory.'' *Behavioral Science,* 1974(a), **19,** 1–15.

Derogatis, L. R., Lipman, R. S., Rickels, K., Uhlenhuth, E. H., and Covi, L. ''The Hopkins Symptom Checklist (HSCL): A measure of primary symptom dimension.'' In *Psychological Measurements in Psychopharmacology,* P. Pichot, ed. Basel: Karger, 1974(b).

Efran, J. S. ''Some personality determinants of memory for success and failure.'' Doctoral dissertation, Ohio State University, 1963.

Ellis, A. *Review of the MMPI; 5th Mental Measurements Yearbook.* Oscar Buros, ed. Highland Park, N.J.: The Gryphon Press, 1959, 166–167.

Engel, G. L. ''A life setting conducive to illness: The given-up–giving-up complex.'' *Bulletin of the Menninger Clinic,* 1968, **32,** 355–365.

Friedman, M. and Rosenman, R. H. *Type A. Behavior and your Heart.* New York: Knopf, 1974.

Gersten, J. C., Frii, S. R., and Lengner, T. S. ''Life dissatisfactions, job dissatisfaction and illness of married men over time. *American Journal of Epidemiology,* 1976, **103,** 333–341.

Gilberstadt, H. and Duker, J. *A Handbook for Clinical and Actuarial MMPI Interpretation.* Philadelphia: Saunders, 1965.

Good, P. and Brantner, J. *A Practical Guide to the MMPI.* Minneapolis: University of Minnesota Press, 1974.

Green, C., Meagher, R. and Millon, T. ''The management of the ''problem'' patient in the milieu setting.'' Paper presented at the Society of Behavioral Medicine Meetings, New York, November 1980.

Head, R. ''The impact of personality on the relationship between life events and depression.'' Doctoral dissertation, University of Miami, 1979.

Holmes, T. H. and Rahe, R. ''The social readjustment rating scale.'' *Journal of Psychosomatic Research,* 1967, **11,** 213.

Jackson, D. N. ''A sequential system for personality scale development.'' In *Current topics in clinical and community psychology* (Vol. 2), C. D. Spielberger, ed. New York: Academic Press, 1970.

Jackson, D. N., and Messick, S. ''Acquiescence and desirability as response determinants in the MMPI.'' *Educational and Psychological Measurement,* 1960, **21,** 771–790.

Jenkins, C. D. ''Psychologic and social precursors of coronary disease.'' *New England Journal of Medicine,* 1976, **284**(6), 307–317.

Jenkins, C.D., Zyzanski, S. J., and Rosenman, R. H. *Jenkins Activity Survey Manual.* New York: The Psychological Corporation, 1979.

Karson, S. and O'Dell, J. *Clinical Use of the 16PF.* Champaign, Ill,: Institute for Personality and Ability Testing, 1976.

Krug, S. E. *Psychological Assessment in Medicine.* Champaign, Ill.: Institute for Personality and Ability Testing, 1977.

Lambley, P. and Silbowitz, M. ''Rotter's internal-external scale and prediction of suicide contemplators among students.'' *Psychological Report,* 1973, **33,** 585–586.

Lazarus, R. ''Cognitive and coping processes in emotion.'' In *Stress and Coping,* A. Monart and R. Lazarus, eds. New York: Columbia University Press, 1977.

Levine, R. ''The impact of personality style upon emotional distress, morale and return to work in two groups of coronary bypass surgery patients.'' Master's thesis, University of Miami, 1980.

Lingoes, J. C. *Review of the MMPI; 6th Mental Measurements Yearbook.* Oscar Buros, ed. Highland Park, N.J.: The Gryphon Press, 1965, 316–317.

Lipsitt, D. R. ''Medical and Psychological Characteristics of 'Crocks.''' *International Journal of Psychiatry in Medicine,* 1970, **1,** 15–25.

Loevinger, J. ''Objective tests as instruments of psychotherapy.'' *Psychological Reports,* 1957, **3,** 635–694.

Lowy, F. H. ''Management of the Persistent Somatizer.'' In *Psychosomatic Medicine: Current Trends and Clinical Applications,* Lipowski, Z. J., Lipsitt, D. R., and Whybrow, P. C., eds. New York: Oxford University Press, 1977.

Lucente, F. E. and Fleck, S. ''A study of hospitalization anxiety in 408 medical and surgical patients.'' *Psychosomatic Medicine,* 1972, **34,** 304–312.

Marks, P. and Seeman, W. *The Actuarial Description of Abnormal Personality.* Baltimore: Williams & Wilkins, 1963.

Marston. ''Compliance with medical regimens as a form of risk taking in patients with myocardial infarctions.'' *Dissertation Abstracts International* 1969, **30,** 2151A–2152A.

Mechanic, D., and Volkart, E. H. ''Stress, Illness behavior and the sick role.'' *American Sociological Review,* 1961, **26,** 51.

Meehl, P. E. and Rosen, A. ''Antecedent probability and the efficiency of psychometric signs, patterns or cutting scores.'' *Psychological Bulletin,* 1955, **52,** 194–216.

Mei-Tal, V., Meyerowitz, S., and Engel, G. L. ''The role of psychological process in a somatic disorder: Multiple sclerosis. 1. The emotions of illness onset and exacerbation.'' *Psychosomatic Medicine,* 1970, **32,** 67–86.

Millon, T. *Modern Psychopathology.* Philadelphia: Saunders, 1969.

Millon, T. *Millon Clinical Multiaxial Inventory Manual,* 2nd ed. Minneapolis: National Computer Systems, Inc., 1982.

Millon, T., Green, C., and Meagher, R. *Millon Behavioral Health Inventory Manual.* Minneapolis: National Computer Systems, Inc., 1982.

Moss, E. ''Biosocial resonation: A conceptual model of the links between social behavior and physical illness.'' In *Psychosomatic Medicine: Current trends and Clinical Application* Lipowski, Z. J., Lipsitt, D. R., and Whybrow, P. C., eds. New York: Oxford University Press, 1977.

Parkes, M., Benjamin, B. and Fitzgerald, R. G. ''Broken heart: A statistical study of increased mortality among widowers.'' *British Medical Journal,* 1969, **1,** 740–743.

Paull, A. and Hislop, I. G. ''Etiologic factors in ulcerative colitis: Birth, death and symbolic equivalents.'' *International Journal of Psychiatry in Medicine,* 1974, **5,** 57–64.

Phares, E. J. Expectancy changes in skill and chance situation.'' *Journal of Abnormal and Social Psychology,* 1957, **54,** 339–342.

Rabkin, J. C. and Struening, E. L. ''Life events, stress and illness.'' *Science,* 1976, **194,** 1013–1020.

Rahe, R. H. ''Subjects' recent life changes and their near future illness susceptibility.'' *Advances in Psychosomatic Medicine,* 1977, **8,** 2–19.

Rosenman, R. H., Friedman, M., Strau, R., Wurm, M., Kostichek, R., Hahn, W., and Werthessen, N. T. ''A predictive study of coronary heart disease: The Western Collaborative Group Study.'' *Journal of the American Medical Association,* 1964, **189,** 15–22.

Rorer, L. G. *Review of the 16PF; 7th Mental Measurements Yearbook.* Oscar Buros, ed. Highland Park, N.J.: The Gryphon Press, 1972.

Rotter, J. B. In *Review of the MMPI, 3rd Mental Measurements Yearbook.* O. Buros, ed. Highland Park, N.J.: Gryphon Press, 1945.

Rotter, J. B. ''Generalized expectancies for internal versus external control of reinforcement.'' *Psychological Monographs,* 1966), **80**(1), 1–28.

Rotter, J. B., Liverant, S., and Crowne, D. P. ''The growth and extinction expectancies in chance-controlled and skilled tests.'' *Journal of Psychology,* 1961, **52,** 161–177.

Sarason, I. G., Johnson, J. H., and Siegel, J. M. ''Assessing the impact of life changes.'' *Journal of Consulting and Clinical Psychology,* 1978, **46,** 932–946.

Schmale, A. H. ''Giving up as a final common pathway to changes in health.'' In *Psychological Aspects of Physical Illness,* Lipowski, Z. J., ed. Basel, Switzerland: Karger, 1972.

Schmidt, H. O. ''Test profiles as a diagnostic aid: The MMPI.'' *Journal of Applied Psychology.* 1945, **29,** 115–131.

Schofield, W. ''Clinical and counseling psychology: Some perspectives.'' *American Psychology,* 1966, **11,** 122–131.

Schwartz, M. S., Osborne, D. and Krupp, N. E. ''Moderating effects of age and sex on the association of medical diagnoses and 1-3/3-1 MMPI profile.'' *Journal of Clinical Psychology,* 1972, **28,** 502–505.

Spielberger, C. D., Gorsuch, R. L. and Lushene, R. *The State-Trait Anxiety Inventory Manual.* Palo Alto, Ca.: Consulting Psychologists Press, 1970.

Stavraky, K. M., Buch, C. N., Lott, J. S., and Wanklin, J. M. "Psychological factors in the outome of human cancer." *Journal of Psychosomatic Research,* 1968, **12,** 251–259.

Wallston, K. A. and Wallston, B. S. "Health locus of control scales." In *Research With the Locus of Control Construct* (Vol. I), H. Lefcourt, ed. New York: Academic Press, 1981.

Wallston, K. A. and Wallston, B. S. "Who is responsible for your health? The construct of health locus of control." In *Social Psychology of Health and Illness,* G. S. Sanders and J. Suls, eds. Hillsdale, N.J.: Erlbaum, 1982.

Wider, A. *The Cornell Medical Index.* New York: The Psychological Corporation, 1948.

Wright, B. A. *Physical Disability:* A Psychological Approach. New York: Harper, 1960.

Wright, H. "Well person screening and executive health supervision." In *Occupational Health Practice,* R. Schilling, ed. London: Butterworths, 1973.

Yunik, S. "The relationship of personality variables and stressful life events to the onset of physical illness." Doctoral dissertation, University of Miami, 1980.

CHAPTER 17

Cost-Effectiveness and Cost-Benefit Analyses of Occupational Health Promotion

Roberta B. Hollander
Joseph J. Lengermann
Nancy M. DeMuth

This chapter discusses the application of cost-effectiveness and cost-benefit analyses to worksite health promotion programs. It explores how these techniques can be used to assess programs and justify the allocation of resources (Stason and Weinstein, 1973). More specifically, this chapter includes definitions; advantages and disadvantages of cost-effectiveness and cost-benefit analyses; steps in applying each technique; the selection of items commonly included as direct and indirect costs, effects, and benefits in these calculations; and problems inherent in cost-effective and cost-benefit analyses relative to other forms of evaluation of worksite health promotion programs (Hannan and Graham, 1978).

INTRODUCTION

In the past decade, we have witnessed a steady increase in the development of health promotion programs within occupational settings. Organizations have been attracted to these programs primarily because of their potential for reducing a broad range of health-related costs currently incurred by employers. As health care costs skyrocketed, employers found themselves taking over a greater share of these costs. From 1950 to 1970, for example, the portion of insurance coverage paid for by employers rose from 35 to 84% (Chadwick, 1979). Employers have become aware that they pay for health-related costs in other ways as well: through high absenteeism and turnover rates, low productivity, accidents, and disability claims from the employees. Fielding (1979) reports, for instance, that the average yearly loss in wages alone from cigarette smoking is three billion dollars.

In order to contain costs many organizations have established onsite health promotion programs. These range from specific interventions for targeted groups to "broad spectrum" approaches. Underlying these programs are the beliefs that the workplace is a viable setting for promoting health and that prevention pays—that is, worksite health promotion programs will yield substantial savings in health care costs to the company (Woods, 1981). Though these programs may represent an initial increase in health-related costs, it is believed that savings will be produced over time in the form of less expensive health activities and healthier employees. Cost savings are assumed to be likely because of such features of worksite programs as the broader use of nonphysician health care providers; centralization of services; the wholesale purchase of supplies and drugs; and the potential for achieving economies of scale (Alderman, 1981; Hannan and Graham, 1978; LaRosa and Ziska, 1981). As Table 17.1 suggests, in addition to these cost savings, it is believed that there may be other advantages to companies from programs. These beliefs may or may not be well-founded. One way to resolve this question is through the application of cost-effectiveness and cost-benefit analyses. To date, few companies have

Table 17.1 Potential Advantages of Worksite Health Promotion Programs

Advantages for Employers
Reduced health insurance costs
Reduced disability and death claims
Reduced treatment costs
Reduced absenteeism
Reduced on-the-job accidents
Reduced turnover rates and replacement costs
Increased productivity
Increased worker morale
Increased worker health and quality of life
Advantages for Employees
Reduced health-related costs
Reduced transportation and waiting time for health
Reduced sick leave
Increased co-worker and employer support for adherence to positive health behaviors
Increased morale based on management's concern for health
Increased satisfaction with health activities
Improved health and quality of life
Advantages for Society
Reduced health costs
Improved health and quality of life
Adoption of health promotion emphasis

committed sufficient resources to carrying out vigorous evaluation involving cost analyses. Moreover, only a few of the potential advantages listed in Table 17.1 are built into these analyses. But given an era of belt tightening and the expressed need by organizations to document program worth according to specific quantifiable measures, worksite health promotion programs will be expected to show that they are cost-effective, cost-beneficial, or otherwise make economic sense. Thus the current acceptance of estimates of potential advantages will increasingly be replaced by efforts at more systematic and precise measurement.

Table 17.1 indicates that worksite health promotion programs can also provide advantages to employees and society. Workers and health professionals therefore also have an interest in cost-effectiveness and cost-benefit analyses. They are aware that such analyses could provide the key rationale for work organizations to enhance their commitment to programs. Analyses of this order could yield data that allow programs to be developed in which each group's interests are more directly met. It is important to recognize that the various groups interested in cost-effectiveness and cost-benefits may prioritize differently the criteria on which data are collected and analyzed. Health professionals and employees think of health as an inherent "good," and program costs are assessed in terms of whether the maximum amount of health is achieved in one program compared to another. It is also asked whether a worksite health promotion program achieves as much health for the cost incurred as another approach. For most work organizations, however, health and health promotion are means of reducing health-related costs which detract from the primary organizational goals of productivity and profit. These issues are further confounded because at this point the empirical links between specific activities—programs, health outcomes, and organizational goals—are not very clear. Ironically, as these links become clearer and as cost-effectiveness and cost-benefit analyses become more sophisticiated, differences between the priorities of organizations, employees, and health professionals are likely to be made more explicit. However, this can well lead not to conflict but to assurance that worksite health promotion programs can be made mutually advantageous and deserving of cooperative commitment.

COST-EFFECTIVENESS ANALYSES OF WORKSITE HEALTH PROMOTION PROGRAMS

Definitions of *cost-effectiveness analysis* typically describe it as a procedure for assessing outputs, results, or objectives attained per dollars spent in different programs or by different strategies. DeFriese and Barry (1982, p. 144) suggest that "cost-effectiveness analysis is a technique designed for the comparison of different approaches to the achievement of the same objective with respect to program costs." Goldschmidt (1976, p. 30) defines cost-effectiveness as "the cost of producing a unit of effect within a given program."

Cost-effectiveness is expressed as a ratio: Total program costs are divided by some measure of "success," such as the number of program participants who have adopted a particular health behavior (e.g., stopped smoking, lost weight, were screened for high blood pressure). The result of these calculations are then compared with related health care costs to the company prior to the introduction of the program, or with alternative interventions (Goldbeck and Kiefhaber, 1981).

Cost-effectiveness analysis can aid in making decisions about how to use resources. For example, a hypertension control program designed to help workers reduce and maintain blood pressure levels may be assessed through cost-effectiveness analysis. Two approaches or two programs designed to meet these same objectives can be compared with regard to the cost of achieving a specified degree of success (DeFriese and Barry, 1982; Fleming, 1982). The worksite health promotion program or strategy that meets its objectives most economically, that is, is cost-effective, is likely to be maintained in contrast to programs or strategies that achieve the same objectives but at greater cost.

There are four major steps involved in calculating cost-effectiveness of worksite programs ("Cost-benefit and cost-effectiveness analysis for health promotion programs," 1982, 3–4): "Step 1: define program (intervention) objectives; Step 2: compute the program's net monetary costs; Step 3: define program outcomes; and Step 4: do a sensitivity analysis." These steps are discussed in the following sections.

Step 1—Define Program (Intervention) Objectives Objectives of worksite health promotion programs (or interventions) derive from the broader purposes or goals of programs. Goals are general statements about what a program seeks to accomplish, such as "to increase the level of information" about a particular health issue, or "to improve health" in the target population by reducing the amount of a behavior, for example, smoking. Program objectives usually are expressed in measurable terms, that is, a percentage (or numerical) change to be achieved within a specified period of time. For example, an objective of a worksite weight control program may be an average weight loss of X pounds per participant in three months. A smoking cessation program's main objective could be to reduce the number of workers who smoke by X% over the course of a year.

Selecting appropriate objectives for worksite programs involves knowledge of the needs and characteristics of the target population obtained through a needs assessment. The extent to which needs are met by worksite health promotion programs is tempered by available resources (e.g., funds and personnel) and potential barriers (e.g., educational level or cultural values of the target population). Therefore, these resources and barriers are considered when objectives are developed. Alternative activities by which objectives can be met, the responsibilities of personnel, a time frame for meeting objectives, and evaluation procedures should be provided for each objective. Program objectives developed in this fashion are more likely to be realistic and achievable.

The development of objectives for worksite health promotion programs and the ability to carry out cost-effectiveness and cost-benefit analyses is dependent on the state-of-the-art of related epidemiological evidence. There is a lack of data linking specific risk factors to disease outcomes through given mechanisms, and, conversely, linking changes in behavior (and some treatments) to reduced health risks. Certainly some evidence is more compelling than other, such as that concerning the relationship between smoking and various health problems. But for many other health issues that worksite health programs address, evidence of association remains weak (Scheffler and Paringer, 1980; Kristein, 1977; 1982). Even though the data often are only suggestive, programs in the work setting premised on the best available evidence are being established.

Step 2—Compute the Program's Net Monetary Costs Several items may be included as *costs* in calculations: total costs of the health promotion program, health-related costs to the company prior to introducing the program, as well as costs at the individual and societal levels. Costs can be direct and indirect, and may include opportunity costs, that is, whatever services or programs the company forgoes when it provides health promotion at the worksite. Costs are calculated in dollar amounts for purposes of comparisons between worksite health promotion programs or strategies.

Direct costs of a worksite program include those costs that are directly incurred in providing the program. Direct costs, Fleming suggests (1982, p. 25), "constitute the resources expended in the planning, implementation and evaluation of program (sic) and are usually attainable from budgets prepared according to market prices of inputs." Direct costs of worksite health promotion programs include salaries for personnel, cost of all services, and expenses for equipment, supplies, and facilities.

Indirect costs are more difficult to define and select. This group includes "expenditures incurred as a result of program participation," and is likely to contain items about which there is considerable question (Fleming, 1982, p. 25). In fact "unresolved debates" over what to include as indirect program costs sometimes result in their being omitted from analyses altogether (Fleming, 1982; Cost-benefit and cost-effectiveness analysis for Health Promotion Programs," 1982). Indirect costs usually include the cost of time lost from work. This can be broken down further into the time it takes to participate in the worksite program and time associated with transportation to and from the program. Costs in time depend on the arrangements for release time which the company makes with employees and where health promotion programs are situated. Clearly, one of the major advantages of health promotion programs at the worksite is that indirect costs are reduced.

Logan et al. (1981) illustrate how costs may be calculated for a worksite program. In a 1976–77 study in Toronto they compared the cost-effectiveness of treating hypertension at the workplace to regular care in the community, i.e., treatment by physicians in private practice. Cost calculations covered all medical care costs, including health system costs and patient costs. In addition to health system costs, costs to patients were also calculated. These involved participants' time, particularly time lost from work. In comparing hypertension treatment at the place of employment to usual care, they found that it was both more effective and cost-effective to carry out treatment programs in the workplace. Although health system costs per patient were higher for the worksite program ($197 vs. $129), worksite care resulted in significantly greater reductions in diastolic blood pressure levels (12.1 vs 6.5 mm Hg). Thus, "the incremental cost-effectiveness ratio of $5.63 per mm Hg for worksite care was less than the base cost-effectiveness ratio of $32.51 per mm Hg for regular care, indicating that the worksite program was substantially more cost-effective" (Logan, 1981, p. 211).

An important additional step in dealing with worksite health promotion program costs is to conceptualize and calculate them for each area of activity. This permits a fuller understanding of the nature of the program, including which components are least or most costly. Decision making is enhanced as a result. For example, Foote and Erfurt (1977, p. 340) found in examining hypertension programs in industrial and community settings that, compared with screening activities, "follow-up is the more expensive component."

Step 3—Define Program Outcomes *Outcomes* or *effects* of worksite health promotion programs may be easier to determine than costs. One of the advantages that cost-effectiveness has over cost-benefit analysis is that program effects do not have to be put into economic terms or monetarized, but rather are assumed to be of value. Effects of programs, then, can be any agreed-upon output or change, such as the number of program participants who stop smoking for a fixed period; average weight loss in a weight control program; or mean reduction in diastolic blood pressures. In more general terms effects can be measured as "additional healthy years of life" or "quality-adjusted life years" (Shepard, 1982, p. 7).

In addition to changes that are anticipated and desired, evaluators need to decide what is the degree or magnitude of change expected and when measures will be taken. These issues are not quite as simple as they may appear at first to be. For example, Green and Lewis (1981) note several ways in which the magnitude of effects can be expressed: as a percentage of the target population who made any change; as the percentage of the target population who experienced a statistically significant change; those who reached a threshold level of change essential to deriving some benefit; and those who experienced some greater level of change. Further, effects can be calculated at the individual level and/or aggregated level.

The timing of measures is crucial in that effects can also be short or long term. Most health promotion programs at the worksite can probably be shown to have some effect if measures are taken immediately, but these effects may drop off rapidly. Other effects may be long-standing. Evaluations are not often designed to measure effects sustained over a long period. What is being suggested here is that effects or outcomes of health promotion programs at the worksite are complex and require creative assessments.

Step 4—Sensitivity Analysis The final step in cost-effectiveness calculation—sensitivity analysis—involves trying to assess the impact of various uncertain costs and effects of worksite health promotion program (Shepard, 1982). Because many of the desired effects of programs are realized at a later point in time, some effort is made to discount or adjust costs and effects back to the present time, as well as to estimate costs where data are missing. For example, a major effect of a hypertension screening program at the workplace, reduced morbidity, accrues over many years, but is paid for in "today's dollars." Therefore, to assess the "true" cost-effectiveness of this program, various discount rates within different time frames are tried out. Typically high and low estimates are made. For example, Stason and Weinstein in the final stage of an analysis of hypertension treatment tested the impact on cost-effectiveness of varying discount rates ranging from 0 to 10% (1977).

Where data were incomplete, Logan et al. (1981) included in their sensitivity analysis of another worksite hypertension program highest possible costs of drugs, time, and so on, for worksite praticipants and lowest possible costs for the group in regular care. This strengthened their confidence in the finding that the worksite program was more cost-effective.

Additional Issues in Cost-Effectiveness Analysis

Program evaluation may fail to show that a health promotion program is cost-effective. The apparent lack of success of a program in this regard may be a function of one or more problems: the program may be poorly conceived or inadequately based in theory; it may be poorly designed; poorly implemented or assessed through inadequate measures (Flay and Best, 1982; Green, 1977). Program objectives, for example, may have been selected with little understanding of the target population. As a result objectives may be inappropriate or unrealistic. Moreover, as Stason and Weinstein (1977) found with regard to hypertension control, if program objectives are appropriate but participants fail to adhere to recommended procedures, the effectiveness and cost-effectiveness of such programs may be seriously compromised. Similarly, although it is fairly inexpensive and effective to carry out hypertension screening programs, if screening is not accompanied by follow-up, then the cost-effectiveness and overall value of these programs may be reduced substantially (Stason and Weinstein, 1977; Scheffler and Paringer, 1980). The question of cost-effectiveness of worksite hypertension screening-programs without subsequent referral and follow-up procedures has been raised in other studies as well. Haynes et al., for example, found in their 1972–73 study of workers at a Canadian steel company that the rate of absenteeism increased significantly for workers who were made aware of their hypertension as a result of the program, which clearly reduced its potential cost-effectiveness (Haynes et al., 1978; Taylor et al., 1981). Alderman, Charlson, and Melcher (1981) found, however, at the Massachusetts Mutual Life

Insurance Company that, although the absenteeism rate among newly aware hypertensives also rose, the increase was not statistically significant, and was more modest than that for Canadian workers.

Other conditions may contribute to worksite health promotion programs being cost-effective at the outset but yielding diminishing returns over time. Programs in their early stages when "innovators" or early adopters can be reached through low-cost methods may be more effective and cost-effective than at a later period when personnel try to reach high-risk or hard-to-reach groups by more costly methods (Green, 1977).

Some programs that are not initially cost-effective or only marginally so, may become cost-effective in a relatively short time. Ruchlin and Alderman (1980) suggest that some costs associated with health promotion porgrams in the work setting are highest during the program's first year and then taper off; others may be lower initially and rise with time. In their cost analyses of hypertension control programs at the workplace the authors found that two major costs in particular fluctuated considerably during a five-year period. The portion of operating costs for revisits to health care providers was highest during the first year, but as blood pressure control was achieved the number of revisits dropped off. The optimal number of five revisits per year for each participant was reached during the fourth or fifth years of care. Drug costs, however, for the same programs tended to be lowest during the first year of treatment, increased greatly during the second year, and rose more moderately through the third and fourth years. These increases in drug costs can most likely be attributed to greater use of drugs over time or a change in drugs prescribed to more costly types. Even in light of fluctuations in two items that figure centrally in the cost of hypertension control programs, Ruchlin and Alderman found that total operating costs for worksite programs were lower than costs of hypertension treatment for both private care and care in hospital outpatient departments.

There are several other possible explanations why there may be a lag in time until a worksite health promotion program is cost-effective. Start up expenditures for some types of health promotion programs can be particularly high. It is expected, however, that as the popularity of a program increases and equipment and other resources are used on a continuing basis, program costs will be equalized.

The ability to centralize efforts (often using a site available in the organization), to make more use of health providers other than physicians, and to serve large populations are essential aspects of worksite health promotion programs that are likely to contribute to cost-effectiveness. As more and more employees use in house services, companies may realize health cost savings by achieving economies of scale. However, for some items the scale of the program (and parent organization) will have little impact on the cost per participant; these are set and cannot be reduced past a certain point to yield additional savings. Other items represent potential savings for companies. Sometimes an item can be viewed as something of a mixed case: initially it contributes to cost savings, but at one point the cost per program participant cannot be reduced further. The wholesale purchase of drugs is one example (Ruchlin and Alderman, 1980).

Health promotion programs most likely to be cost effective are those established by large companies, which are self-insured for health care costs and which come to achieve economies of scale. Smaller companies tend to have greater difficulty in attaining cost-effectiveness—the number of participants in these organizations simply may not be adequate to support the program (Chadwick, 1979; Hannan and Graham, 1978). By the same token programs with more individualized approaches may be effective in meeting objectives, but too costly to be cost-effective (Green, 1977).

COST-BENEFIT ANALYSES OF WORKSITE HEALTH PROMOTION PORGRAMS

Most steps in cost-benefit analysis look very much like cost-effectiveness. Cost-benefit analysis also builds around objectives, costs, and outcomes (benefits). However, the critical difference comes in Step 3 in the monetarization of benefits into the same medium as used in measuring costs, namely dollars. As a consequence some of the problems associated with cost-effectiveness analyses of worksite health promotion programs are magnified many times over. The value of outcomes that are abstract and fluid is not easily monetarized. Cost-benefit analyses often involve some very questionable assumptions about the relative monetary worth of intangible issues such as the value of life, the quality of life, or suffering averted. Limitations in the availability of sound epidemiological data that were earlier discussed as a problem in cost-effectiveness analyses limit even more severely our ability to do cost-benefit analyses of worksite programs. Not only must benefits be of value but the value must be quantifiable into definite dollar amounts instead of being assumed to be of general value. It is this feature that contributes to cost-benefit analysis being applied very infrequently to health promotion programs in the worksite.

As indicated, the estimation of costs is done similarly as in cost-effectiveness, taking into account both direct and indirect costs as well as allowing for opportunity costs. However, benefits (effects) are considered differently because they are monetarized instead of being expressed in terms of substantive outcome units. These monetarizations should be done for all categories of benefits, such as direct and indirect, short-term and long-term. *Direct benefits* are considered to be the reduction in costs or costs averted as a result of the program (Tolpin, 1980; Fleming, 1982). More specifically, direct program benefits involve savings in costs for health and medical services. Though this initially seems relatively simple, in fact there is some uncertainty about what portion of health and medical costs should be included in cost-benefit calculations (Kristein, 1982). Direct benefits may include cost savings for any aspect of health services: prevention efforts, screening, treatment, research-related activities, supplies (including drugs, etc.) (Bootman et al., 1979).

Indirect benefits of worksite health promotion programs are even more difficult to assess. Bootman and associates (1979, p. 136) refer to indirect benefits as "the result of the avoidance of earnings and productivity losses that would have been borne without the health program in question." They refer to a method used by Rice (Bootman et al., 1979) by which indirect benefits are measured that includes consideration of "wage and productivity losses resulting from illness, disability and death, based on age and sex for major causal categories of morbidity and mortality." This method of calculating indirect benefits of worksite health promotion programs introduces other problems in that it values lives differentially based on age and sex variables. As a result, employees who earn less (and have less earning potential), such as women, blacks, and others, are likely to be undervalued in these calculations independent of their risks and levels of morbidity (and mortality) (Schleffler and Paringer, 1980, p. 475).

The problem with calculating indirect benefits is further complicated by uncertainty over what discount rate should be applied. Given that most benefits of worksite health promotion programs will be realized many years later, calculating benefits at present rates results in underestimating them (Kristein, 1982). The fact that many benefits of these programs are intangible, thus difficult to calculate in economic terms, and accrue at a later point in time may make it particularly difficult for companies who wish to use cost-benefit techniques now to justify allocating resources for onsite health promotion programs. As a result, many authors concerned with worksite programs agree that widespread use of cost-benefit analysis in this area is probably premature. Green (1977, p. 160), for instance, notes that "we are so far from having adequate data to compute comparable ratios that there will be few applications of this criterion as an administrative decision tool for the near future." However, if decisions could be reached about the dollar-value equivalents of benefits, the technique offers the advantage of allowing direct comparisons to be made between program benefits and the costs of benefits. As Rossi and Freeman (1982, pp. 273, 275) point out, even if some of the monetarized values and conclusions are questionable, and even if benefits and costs tend to be defined differently from different perspectives, a main strength of the approach "lies in its discipline with which it forces the evaluator, policy maker, planner, and manager to articulate economic considerations that might otherwise remain implicit or unstated."

Summary

In some respects, cost-effectiveness and cost-benefit analyses of worksite health promotion programs may be seen as simply special cases of the broader category of health and social program evaluation. To this extent, these analyses must contend with the standard questions, caveats, and limitations inherent in other forms of program evaluation, including:

1. Program administrators, company managers, and others tend to be shortsighted with respect to the need for evaluation. They often fail to recog-

nize the importance of including *a priori* a cost-effectiveness and/or cost-benefit component as part of the broader program planning, implementation, and evaluation cycle. When cost analysis is included, it is likely to consist only of simple accounting procedures.

2. Evaluators, program administrators, employees, and management represent competing perspectives with regard to the need for worksite health promotion programs altogether and related evaluative efforts. These perspectives influence the process at every point, including the uses to which findings are put.
3. It is difficult to carry out rigorous evaluation on programs established within natural settings. Issues central to research of this order that affect control and statistical power are difficult to resolve in these settings (Flay and Best, 1982; Green and Figa-Talamanca, 1974).
4. Finally, it is very difficult to assess quantitatively abstract elements of programs. Persons involved in worksite health promotion programs often express a feeling that they know something positive is taking place, but these feelings may be very hard to substantiate through cost-effectiveness and cost-benefit analyses.

C. C. Wright, director of Health Services for the Xerox Corporation notes for example that while formal studies indicated some important changes had taken place for participants in a worksite exercise program, informal indicators may be the most telling.

> In short, the Physical Fitness Programs are considered to be effective in cost containment because so many individuals believe they are (Wright, 1982, p. 967).

There are limitations inherent in all evaluation techniques. One way to offset some of the difficulties involved in applying the cost analytic procedures described here is to combine these with other forms of evaluation. By doing so, complex processes and outcomes of worksite programs may be open to more understanding. Ultimately, as DeFriese and Barry note (1982, p. 145), decisions about worksite health promotion programs "would seem to require a good measure of informed judgement in addition to the results of cost-benefit/effectiveness studies."

Cost-effectiveness and cost-benefit can bring to the program a degree of order, a framework. Even where results of such analyses are open to doubt, the fact of including these techniques forces some order on the process (DeFriese and Barry, 1982). It encourages those involved to regard evaluation as essential, to develop objectives that are realistic and measurable (through needs assessments of the target population), to select activities by which objectives can be met, and to specify indicators and measures of program success.

In the final analysis even programs that are demonstrated to be cost-effective and/or beneficial must also be clinically important and socially meaningful (Green and Lewis, 1981). Ideally, worksite health promotion programs increasingly will be designed to meet all these criteria.

REFERENCES

Alderman, M. H. ''Improved compliance with antihypertensive therapy: lessons from the work site.'' *Cardiovascular Reviews and Reports,* September 1981, **2**(9), 909–917.

Alderman, M. H., Charlson, M. E., and Melcher, L. A. ''Labeling and absenteeism: the Massachusetts Mutual experience.'' *Clinical and Investigative Medicine,* 1981, **4**(3/4), 165–171.

Alderman, M. H. and Davis, T. K. ''Blood pressure control programs on and off the worksite.'' *Journal of Occupational Medicine,* March 1980, **22**(3), 167–170.

Alderman, M. H., Green, L. W. and Flynn, B. S. ''Hypertension control programs in occupational settings.'' *Public Health Reports,* March–April 1980, **95**(2), 158–163.

Alderman, M. H. and Miller, K. F. ''Blood pressure control: the effect of facilitated access to treatment.'' *Clinical Science and Molecular Medicine,* 1978 **55** suppl., 349S–351S.

Alderman, M. H. and Schoenbaum, B. A. ''Detection and treatment of hypertension at the worksite.'' *New England Journal of Medicine.* July 10, 1975, **293**(2), 65–68.

Anderson, D. R. and Fleming, P. L. ''Estimating cost savings from the Staywell program: a cost impact model.'' Control Data Corporation, 1982, Hrr 421–482.

Boden, L. I. ''Cost-benefit analysis: caveat emptor.'' *American Journal of Public Health,* 1979, **69**(12), 1210–1211.

Bootman, J. L., Rowland, C., and Wertheimer, A. I. ''Cost-benefit analysis: a research tool for evaluating innovative health programs.'' *Evaluation and the Health Professions,* 1979, **2**(2), 129–154.

Brennan, A. J. J. ''Health promotion in business: caveats for success.'' *Journal of Occupational Medicine,* September 1981, **23**(9), 642–659.

Chadwick, J. H. ''Costs and cash benefits of heart disease programs at work.'' In *State of the art papers.* National conference on health promotion programs in occupational settings. Office of Health Information and Health Planning, U.S. Department of Health, Education, and Welfare, Public Health Service, 1979.

Christianson, J. B. and Bender, S. G. Benefit-cost analysis and medical care delivery system change: an application to rural hospital closure.'' *Evaluation Review,* 1982 **6**(4), 481–504.

Cost-benefit and cost-effectiveness analysis for health promotion programs. *Baseline,* September 1982, **1**(2), 1–5.

Dedmon, R. E. et al. ''Employees as health educators: a reality at Kimberly-Clark.'' *Occupational Health and Safety.* April 1980.

Dedmon, R. E., Smoczyk, C. M., and Konkol, P. D. ''Kimberly-Clark's health management program: results and prospects.'' October 1981.

DeFriese, G. H. and P. Z. Barry. ''Questions about costs, benefits, and the effectivenesss of health promotion programs.'' *Mobius,* July 1982, **2**(3), 142–146.

Erfurt, J. C. and Foote, A. ''Final Report: Hypertension control in the work setting.'' University of Michigan-Ford Motor Company demonstration program, June 1982, Contract No. NO1-HV-8-2913.

Fielding, J. E. ''Effectiveness of employee health improvement programs, *Journal of Occupational Medicine,* November 1982, **24**(11), 907–916.

Fielding, J. E. ''Preventive medicine and the bottom line.'' *Journal of Occupational Medicine,* February 1979, **21**(2), 79–88.

Flay, B. R. and Best, J. A. ''Overcoming design problems in evaluating health behavior programs.'' *Evaluation and the Health Professions,* 1982, **5**(1), 43–69.

Fleming, P. L. ''Cost effectiveness/Cost benefit analysis strategies.'' *The Community Nutritionist,* March–April 1982, **1**(4), 23–27.

Foote, A. and Erfurt, J. C. Controlling hypertension: a cost-effective model. *Preventive Medicine,* 1977, **6,** 319–343.

Goldbeck, W. B. and Kiefhaber, A. K. ''Wellness: the new employee benefit.'' *Group Practice Journal,* March 1981, **30**(3), 20–21, 24–27.

Goldschmidt, P. G. ''A cost-effective model for evaluating health care programs: applications to drug abuse treatment.'' *Inquiry,* 1976, **XII**(1), 29–47.

Green, L. W. ''Evaluation and measurement: some dilemmas for health education.'' *American Journal of Public Health,* February 1977, **67**(2), 155–161.

Green, L.W. and Figa-Talamanca, I. ''Suggested designs for evaluation of patient education programs.'' *Health Education Monographs,* 1974, **2**(1), 54–71.

Green, L. W. and Lewis, F. M. ''Issues in relating evaluation to theory, policy, and practice in continuing education and health education.'' *Mobius,* April 1981, **1**(2), 46–58.

Hannan, E. L. and Graham, J. K. ''A cost-benefit study of a hypertension screening and treatment program at the work setting.'' *Inquiry,* December 1978, **XV**(4), 345–358.

Haynes, R. B. et al. ''Increased absenteeism from work after detection and labeling of hypertensive patients.'' *New England Journal of Medicine,* October 5, 1978, **299**(14), 741–744.

''High blood pressure control in the work setting: issues, models, resources.'' Proceedings of the National Conference, October 14, 1976, Washington, D.C.

Kristein, M. M. ''The economics of health promotion at the workplace.'' *Health Education Quarterly,* **9,** supplement, 1982, 27–36.

Kristein, M. M. ''Economic issues in prevention.'' *Preventive Medicine,* 1977, **6,** 252–264.

La Rosa, J. H. and Ziska, D. ''Lowering employee blood pressure and costs.'' *Nation's Business,* December 1981, 72, 74.

Lave, J. R. and Lave, L. B. ''Cost-benefit concepts in health: examination of some prevention efforts.'' *Preventive Medicine,* 1978, **7,** 414–423.

Logan, A.G. et al. Cost-effectiveness of a worksite hypertension treatment program. *Hypertension,* March-April 1981, **3**(2), 211–218.

Logan, A. G. et al. ''Work-site treatment of hypertension by specially trained nurses.'' *Lancet,* December 1, 1979, **II**(8153), 1175–1178.

Morris, J. N. ''Epidemiology and prevention.'' *Millbank Memorial Fund Quarterly/Health and Society,* 1982, **60**(1), 1–16.

Naditch, M. P. ''The Staywell Program: health enhancement at work,'' 1982.

Parkinson, R. S. et al. *Managing health promotion in the workplace: Guidelines for implementation and evaluation.* Palo, Alto, Ca.: Mayfield, 1982.

Preceedings of the National Conference on Health Promotion in Occupational Settings. 1979, U. S. Department of Health, Education and Welfare.

Romans, J. T. "The economic evaluation of mental health programs." In *Program Evaluation in the Health Fields* (Vol. II) H. C. Schulberg and F. Baker, eds. N.Y.: Human Sciences Press, 1979, 238–250.

Rossi, P. and Freeman, H., *Evaluation: A Systematic Approach,* Second Edition, Beverly Hills, California: Sage, 1982.

Ruchlin, H. S. and Alderman, M. H. "Cost of hypertension control at the workplace." *Journal of Occupational Medicine,* December 1980, **22**(12), 795–800.

Ruchlin, H. S. and Alderman, M. H. *Cost of worksite hypertension treatment,* November 1980, U. S. Department of Health and Human Services, Public Health Service. National Institutes of Health, NIH Publication No. 81-2115.

Scheffler, R. M. and Paringer, L. "A review of the economic evidence on prevention." *Medical Care,* May 1980, **XVIII**(5), 473–484.

Shepard, D. S. et al. "Cost-effectiveness of interventions to improve compliance with antihypertensive therapy." Presented at the National Conference on High Blood Pressure Control, Washington, D.C., April 4–6, 1979.

Shepard, D. S. "Cost-effectiveness of preventive health programs." Presented at Conference on Research on Cost-Effectiveness and Cost-Offset of Mental Health Programs, Washington, D.C., June 21–22, 1982.

Shepard, D. S. "Incentives for not smoking: experience at the Speedcall Corporation, First Report." Presented in part at the Corporate Commitment to Health: First Executive Conference, Washington, D.C., June 9–10, 1980.

Stason, W. B. and Weinstein, M. C. "Allocation of resources to manage hypertension." *New England Journal of Medicine,* March 31, 1977, **296**(13), 732–739.

Taylor, D. W. et al. "Long-term follow-up of absenteeism among working men following the detection and treatment of their hypertension." *Clinical and Investigative Medicine,* 1981, **4**(3/4), 173–177.

Tolpin, H. G. "Economics of health care." *Journal of the American Dietetic Association,* March 1980, **76,** 217–222.

U. S. Department of Health and Human Services, Health Education Branch, Office of Prevention, Education, and Control; National Heart, Lung, and Blood Institute. *Cardiovascular primer for the workplace.* January 1981, NIH Publication No. 81-2210.

Woods, B. and Hamilton, G. "Case study in HBP control: General Motors Corporation." *Occupational Health Nursing.* November 1981, **29**(11), 43–45.

Wright, C. C. "Cost containment through health promotion programs." *Journal of Occupational Medicine,* December 1982, **24**(12), 965–968.

Yates, B. T. "The theory and practice of cost-utility, cost-effectiveness, and cost-benefit analyses in behavioral medicine: toward delivering more health care for less money." In *The Comprehensive Handbook of Behavioral Medicine* (Vol. 3), J. Ferguson and C. B. Taylor, eds. Jamaica, N.Y.: S & P Scientific, 1980, 165–205.

CHAPTER 18

Future Directions in Occupational Health Promotion

Robert H. L. Feldman
George S. Everly, Jr.

A wide range of programs and issues concerning occupational health promotion have been discussed and demonstrated in previous chapters. The support and interest of management and labor to these programs indicates an increased commitment to the maintenance of health promotion programs in work settings.

The effectiveness of these programs to make meaningful changes in the health behavior and health status of workers has been demonstrated in programs such as smoking cessation programs, weight control programs, and blood pressure control programs (Alderman et al., 1982). Preliminary findings on cost-effectiveness indicate that occupational health promotion programs are an effective means of reducing health costs. Since cost-effectiveness results are beginning to accumulate, more studies are needed before definite conclusions can be reached.

Health professionals need to be cautious about overselling occupational health promotion programs. Results of programs are encouraging but not conclusive, and more evaluations are needed. Also, health professionals in their quest to effect health behavior change should avoid "victim blaming." Personal health behavior is due to a complex set of factors and individuals are not to blame for their unhealthy life-styles. In addition, health professionals involved in work settings need to be especially aware of work hazards in order to ensure that employees work in a safe and healthy environment. For example, a smoking cessation program for asbestos workers would have minimum impact if at the same time exposure to asbestos fibers were not reduced to a safe level.

FUTURE DIRECTIONS

Other Populations

The future of occupational health promotion programs is encouraging. Programs are expanding to meet the needs of groups who previously have not received programs.

For example, American Telephone and Telegraph (AT&T) has been one of the few corporations that has addressed a major women's health issue: breast cancer and breast self-examination (Parkinson, Denniston, Baugh, Dunn and Schwartz, 1982). Breast cancer is seen as the most serious health concern of women and AT&T developed a program to increase monthly breast self-examination among its women employees. Other women's health issues include cervical cancer, safe and healthy working conditions for pregnant women, and adequate child care facilities. As women become more organized and establish political clout, more resources will be devoted to women health issues in the workplace.

Since the health and well-being of workers are related in many cases to the health and well-being of their families, occupational health promotion programs are expanding to include workers' families. Corporations are selectively opening their health promotion programs to spouses and children of their employees. Physical fitness programs and health screening programs are being made available to workers and their families. As part of a comprehensive occupational health promotion program some corporations have expanded their programs to involve members of the greater community. This outreach into the community is a possible future direction that programs may take.

As affirmative action programs and other government policies have encouraged the employment of disabled workers, industries have developed programs to meet the needs of this population. Though few occupational health promotion programs are specifically designed for disabled workers, this growing segment of the work population is demanding programs of its own.

Another growing segment of the workforce is older workers. Since older workers utilize health care facilities at greater rates than younger workers, health promotion programs are especially essential for this group. Programs developed for older workers and outreach programs for retired workers are new dimensions of occupational health promotion programs.

Since most corporations are in multicultural multiethnic settings, health programs that target workers from minority and non-English-speaking groups are reaching out to an important segment of the workforce. Health problems may differ for various ethnic groups, therefore requiring different programs for different types of workers.

Expansion of Services

As various populations are being considered for occupational health promotion, new and expanded services are also being offered. Health risk assessments, and more sophisticated screening programs are being offered. Programs on self-care, first aid, CPR, and the clearing of obstructed airways are increasingly being offered. Since back injuries are one of the most common disabling injuries in industry, programs in the care of the back are expanding in industry. In addition, psychological, mental health, and stress management (including the management of stress both on the job and off the job) programs are increasing in popularity.

Coordination of Services

A trend that is seen in occupational health promotion programs is the coordination of services between corporate-run programs and other groups: Cooperation between labor and management (Foote and Erfurt, 1976) was instrumental to the success of a worksite hypertension program. Collaboration between industry and universities—University of Michigan and Ford Motor Company (Alderman, Green, and Flynn, 1982)—demonstrates the utility of joint ventures. Government-industry cooperation can also produce beneficial results. In addition, collaboration between industry and health care providers in the larger community (Murphy, 1976) has shown to improve worker health.

Technological Advances

The age of the computer and microcomputer has also affected occupational health promotion. Health risk assessments generated on computers can yield accurate appraisals of individual health risks as well as appropriate risk reduction strategies (Ellis and Raines, 1983). Health screening programs and education interventions can be individualized by utilizing computer-based programs (Naditch, 1983). For example, individualized physical fitness programs have been developed utilizing a computer-based program (Felts, Feldman, and Dotson, 1983).

THE HEALTHY WORK ENVIRONMENT AND THE CULTURE OF HEALTH

As occupational health promotion programs have expanded throughout industry, interest has increased about how to make the work environment conducive to a healthy life-style (see Table 18.1). For example, to ensure a clean and smoke-free environment some companies have prohibited smoking on their premises. To reduce the stresses of work due to a rigid organizational structure, some government agencies have instituted ''flexitime.'' Flexitime allows workers flexibility in choosing their work hours. For example, rather than having all employees work from nine to five, workers have the option to begin work between the hours of seven and eleven and to leave work between the hours of three and seven (as long as they work an eight-hour day.) The time from eleven to three is common to all employees, so that meetings and other common activities can be arranged. Flexitime has been found to increase worker satisfaction by reducing one of the aspects of occupational stress.

To encourage more nutritious food habits, some organizations are including

Table 18.1 The Health-Conscious Work Environment

1. Smoking:
   Prohibited at the workplace
2. Alcohol use:
   Discouraged, nonalcoholic beverages available
3. Nutrition:
   a. Meals: salad and fruit bars available
   b. Snacks: nutritious snacks available, nutritious foods available in vending machines
   c. Caffeine: noncaffeine beverages available
   d. Other: information on the nutritive value of foods is available
4. Exercise:
   Exercise time
   Miniexercise breaks
   Equipment and facilities available
5. Reducing occupational stress:
   Flexitime
   Four-day work week
   Expanded vacation time
   Liberal maternity and paternity leave
   Job sharing
   Extended child care
   Nondiscriminatory practices
   Worker participation and consultation
6. Other:
   Publicizing health messages
   Requiring and encouraging physical examinations
   Assessibility of health care providers
   Availability of health educators and health education materials

juice, fresh fruits, and bran muffins as morning snacks (Cunningham, 1982). Company cafeterias are offering fruit and salad bars and low-calorie, low-fat, and low-sodium food alternatives. To inform employees of the nutrient content of the food they eat, some company cafeterias are posting nutrition information regarding the food being served. The nutrition information includes the number of calories, amount of fat, cholesterol, sugar, sodium, and percentage of U.S. Recommended Daily Allowances of selected vitamins, calcium, and iron.

Good nutrition can be supported as well by offering nutritious snacks in vending machines. More vending machine companies are including juice as well as soda, low-fat yogurt as well as ice cream, unsalted nuts as well as potato chips, and so forth. An alternative to snacking that offers a break from work is exercise. Rather than taking a ten-minute donut-and-coffee break, some health-conscious workers are taking ten-minute exercise breaks. A walk around the block, the use of a stationary bicycle, and ten-minute in-place exercises are alternatives to high-calorie, high-fat snacks.

A way that companies can promote a health-conscious work setting is by expanding the lunch break. Bonnie Bell Cosmetics Company expanded their lunch break from a half-hour to an hour so that employees could have time to exercise and have a healthy lunch meal. To promote exercise among its workers, Bonnie Bell installed an exercise room with a shower and locker facilitiy and encouraged its employees to utilize a nearby park for jogging, running, and biking. The value of health and fitness is so strong that office employees often wear their warm-up clothes at their desks. In addition, Bonnie Bell Cosmetics' commitment to fitness has lead it to sponsor a series of long-distance women's races. Due to these changes over a five-year period, Bonnie Bell reports no increase in health insurance costs, a reduction in absenteeism, increases in productivity and morale, and an improvement in the general well-being of its employees (Cunningham, 1982).

Companies that support a health consciousness or culture of health may reap additional benefits. Attitudes and values concerning drinking and driving are changing in the United States. Employees attending cocktail parties and "happy hours" are frequently choosing nonalcoholic beverages. Company parties and social events are serving alternatives to alcoholic and caffeine beverages such as apple cider, juices, and herb tea.

Since occupational stress plays an important role in the health and well-being of workers, a number of changes in the workplace can minimize this stress. In addition to flexitime and 4-day (10 hours a day) work weeks, some companies have expanded vacation time, including offering both winter and summer vacations. To reduce the added burden of combining child-rearing with employment, progressive companies have instituted liberal maternity leave (and paternity leave, e.g., New York City school system), job sharing, and extended child care. Laws and a climate of tolerance for diversity have lead to less discrimination in corporate settings against individuals of different gender, ethnicity, physical condition, or life-style. These reductions in discriminatory practices have had a positive effect in decreasing stress in the workplace.

Since little or no participation in decision making at work is an important source of occupational stress (Cooper and Marshall, 1976), increased involvement of employees in decisions that affect their work has been suggested. Consultation with employees, participation in decision making and democratization of workplace practices should reduce many of these sources of occupational stress.

Additional means of promoting within the work setting a culture of health are publicizing health messages (e.g., posters, signs, notices with pay checks), requiring or encouraging periodic physical examinations (e.g., medical, dental, eye, gynecological), making health care providers (e.g., physicians, nurses, dentists, physical therapists) accessible for consultation either at the worksite or in nearby offices, and last, having health educators and health education materials readily available at the workplace.

The enthusiasm and support of corporate mangement and workers to health promotion programs has been accompanied by changes in norms and values concerning health living. New cultural patterns are emerging regarding a healthy lifestyle and this is reflected in the making of the workplace a healthy place to be. By promoting positive health practices managers and workers can benefit alike.

Concluding Remarks

> To be adopted for widespread use, surgical techniques must be proven safe. Drugs must be proven both safe and effective. But preventive programs must be proven not only safe and effective, but also highly cost-effective
>
> William Foege, in his address to the APA Public Policy Forum, Jan. 20, 1983.

In his treatise on human health, *The Survival of the Wisest,* Jonas Salk argues that we have left the era in which humanity was threatened by microbial disease, only to enter an era where the greatest threat to human health is humankind itself. In the same breath, however, Salk tells us that we can exercise our power of choice and self-direct our future by eliminating threats to health such as pollution, overcrowding, and the exhaustion of natural resources. Furthermore, we can help ourselves by beginning to promote health through the encouragement of health promoting behaviors.

In order for business and industry, as well as society at large, to begin to seriously encourage health promotion, a perceived reinforcement must be available. Are reinforcers such as altruism, or self-preservation, sufficient to do the job? Or as William Foege suggests, must there be an economic incentive as well? The answer to this question is unclear at this time. However, should it become clear that economic incentives are necessary to support the trend that has become known as health promotion, then let us seek to document such benefits. But what if we should find that some health promotion programs simply cannot become cost-effective? Do we then in some way provide economic subsidy? The issue is more than an economic one. According to Salk, the issue is one of survival.

REFERENCES

Alderman, M., Green, L. W., and Flynn, B. S. ''Hypertension control programs in occupational settings.'' In *Managing health promotion in the workplace: Guidelines for implementation and evaluation,* R. S. Parkinson and Associates, Palo Alto, Ca., 1982.

Cooper, C. L. and Marshall, J. ''Occupational sources of stress: A review of the literature relating to coronary heart disease and mental ill health.'' *Journal of Occupational Psychology,* 1976 **49,** 11–28.

Cunningham, R. M., Jr. *Wellness at Work: A Report on Health and Fitness Programs for Employees of Business and Industry*. Chicago: Inquiry (Blue Cross Blue Shield Association), 1982.

Ellis, L. B. M. and Raines, J. R. ''Health risk appraisal: A tool for health education.'' *Health Education,* 1983, **14**(6), 30–34.

Felts, W. M., Feldman, R. H. L., and Dotson, C. O. ''Exercise, health and the microcomputer.'' *Health Education,* 1983, **14**(6), 39–42.

Foote, A. and Erfurt, J. ''A model system for high blood pressure control in the work setting.'' In *High Blood Pressure Control in the Work Setting.* National High Blood Pressure Education Program, Washington, D.C., October 1976.

Murphy, A. F. ''The Burlington Industries industrial hypertension program.'' In *High Blood Pressure Control in the Work Setting: Issues, models, resources.* National High Blood Pressure Education Program, Washington, D.C., October 1976.

Naditch, M. ''The application of advanced technology to behavioral medicine in the workplace: A case study.'' Presented at the meeting of the Society of Behavioral Medicine, Baltimore, March 1983.

Parkinson, R. S., Denniston, R. W., Baugh, T., Dunn, J. P., and Schwartz, T. L. ''Breast cancer: Health education in the workplace.'' *Health Education,* 1982, **9** (Suppl.), 61–72.

Salk, J. *Survival of the Wisest.* New York: Harper and Row, 1973.

Appendix Three Examples of Multidimensional Occupational Health Promotion Programs

In the Appendix, we have provided the reader with three examples of ongoing, multidimensional occupational health promotion programs. In our opinion, these programs represent differing, yet useful examples of how health promotion efforts may be realized in the workplace. We will highlight Control Data's Staywell Program, IBM's Health Education Program, and Johnson & Johnson's Live for Life Program.

Appendix Control Data's STAYWELL

Murray P. Naditch

PROGRAM DESCRIPTION

The Control Data STAYWELL program is based on the premise and evidence that life-style and behavior have an important impact on health and that individuals can be helped to change those harmful behaviors.

The program was initiated in 1979 and is currently being delivered to more than 22,000 Control Data employees and their spouses in more than 80 plants as well as to employees of a number of other companies. Control Data offers the program as a free corporation benefit to all of its employees and their spouses, and also sells the program to other companies for their own use.

The focus of the program is on long-term changes in behavior. Although there are major awareness and educational aspects of the program, the key emphasis in the program is on assisting people in creating the skills that they need to change their behaviors and on creating an environment in the work place conducive to the initiation and maintenance of positive life-style behaviors. The program focuses on five key risk areas: smoking cessation, weight control, nutritional practices related to reduction in cholesterol, salt and sugar, fitness behavior, and stress management.

PROGRAM DELIVERY

STAYWELL is initiated at each new plant with an extensive presite preparation phase that begins months before employees are introduced to the program. A key element in that phase is an extensive communication program for managers, consisting of special management orientations and individual contact by key members of the STAYWELL team. The objective is not only to help management understand the purpose and nature of the program, but also to enlist an active management support. Every effort is made to enlist managers as program participants so that they can serve as role models for employees. Implementation of the program is planned in order to minimize any disruption of work schedules and to accommodate needs of each facility.

This period of intense preparation is followed by a promotion and orientation program for employees. Employees receive two mailings informing them about the program. The mailings are followed by special orientation sessions during which employees and spouses learn about the program and have the opportunity to enroll. Employee enrollment ranges from 65 to 95% of the work force. Enrollment is higher at sites where people do less traveling, and is higher among people with at least a high-school education. Enrollment has not been linearly related to education, but is lowest among employees with less than a high-school education. Enrollment has also been higher among employees over 45, although enrollment files off in the over-60-years-old group. Females are slightly more likely to enroll than males, and employees who earn $30,000 or more a year are more likely to enroll than those making less than $30,000.

Employees who do enroll are scheduled for a health screening session. At that screening a professional health team collects data on each participant including blood pressure, height, weight, health history, and life-style and a small sample of blood. Data from the health screening is used to generate a confidential health risk profile.

The Health Risk Profile is a computer-generated analysis of each participant's most significant health risks. All participants receive a report that compares their chronological age with their risk age (how old they really are in terms of risk) and their achievable age (how old they could be if they decide to make the appropriate changes in their health-related behaviors).

Control Data has a very strict confidentiality policy. No one at the company has access to any individual data except a physician who reviews the data for abnormalities. Individual participants decide when they register for the program whether they would like any findings suggesting potential abnormality sent to their personal physician. Health Risk Profiles are interpreted for employees at group interpretation workshops. Workshops help individuals identify their health risks and suggest specific actions.

After completing the interpretation workshops, STAYWELL participants can enroll in a variety of activities related to their individual needs. These activities include: Life-style change courses, support groups, and a variety of other group activities.

Life-style Change courses teach employees the skills they need to change lifelong hahits such as smoking and improper eating habits. There are Life-style Change courses on smoking cessation, stress management, weight control, nutrition, and fitness.

Courses are taught by health educators, psychologists, exercise physiologists, and other health professionals, trained by CDC to deliver the program.

A unique part of the Control Data STAYWELL program is the social support activities offered through the employee participation groups. These groups are composed of employees and spouses who are interested in taking a more active leadership role in the program. The purpose of these groups is to change the worksite environment in a manner that will create norms and expectations about more positive health-related activities on the part of employees.

Participation group activities can take the form of action teams in which groups of employees work with the company to change some part of the work environment to make it more conducive to positive health-related behaviors.

Employees have formed action teams to change food in vending machines so that there are more nutritious selections, have initiated low-calorie cooking classes, running, walking, hiking, bicycling, and cross country ski clubs, aerobic dance classes, and a large number of other health-related activities.

Other employees have joined support groups. Support groups are formed at the end of Life-style Change courses. These groups assist members in implementing, continuing, and maintaining the difficult job of changing health-related behaviors. Support group activities include group sharing of experiences, innoculation against failure techniques, and other activities whose purpose is to provide a supportive, sympathetic environment in which to continue to initiate or maintain hard-won changes in health behavior habits.

There has been a strong positive response on the part of employees to participation group activities. More people have become involved in participation group activities than have enrolled in courses.

Participation group activities are usually very visible at the worksite. These activities are attractive to employees, increase awareness, establish positive health-related norms, and attract employees into the program who may not have been attracted into the program solely because of a desire to modify health behaviors. Participation groups give people an opportunity to socially interact with one another, extend opportunities for program entry, and build program momentum.

Orientation and screening sessions are on company time. Interpretation workshops, courses, and participation group activities are on the employee's time. Most classes and activities take place during the lunch break, although activities are also offered before or after work.

PROGRAM EVALUATION

A basic premise in implementing the program was that the program would be evolutionary and based on empirical data generated by the evaluation. Evaluation has included formative as well as summative components. The formative component has focused on how well the program has been functioning while the summative component has focused on changes in attitudes, health-related behaviors, and health care costs.

Initial formative data from the pilot study of the program in 1979 indicated a significant problem in management support at the lower-management level. A special orientation for lower-level managers was instituted in 1980 and that problem subsequently disappeared.

The initial course interventions developed in 1979 and 1980, although academically sound, were not initially well-received by employees. Employees felt that these courses were too academic and they wanted a more "hands-on" approach that would give them direct advice about what to do in changing their behaviors. All of the course interventions were extensively revised in 1980 and 1981, and the new courses had higher completion rates, resulted in more knowledge acquisition, and resulted in more behavior change in each of the targeted areas. These results will be summarized below.

Formative data also indicated that the initial set of courses were not working well for production workers. One major problem was logistic. Courses required an hour of employee time for each session, and blue-collar workers for the most part had only one half-hour for lunch, and were not willing to stay before and after work to take the courses. A new set of courses focusing on the needs of production workers, with 20–30 minute session lengths, were instituted in 1982 and 1983.

An additional set of courses based on a self-instructional correspondence model was also implemented in 1983. These courses were targeted at people who were either too busy to continue to schedule course sessions or were hesitant to take group-administered courses.

Substantive evaluation data indicate that there is significant improvement in behavior after the weight control, smoking cessation, fitness, stress management, and nutrition courses. Approximately 50–62% of employees completing these courses reported significant behavioral change lasting for at least 12 months. The strongest effects are in the area of smoking cessation where 58% of the respondents indicated some change of habit and 35% of the respondents continued not to smoke 12 months after the course was completed. Evaluation data also indicate that health screening and assessment of risk factors themselves do not significantly impact behaviors. Life-style behaviors are only impacted as a result of participation in either courses or other group-related activities.

A major premise underlying the program is that long-term health behavior change will be facilitated in a normatively conducive environment. Normative change was measured by asking respondents the extent to which they perceived that *other* employees at the worksite were making positive changes in specific health-related behaviors. Three-year trend data indicate strong positive normative changes in the areas of physical fitness, weight, smoking, stress, nutrition, and general energy level.

The three-year trend data also indicate that STAYWELL participants observe more normative change than do nonparticipants at STAYWELL sites or people at matched control sites. Nonparticipants at STAYWELL sites, however, observe consistently more normative change than did people at matched control sites in all three years. Normative change data become stronger in the second and third year of data collection.

A number of items were included in order to test for response bias in the normative data. These data indicated that the normative change was only in the areas in which the STAYWELL intervention was focused, and not in other socially desirable areas such as coffee consumption or alcohol problems.

STAYWELL PLATO: A COMPUTER-MANAGED HEALTH BEHAVIOR CHANGE PROGRAM

Control Data has developed a computer-managed program of health behavior change. That program uses computer technology to address problems that resist solution in traditional clinical approaches. The program is highly individualized. Each person is matched with the intervention most likely to be effective for that person. The program is also modified while the person is in it as a function of behavior change and compliance. The efficiency of matching and reassignment is evaluated by a mathematical model that enables the program to make more decisions as more people complete the program.

Increased recognition of the importance of individual differences in response to health behavior change programs is an often cited area of need inquiry in behavioral medicine. Clinically and academically based programs have difficulty in responding to the need for more individually tailored programs because samples are usually too small to perform the complex multivariate analysis required to determine which interventions work with what kinds of people to what effect.

The STAYWELL program has a large enough population to establish the basic parameters required for an individualized program. Program users complete a behavioral profile prior to entry into the program. This behavior profile contains a listing of the major variables that have been hypothesized to relate to individual differences in response to programs in the clinical literature. For example in the weight control area, variables such as knowledge about nutrition, the degree of social support at home, the degree to which an individual is overweight, the number of programs a person has been in previously, sex, and other demographic characteristics are included in the profile.

In the initial iteration of the program, subjects are randomized across a number of intervention approaches. When a sufficient sample of people have run through the program, the individual difference variables are examined to determine their efficacy in predicting outcomes at program end and 12 months after program completion. Individual difference variables that are useful predictors remain in the model and those that do not account for significant variance are deleted. Variables whose main or interactive effects accounted for significant variance are used to match individuals to program paths in the next intervention. This procedure is repeated with each iteration (approximately every time 2500 people complete the program), so that the system is able to make increasingly accurate predictions about the effects of matching people to program paths.

Each program path includes branches so that individuals who are not doing well may (1) move to an alternative intervention, (2) have the intervention they are enriched with adjunctive material, or (3) repeat certain aspects of the intervention that they are in. Each branch point is treated and tested as an alternative experimental intervention. The efficiency of branch points are evaluated and reconsidered with each new cohort of people comprising one of the iterations in the evaluation process.

Using this approach the computer-managed model can be considered as a paradigm for the evolution of empirically based theory that could yield cumulative scientific knowledge of this area. This method is consistent with the structural equation-modeling paradigm using theory construction proposed by Blalock (1969).

The computer-managed program is individualized further by having a friendly, supportive tone, using the subject's name, remembering statements made by the subject earlier in the program, providing quantifiable and graphic feedback on the user's progress, facilitating comparison of the individual progress with other similar people in the program, and allowing users a wide latitude of choice.

The emphasis is on behavior change rather than limited to education. Each user receives only the information relevant to their specific problem. Users have opportunities to use new information in computer-managed simulation situations, make specific goals and objectives for behavior during the week following the session, and receive individualized computer generated feedback on their success or lack of success in applying the techniques between lessons. Programs for smoking cessation, weight control, stress management, blood pressure management, and fitness were implemented at Control Data and were made available to other companies for sale in 1983.

STAYWELL IN OTHER COMPANIES

Control Data is currently selling the STAYWELL program to other companies. In some cases, Control Data has implemented the entire program, and in other cases Control Data has worked with companies and their employees to implement the whole or parts of the program. This has provided a unique opportunity to assess the efficacy of a variety of program components as well as to test hypotheses about different procedures.

Initial evaluations data in comparing the STAYWELL program at Control Data to the STAYWELL program at other companies has indicated that some employee payment for the program may have beneficial results. Control Data offers the program as a free corporate benefit to employees and spouses. A matched sample of Control Data employees and employees in a STAYWELL program at another company that charged employees a modest fee for program participation ($2 per month) indicated that although initial enrollment was lower, course completion and behavior change was higher in the situation in which there was a token charge for participation.

REFERENCES

Blalock, H. M. *Theory Construction: From Verbal to Mathematical Formulation.* Englewood Cliffs, N.J.: Prentice-Hall, 1969.

Appendix IBM Health Enhancement Activities

Kent W. Peterson

In recent years, medical research has found increasing evidence that life-style has a major influence on health. In fact, more than 50% of premature deaths among Americans is thought to be caused, not by hereditary or environmental factors, but by life-styles that include heavy cigarette smoking, insufficient exercise, and poor diets.

The growing interest in good health among our employees, as well as increases in health care costs, has led IBM to develop a voluntary Health Education Program to help individuals improve skills related to maintaining good health.

The IBM Health Education Program includes nine comprehensive courses scheduled before or after working hours. These comprehensive courses last from 2 to 12 weeks and include Exercise, Smoking Cessation, Stress Management, Weight Management, Healthy Back, First Aid, CPR and Obstructed Airway Maneuver, Driver Improvement, and Water Safety. Mini-courses, which generally consist of an hour-and-a-half class, are also offered. They cover Health and Nutrition and Risk Factors Management, as well as the nine subject areas listed above.

The courses are taught by qualified instructors affiliated with community organizations such as the "Ys," American Red Cross, American Lung Association, and hospitals and colleges. All regular and regular part-time IBM employees, retirees, spouses and dependent children ages 15–22, are eligible to enroll.

Any course offered by IBM specifically for its employees and their families is paid for by the company. However, individuals may also enroll in similar courses provided by community or commercial organizations on a tuition assistance basis. Maximum tuition assistance amounts and eligibility criteria have been established for each course.

Since the start of the program in 1981, over 4000 courses have been conducted that include over 100,000 enrollments. In addition, about 8000 people have participated through tuition assistance.

COMPREHENSIVE COURSES

Comprehensive courses that provide detailed information and an opportunity to practice, when appropriate, are conducted over a period from 2 to 12 weeks.

Exercise: A course designed for those who have not been physically active. The program covers the benefits and hazards of exercise as well as aerobic activities and the technique for monitoring pulse rate.

Smoking Cessation: A course designed to help smokers understand their habit and to learn new behavior patterns in order to stop smoking.

Stress Management: A course designed to teach the sources of excessive stress and approaches for achieving mental and physical relaxation. Techniques for channeling and managing stress more productively are covered.

Weight Management: A course designed to assist in developing specific approaches for changing eating patterns to achieve permanent weight loss. Information on nutrition and the value of exercise is included.

Healthy Back: A course designed to assist people who have low back pain. The program includes appropriate exercises and is medically approved.

First Aid: A course that provides certification in emergency first aid procedures.

CPR and Obstructed Airway Maneuver: A course that provides certification in emergency procedures for respiratory failure and cardiac arrest.

Driver Improvement: A course that teaches the principles and techniques involved in avoiding accidents, responsibility of the driver, traffic laws, and defensive and emergency driving techniques.

Water Safety: A course designed to teach accident prevention, water survival skills, and rescue techniques for swimmers and nonswimmers.

MINICOURSES

Minicourses generally consist of an hour-and-a-half class that provides an introduction to basic concepts with a brief overview of the subject. Minicourses provide general principles and guidelines.

The nine comprehensive course subjects (listed above) may also be covered in minicourses. In addition, the following two subjects may be offered.

Health and Nutrition: An introduction that provides nutritional information and ideas regarding diets.

Risk Factors Management: An introduction designed to provide information on risk factors (i.e., smoking, obesity, hypertension, stress).

IBM VOLUNTARY HEALTH SCREENING EXAMINATION PROGRAM

The IBM Voluntary Health Screening Examination Program, initiated in 1968, assists employees in maintaining a planned program of medical checkups at no cost to the employee.

The health screening examination is completely voluntary and is available to employees at five-year intervals beginning at age 35. The results are strictly confidential and are provided only to the employee and his/her personal physician at the employee's request.

Examinations are conducted at the IBM medical department or at the office of an IBM-designated physician or medical service organization. Over 70% of eligible employees in the U.S. and Canada have participated in the program since its inception through 1983—over 200,000 exams have been given. More than 10% of the participants have had at least one previously undetected medical condition or significant risk factor identified.

IBM RECREATIONAL/FITNESS FACILITIES

IBM provides recreational/fitness facilities for use by employees and their families at a number of major locations. Currently, there are 30 recreational facilities. The size and type of activity depends upon the interests of employees and the availability of recreational opportunities in the community.

Nationwide, IBM has approximately 122 tennis courts, 59 athletic fields, 23 basketball courts, and 12 outdoor jogging tracks.

Appendix The Johnson & Johnson LIVE FOR LIFE Program

Curtis S. Wilbur

INTRODUCTION

The Johnson & Johnson LIVE FOR LIFE Program is a comprehensive health promotion effort intended ultimately for all Johnson & Johnson employees worldwide. The LIVE FOR LIFE Program is specifically designed to encourage employees to follow life-styles that will result in good health. The program is based upon the assumptions that

1. Life-style activities such as eating, exercise, smoking, and stress management contribute substantially to an individual's health status.
2. Life-style activities that support good health can be successfully promoted at the work setting.

During the initial phase, Johnson & Johnson is committed to a careful evaluation of the LIVE FOR LIFE Program in terms of its impact on employee health and its overall cost-benefit to the corporation.

The LIFE FOR LIFE Program began in early 1979 with two primary goals:

1. To provide the means for Johnson & Johnson employees to become among the healthiest employees in the world.
2. To determine the degree to which the program is cost-effective.

Program objectives include improvements in nutrition, weight control, stress management, fitness, smoking cessation, and health knowledge. Moreover, the proper utilization of medical interventions such as high blood pressure control and the employee assistance program is strongly encouraged. It is anticipated that such improvements will lead to positive changes in employee morale, relations with fellow employees, company perception, job satisfaction and productivity, as well as reductions in absenteeism, accidents, medical claims, and total illness care costs.

Responsibility for the LIVE FOR LIFE Program rests with LIVE FOR LIFE staff at the corporate level. As of December 1983, the LIVE FOR LIFE staff serves about 22,000 employees in active programs at 37 separate Johnson & Johnson locations throughout the United States, Puerto Rico, and Canada. By the end of 1986, LIVE FOR LIFE will be available to all Johnson & Johnson employees worldwide (approximately 75,000 employees).

LIVE FOR LIFE is primarily a service organization. Its mission is to provide to Johnson & Johnson employees, their families, and the community the direction and resources that will result in healthier life-styles and help contain illness care costs. LIVE FOR LIFE supplies participating companies with the consulting expertise, training, core program components, professional services, and promotional materials necessary for program success. LIVE FOR LIFE staff are also responsible for program development and evaluation.

It is important to recognize that Johnson & Johnson is a highly decentralized group of companies, each of which operates in a very independent fashion. Acceptance of and full commitment to the program by a company's senior management is essential to ensure the financial and time commitment to support the program.

THE PROGRAM

The LIVE FOR LIFE program works in the following way:

Sales Presentation to Company Management

The LIVE FOR LIFE program begins with a presentation to a company's management board. The management board is asked to make three commitments: financial resources, management time, and responsiveness to employee requests for the creation of a worksite environment more supportive of good health.

Selection of an Administrator and Volunteer Leaders

A management board member, often the vice president of personnel is assigned operational responsibility for LIVE FOR LIFE. He/she hires a LIVE FOR LIFE administrator and recruits a small group of talented middle managers and labor leaders to serve as a voluntary task force. Both the LIVE FOR LIFE administrator and the task force receive extensive training by the LIVE FOR LIFE staff. Following training they are responsible for managing their company's LIVE FOR LIFE program.

Health Screen

For most employees, the LIVE FOR LIFE process begins with participation in the health screen. The health screen is promoted as a unique opportunity for all employees to find out how healthy they are and to learn a tremendous amount about health in the process. The screening is offered on company time and takes about 1 hour to complete. A wide range of health, life-style, and attitudinal measures are collected during the health screen, including *biometric* (e.g., blood lipids, blood pressure, body fat, weight, and estimated maximum oxygen uptake), *behavioral* (e.g., smoking, alcohol use, physical activity, nutrition, healthy heart behavior pattern, job performance, and human relations) and *attitudinal* measures (e.g., general well-being, job satisfaction, company perception, and health attitudes).

Life-style Seminar

The LIVE FOR LIFE process continues with a life-style seminar that is usually scheduled about six weeks after an employee completes the health screen. The life-style seminar is the primary vehicle for introducing employees to the program. Personal responsibility for health is emphasized, especially within the context of a program in which the company has committed itself to creating a work environment rich in opportunities to improve and maintain personal health. Employee health screen results are returned at this point via an attractive document called a life-style profile. The lifestyle profile is presented as a way to determine "how healthy you are" and provides the basis for taking action to improve health through LIVE FOR LIFE participation. Upcoming LIVE FOR LIFE health enhancement opportunities are reviewed and promoted. The lifestyle seminar is offered on company time to groups of about 50 employees at a time and takes about three hours.

Lifestyle Improvement Programs

An essential component of the LIVE FOR LIFE process is the ongoing opportunity for employees to improve or maintain their health through participation in formal programs. LIVE FOR LIFE action programs are offered regularly in smoking cessation, weight control, stress management, nutrition, exercise, and high blood pressure control. Action programs are distributed in a variety of formats, including groups, individual consultation, self-help kits, and the telephone. In addition to action programs, LIVE FOR LIFE regularly offers a wide range of shorter educational and promotional programs built around such themes as breast self-examination, biofeedback, nutrition, blood pressure, and carbon monoxide analysis for smokers. First year participation targets include

Weight Control	25% of all overweight employees
Smoking Cessation	20% of all employees who smoke
Stress Management	20% of all employees
Exercise	40% of all employees
Blood Pressure	75% of all hypertensives

Programs to Sustain Participation

The introduction of a novel health promotion program like LIVE FOR LIFE invariably creates widespread employee interest and involvement. The challenge is to sustain high program participation after the novelty wears off and initial enthusiasm subsides. Company leaders, in consultation with the LIVE FOR LIFE corporate staff, begin planning ways to sustain participation during the early phases of program development and execution. One popular program is an incentive system. In business, regular feedback to employees on their performance has been shown time and again to be a critical aspect of sustaining and improving performance over time. This basic management principle also applies to health. Employees participate in life-style improvement programs to the extent that they perceive positive consequences (feedback) for doing so. Many employees find a structured incentive system to be a powerful source of positive consequences. Participating in various LIVE FOR LIFE activities can earn "points," and nonmonetary incentives like program T-shirts, key chains, and umbrellas can be awarded to those employees who earn a predetermined number of "points."

Another method for sustaining program involvement is regular follow-up. The LIVE FOR LIFE program includes a system whereby all employees who participated in life-style improvement programs are contacted by phone or mail at regular intervals after the completion of that program. The primary purpose is to provide participants with a minimum structure through which they can report progress in sustaining the life-style improvements (e.g., weight loss, smoking cessation, regular exercise) that were initially achieved during the structured action program. The follow-up system also served to reinforce the key life-style activities considered essential to sustained life-style improvement (e.g., regular weighing and counting calories in weight control, regular practice of stress management skills). Moreover, regular follow-up provides program management with valuable information on effectiveness over time.

Creating a Healthful Environment

The primary component of a LIVE FOR LIFE program is the creation of a work environment that supports and encourages positive health practices among the greatest number of employees possible. Offering regular, convenient, and attractive life-style improvement programs will go a long way toward creating a positive social environment based on high employee participation. Other aspects of the work environment are targets for improvement as well. Examples of environmental improvements in key areas include

Fitness

- Shower and locker facilities on site.
- Exercise facilities either on site or rented from local organizations.

Weight Control/Nutrition

- Scales in restrooms.
- Availability of convenient nutrition information where food is sold.
- Availability of nutritious foods in the company cafeteria and vending machines.

Stress Management

- Employee assistance program to provide professional treatment and referral services to troubled employees.
- Availability of management training programs designed to improve supervisor-supervisee relations.
- Flextime.
- Carpooling.
- Self-administered blood pressure equipment.

Smoking Cessation

- A clearly stated smoking policy.
- Availability of LIVE FOR LIFE "Thank You For Not Smoking" signs.

Publicity

- LIVE FOR LIFE newsletter.
- LIVE FOR LIFE bulletin board and information display area.
- Comprehensive recruitment brochure.
- Health Fairs.
- Poster displays for upcoming programs.

THE EVALUATION

A two year epidemiological study is in progress to evaluate the Johnson & Johnson LIVE FOR LIFE Program, a comprehensive health promotion effort. In LIVE FOR LIFE, all employees have the opportunity to participate at the work setting in such health enhancement programs as exercise, smoking cessation, stress management, nutrition, and weight control. At the same time, voluntary employee task forces are assigned responsibility for creating a work environment that supports positive health practices. Examples of environmental improvements include convenient exercise facilities, lower-calorie foods in the cafeteria and vending machines, designated nonsmoking areas, convenient weight scales, and consistent program publicity.

Design

Ideally, a randomized prospective controlled trial of the LIVE FOR LIFE Program would be the most convincing test of the intervention. For a number of reasons, this design was not possible. When using an educational campaign and facilities for life-style improvement over an entire site or factory, one cannot keep individuals randomly assigned to a control group from being exposed to the intervention. Random assignment of company sites also proved impossible.

Instead, a "quasi-experimental" research strategy was considered feasible and very suitable for evaluation purposes. The LIVE FOR LIFE Program is evaluated employing what Campbell and Stanley (1) term a NonEquivalent Control Groups Design, as described below:

Treatment Group	O	X	O
- - - - - -	- - - -	- - - -	- - - -
Control Group	O		O

The Os refer to observations or Health Screens and the X represents the intervention or LIVE FOR LIFE Program. The hatched line indicates that the groups are not randomized. Thus, the treatment group represents four companies where the LIVE FOR LIFE Program was introduced from September 1979, to May 1980. The control group contains five sites with no intervention except the Health Screen and the return of health and life-style information to participants.

The evaluation time period is two years. The design as applied to the LIVE FOR LIFE Program can be schematized as follows:

Treatment Group	O_1 ——	O_2 —— X	O_3
- - - - - -	- - - -	- - - -	- - - -
Control Group	O_1	O_2	O_3

The O_1, O_2, and O_3 observations or health screens occur at yearly intervals. Thus, for the treatment group the O_1 health screen occurred before the LIVE FOR LIFE program began. The O_2 health screen occurs a year into the program, and the O_3 Health Screen takes place at the end of the second year of the Program. Similarly spaced health screens in the absence of the LIVE FOR LIFE Programs are planned for the control sites.

The Population

At baseline, the treatment group consisted of approximately 2100 employees at four Johnson & Johnson facilities, and the control group consisted of approximately 2000 employees at five additional Johnson & Johnson companies. All treatment and control groups are located in the same geographical region (central New Jersey and northeastern Pennsylvania).

As of July 1, 1980, the distribution of the initial study sample by type of plant was as follows:

	Treatment	**Control**
Screened	1,606	1,440
Not screened[a]	533	572
Total	2,139	2,012
Screened as % of Total	75.1%	71.6%

[a]Preliminary estimates by use of a nonrespondent survey. Approximately 17% of the individuals selected in the nonrespondent survey were either not employed or ineligible for the study at baseline.

Sampling Procedures

Health and life-style measures are collected on all employees who participate in the LIVE FOR LIFE Health Screen at baseline, and year one and two. Variables were described earlier.

All employees at the treatment and control companies are actively encouraged to participate. The primary study focus is on the cohort of employees in the treatment and control sites who participate in the health screen at baseline and at year two. This cohort will steadily shrink in size over the two-year evaluation period due to an estimated 15% employee turnover rate and an estimated 70% health screen recapture rate. Cross-sectional comparisons are also planned, especially for management use.

Employees at treatment and control sites who did not participate in the Health Screen were designated as nonresponders. A nonrespondent sampling methodology was developed and executed by Research Triangle Institute (RTI). The purpose was to make possible comparisons across all employees in all plants, and to determine if responders differed from nonresponders.

Preliminary Findings

Preliminary results on the cohort of employees attending both the baseline and year one health screen (excluding one treatment company since its year one Health Screen data is not completely collected and analyzed) are very exciting. The evidence clearly suggests that the treatment cohort, in comparison with controls, consistently show greater improvements in the major health and life-style areas addressed by LIVE FOR LIFE:

Health Screen Measure	% Change Baseline—One Year: Treatment (N=737)		Control (N=680)
Fitness			
Aerobic Calories/kg/Week	43%	**	6%
Weight Control			
% Above Ideal Weight	−1%	**	6%
Smoking Cessation			
% Current Smokers	−15%	*	−4%
Stress Management			
General Well-Being	5%	**	2%
% With Elevated Blood Pressure (≥140/90)	−32%		−9%
Employee Attitudes			
Self-Reported Sick Days	−9%	*	14%
Satisfaction With Working Conditions	3%	**	−7%
Satisfaction With Personal Relations At Work	1%	**	−3%
Ability to Handle Job Strain	0%	**	−2%
Job Involvement	2%		0%
Commitment to the Organization	0%		−2%
Job Self-Esteem	0%	*	−2%
Satisfaction with Growth Opportunities	−1%		−3%

* = Significant at the 5% level.
** = Significant at the 1% level.

The very preliminary baseline-one year comparison data available at this point strongly supports the contention that the LIVE FOR LIFE program is capable of achieving significant and meaningful improvements in the health and life-styles of Johnson & Johnson employees. The findings are consistent across the range of core program areas. Although these preliminary results are very encouraging, full documentation of the total sustained impact of the LIVE FOR LIFE Program awaits full completion.

Preliminary work is also underway to measure the cost-benefit of the program. Since Johnson & Johnson is self-insured for illness care costs, any changes in the

[1]Campbell, D. T., and Stanley, J. C. *Experimental and Quasi-Experimental Designs for Research.* Chicago: Rand McNally, 1963.

number of dollar amount of illness care claims attributable to a positive health program are of considerable interest as a measure of program benefit. Other potential benefit measures include absenteeism, turnover rates, accident rates, and a host of employee and management attitudes toward themselves, their work, and one another. It is felt that this is pioneering work in an area where potential benefits have been difficult to measure with existing systems and methods.

Authors

George S. Everly, Jr., Ph.D. is associate professor of psychology and director of the Psychophysiology and Health Psychology Research Laboratory at Loyola College in Maryland. He is also a psychologist in private practice in Towson, Maryland and chief consultant to the Biofeedback Laboratory at Crownsville State Hospital. Dr. Everly received his doctorate from the University of Maryland at College Park. He is the author/coauthor of five additional textbooks in the areas of behavioral medicine, health education, and clinical psychology.

Robert H. L. Feldman, Ph.D. is a health psychologist, associate professor, and director of the program in health behavior in the Department of Health Education at the University of Maryland at College Park. Dr. Feldman was previously on the faculty of the Johns Hopkins University, School of Hygiene and Public Health, in the Division of Health Education and the Division of Occupational Medicine. He holds a Ph.D. in Social Psychology from Syracuse University and completed a Post-Doctorate in Health Psychology from the University of Connecticut Medical-Dental School.

Contributors

Arthur Anderson, Ph.D., is associate professor of sociology at Fairfield University, Fairfield, Connecticut.

Samuel Berkowitz, Ph.D., is clinical director of the Psychological Sciences Institute, Baltimore, Maryland.

Nancy M. DeMuth, Ph.D., M.B.A., is an associate in Operations Research at the Johns Hopkins University, School of Hygiene and Public Health, Baltimore, Maryland.

Calvin F. Fuhrmann, M.D., is chief of pulmonary medicine at South Baltimore General Hospital, Baltimore, Maryland.

Catherine J. Green, Ph.D., is associate professor of psychology, University of Miami, Coral Gables, Florida.

Roberta B. Hollander, Ph.D., M.P.H., is assistant professor of Health Education at the University of Maryland, College Park.

Joseph J. Lengermann, Ph.D., is associate professor of sociology at the University of Maryland, College Park.

Carole B. Lewis, Ph.D., R.P.T., is co-director of Physical Therapy Services of Washington, D.C.

Donald E. Morisky, Sc.D., M.S.P.H., is assistant professor of Behavioral Sciences and Health Education at the School of Public Health, University of California, Los Angeles.

Murray P. Naditch, Ph.D., is director of Advanced Programs, Healthcare Services for Control Data in Minneapolis, Minnesota.

Eileen C. Newman, M.A., is in the department of psychology, University of Miami, Coral Gables, Florida.

Kent W. Peterson, M.D., is corporate manager for Preventive Medicine Programs at International Business Machines Corporation in Purchase, New York.

Melvin Sandler, M.S.S.W., is Employee Assistance Program representative of United Airlines Medical Department in New York City.

Kathryn Nickelsberg Schaefer, R.P.T., is co-director of Physical Therapy Services of Washington, D.C.

Steven A. Sobelman, Ph.D., is associate professor of psychology, Loyola College in Maryland, Baltimore, Maryland.

Sharlene M. Weiss, R.N., Ph.D., is a psychologist at the Behavioral Medicine Center of Washington, D.C., Chevy Chase, Maryland.

Donald Weller, Ph.D., is director of the Dundalk Community College Fitness Center, Baltimore, Maryland.

Curtis S. Wilbur, Ph.D., is program director of the LIVE FOR LIFE program at Johnson & Johnson in New Brunswick, New Jersey.

Index

Abrams, D., 81
Accidental deaths, 5
Activity, physical, *see* Exercise; Physical fitness programming
Adams, J., 65
Adams-Webber, J., 271
Adaptive lipogenesis, 77–78
Adcock, C. J., 278
Adler, N., 10
Administrative/organizational approaches, compliance with programs and, 41
Adronaux, John, 21
Aerobic exercise, *see* Physical fitness programming
AFA (union), 164
AFL-CIO, 148, 157
Ainsworth, T. H., 150, 164
Ajzen, I., 235, 236
Akabas, S., 156
Alberti, R., 65
ALCOA Corporation, 211
Alcohol abuse, 5, 147–166, 305
  Employee Assistance Programs, 148, 153–166
  guidelines for program development, 154–163
    foundation building, 155–161
    implementation, 161–163
  introduction, 147–148
  Occupational Alcoholism Programs, 150–152
  resource guide for Employee Assistance Programs, 166
  scope of the problem, 148–150
  setting of goals in program planning, 235
  summary, 163–164
  United Airlines program for, 164–166
Alderman, M. H., 33, 36, 39, 43, 115, 119, 121, 122, 133, 288, 293, 294, 295, 301, 303
ALPA (union), 164
American Association of Fitness Directors in Business and Industry, 29
American College of Sports Medicine, 131, 133, 135, 136
American Cyanamid Company, 26, 27
American Federation of State, County and Municipal Employees, District Council 37 of, 155, 161
American Heart Association, 136
American Institute of Certified Public Accountants, 5
American Lung Association, 92, 93, 103
American Medical Association, 136

American Psychological Association, 4, 14, 251, 266
Task Force on Health Research, 4
American Telephone & Telegraph (AT&T), 302
Anaerobic exercise, 129
Anderson, Arthur L., 17–31
Anderson, Gunnar, 208, 209
Andrew, J. M., 280
Androgynous attitudes, 177–178
Anger, W. K., 188, 193, 195
Anthropometric assessments, 79–80, 138
Approximate methods for assessing program effectiveness, 246–247
Ardell, D., 6
Arnold, M., 55
Aronow, W. S., 94*n*
Arteriosclerosis, 130
Asclepius, god, 17–18
Ashford, N. A., 25
Astrand Ergometer Test, 135
Astrup, P., 94*n*
AT&T, 302
Athanasiou, R., 176
Automobile insurance, 193
Aversion strategies of cessation of smoking programs, 101, 105–109

Back injury prevention programs, 208–219, 302
guidelines for program development, 213–218
introduction, 208–209
research review on, 210–212
resource guide for, 219
scope of the problem, 209–210
summary, 218–219
Back School, 212
Bagley, R. W., 36
Bailey, 155, 163
Baker, S. P., 189*n*
Barnes, L., 141, 142
Barrera, F., 75
Barry, P. Z., 289, 290, 297
Basal metabolic rate, 77*n*, 86
Basmajian, J., 67
Baugh, T., 302
Beck, A. T., 273
Beck Depression Inventory, 273–274, 281
Becker, M. H., 36, 37
Behavioral contracting, *see* Contracting
Behavioral perspective on planning health promotion intervention and evaluation, 251–264
specific behavioral procedures, 257–263
behavioral contracting, 258–261
intermittent reinforcement, 263
negative reinforcement, 258, 259
positive reinforcement, 257–258, 259
shaping, 261–263
stimulus control, 258
token economies, 263
steps in implementing behavioral intervention and evaluation, 252–257
define the target behavior, 252
establish a terminal goal, 255–256
follow-up assessment and evaluation, 257
intervention, 256–257
obtain a baseline, 252–255
select the measurement procedure/technique, 252
summary, 263–264
Behavioral Therapy: Application and Outcome (O'Leary and Wilson), 263
Behavior modification, 235
occupational safety and health programs utilizing, 188–189, 192–194, 195, 200–201
outcome measures in program planning, 235–236
planning, implementing, and evaluating, *see* Behavioral perspective on planning health promotion intervention and evaluation
setting goals and objectives in program planning, 235
weight reduction programs based on, 80, 81–88
Bell, R. A., 233
Bellack, A., 80, 82, 85, 86
Bellinger, S., 156
Belloc, N., 9
Bem, Sandra, 177, 178
Benedict, D. S., 157, 159, 162
Benson, H., 59, 65
Benton, A. L., 278
Berger, P. L., 17
Berkman, L., 10
Berkman, P. L., 280
Berkowitz, S., 251
Bernacca, G., 98
Bernard, J., 173
Bernstein, D., 97, 100, 102
Bertera, R. L., 120
Berry, C. A., 153

Best, J. A., 40, 293, 297
Beutler, L., 59
Bhalla, V., 60, 225
Biofeedback training, 59–60
Bird, C., 170
Björnstrop, P., 77
Blair, 210
Blair, S. N., 35, 38, 140, 141
Blaney, N., 97, 100, 101, 102, 105
Block, 277
Blood pressure readings, 114. *See also* High blood pressure
Blue Cross of Western Pennsylvania, 211
Body composition, measurement of, 137–138
Body strength, measurement of, 139
Bohen, H. H., 183
Bone, R., 95, 101, 103
Bonnie Bell Cosmetics, 41, 305
"Booster sessions," 87
Bootman, J. L., 295, 296
Boruch, R. F., 242
Bowler, M. H., 122
Boyer, J., 130, 141
Brand, R. J., 275
Brantner, J., 277
Braver, E., 190, 191
Bray, G. A., 77
Breast cancer, 302
Brennan, A., 130
Breslow, L., 9
Brown, Bertram, 148
Brown, J., 139
Brown, R., 95
Brownell, K., 76, 77, 80, 86
Brownell, L., 75, 78, 80, 81, 82
Brucker, C., 40, 41
Brunner, B. C., 222
Bryson, J. B., 172
Bryson, R., 172
Buch, C. N., 280
Budgeting for occupational health promotion programs, 237
Buffone, G., 130
Bureau of National Affairs, 25
Burgess, J., 130
Burlington Industries, 43
"Burnout," 52
Business Week, 156, 159, 160
Butcher, J. N., 278
Byers, W., 156, 158, 159, 162
Caffeine consumption, 222
California Personality Inventory, 281
Calliet, R., 210, 213
Campbell, D. T., 240, 241, 243, 245, 246
Campbell Soup Company, 38
Canada Life Assurance Company, 40–41
"Can Companies Kill," 49–50
Cannon, W., 55
Capitalism and occupational health promotion, 26–30
Caplan, R., 51, 52
Carbon monoxide from smoking, 94, 98
Cardi, M., 271
Cardiac rehabilitation and exercise, 131
Cardiovascular disease: exercise and, 130
obesity and, 75
see also Heart disease
Carey, R. G., 239
Carlgren, G., 77
Carrington, Patricia, 61, 222
Cassel, J., 55
Castelli, W., 75
Catalyst, 170, 176, 177
Cattell, R. B., 275, 276
Cautela, J., 252
Chadwick, J. H., 287, 295
Chaffin, Don, 209
Charlesworth, E., 59
Charlson, M. E., 293
Chavat, J., 128
Chicago Heart Association, 33
Child care for dual-career couples, 173, 175, 181–183, 184, 305
Chin, C., 135
Chowdhury, P., 95, 101, 103
Christianson, R., 75
Chwalow, A. J., 39
Circumference assessment, 79–80, 138
Clark, D., 74, 129
Clayton, G. D., 191
Cleary, P., 97
Coal Mine and Safety Act of 1969, 25
Cobb, S., 280
Cognitive/behavioral stress management program, 60
Cohen, A., 193, 195, 196
Cohen, R., 10
Cohen, R. Y., 235
"Cold turkey" approach to stopping smoking, 101–102, 103, 105
Colley, J. R., 74

Collings, G. H., Jr., 41
Communication:
  occupational safety and health and effective, 194–195
  skills in, 179–180
Compliance with occupational health promotion, 33–43
  factors that affect, 35–37
    physical fitness programs, 132–133
  measuring, 34–35
  methods to enhance, 37–43
  summary, 43
Comstock, G. W., 280
Conflict resolution skills, 180–181
Congestive heart failure, 113
Contracting, 40, 258–261
  for cessation of smoking programs, 109
  for weight reduction programs, 86, 87
Control Data STAYWELL program, 310–316
Controlled smoking program, 100
Cook, D., 38
Cook, S. W., 240, 245
Coolidge, Calvin, 30
Cooper, C., 53
Cooper, C. L., 175, 305
Cooper, Kenneth, 65, 131*n,* 133, 135
Cooper's 12 minute run-walk, 135
Cooperative Commission on the Study of Alcoholism, 148
Cost-effectiveness and cost-benefit analyses of occupational health programs, 287–298
  cost-benefit analyses, 295–296
  cost-effectiveness analyses, 289–295
    additional issues in, 293–295
    computing the program's net monetary costs, 291–292
    defining program (intervention) objectives, 290–291
    defining program outcomes, 292
    sensitivity analysis, 293
  introduction, 287–289
  summary, 296–298
Criqui, M., 130
Crisera, R. A., 194
Crow, M., 65
Crowne, D. P., 270
Crown-Marlowe Social Desirability correlations, 269
Cunningham, R., 33, 130, 234, 304
Cunnison, J., 173
Dahlstrom, L. E., 277, 278
Dahlstrom, W. G., 277, 278
Danskin, D., 65
Daum, S., 189, 190, 198
Davidson, A. R., 236
Davis, B., 82
Davis, T. K., 119, 121
Day care, 173, 175, 184. *See also* Child care for dual-career couples
Deeds, S. G., 36, 200
DeFriese, G. H., 289, 290, 297
Degenerative diseases, 4, 8
DeMuth, Nancy M., 287–298
Denniston, R. W., 302
Denny, D., 197
Derogatis, L. R., 271, 272
Derryberry, Mayhew, 252
Diabetes:
  exercise and, 130
  obesity and, 75
Diary for cessation of smoking program, 59–60
Dieting, weight reduction programs based on, 80, 85
Disabled workers, 302
Disabling injuries, 5
Dishman, R., 132
Doctor, corporate, 28
Donoghue, S., 130
Donovan, Ed, 61
Dotson, C. O., 303
Douglass, D., 224, 226
Douglass, M., 224, 226
Drachman, R. H., 36
Drash, G., 75
Draw-A-Person projective test, 82
Drucker, Peter, 223
Dual-career couples, *see* Working couples, health promotion for the
Dubbert, P., 130, 132, 139
Dubos, Rene, 9, 17–18
Duker, J., 277
Dunbar, J. M., 40, 41
Dunn, H. L., 6
Dunn, J. P., 302
duPont de Nemours & Co., E. I., 153–154
Dusek, D., 82, 85
Dwore, Richard, 9, 14
Dwyer, J., 85
Dynameter test, 139

Eber, H. W., 275, 276
Economic assessment of occupational health programs, 247–248, 287–298
Economics of Industrial Health (Follman), 26
Edwards, W., 233
Egdahl, Richard, 12
Eggum, P., 157, 158, 159
Elderly, 302
Elementary Principles of Behavior (Whaley and Malott), 263
Ellis, A., 65, 278
Ellis, L. B. M., 303
Emmons, M., 65
Employee Assistance Programs (EAP's), 148, 153–166
  confidentiality of, 156–157
  goals and measurement issues, 157–159
  implementation of, 161–163
  insurance issues, 157
  location of, 155
  policy statements, 155–156
  resource development, 160–161
  resource guide for, 166
  staffing of, 159–160
  summary, 163–164
  of United Airlines, 164–166
Engel, G. L., 280
Environment and health, 7, 8, 9
Epictetus, 55
Epidemiologists, 199
Epstein, L., 130
Equitable Life Assurance Society, 153
Erfurt, J., 43, 119, 155, 157, 159, 161, 162, 292, 303
Ergonomics, 199
Erikson, E. H., 22–23
Eshelman, F. N., 39
Eustress, 51, 56
Evaluation of occupational health promotion programs, *see* Occupational health promotion, planning and evaluation of programs in
Everly, George S., Jr., 3–15, 49–69, 75–88, 92–111, 127–143, 177, 178, 221–228, 252, 301–306
Everly Behavioral Survey, 82
Exercise, 222, 305
  cessation of smoking programs using, 109
  obesity and, 78
  weight reduction programs based on, 80, 86, 87
  see also Physical fitness programming

Families of employees, health services for, 302
Family Service Association of America, 160
Fardy, P., 131
Farmer, R., 53
Fasting programs, 80, 81
Feedback from program participants, 239
Feldman, Robert H. L., 33–43, 188–205, 233–249, 301–306
Felts, W. M., 303
Ferguson, J., 82, 85
Ferguson, T., 173
Fielding, J. E., 287
Figa-Talamanca, I., 297
Fischhoff, B., 193
Fishbein, M. D., 235, 236
Fisher, I., 24
Fisk, E. L., 24
Fitness director, corporate, 28–29
Fitzgerald, R. G., 280
Fitzler, 210
Fitzloff, J., 39
Fixx, Jim, 41
Flay, B. R., 293, 297
Fleck, S., 281
Fleming, P. L., 290, 291, 295
Flexibility, measurement of body, 139
Flexitime, 183, 303, 305
Florida Office of Mental Health, 148
Flynn, B. S., 39, 303
Foege, William, 306
Folkins, C., 130
Follick, M., 81
Follman, Joseph F., Jr., 26
Food and Drug Administration, 110
Foot, A., 43
Foote, A., 119, 155, 157, 159, 161, 162, 292, 303
Ford Motor Company, 159, 303
Foreyt, J. P., 40, 41
Forman, S., 60
Foundations of occupational health promotion, 1–43
  assessment and enhancement of health compliance in the workplace, 33–43
  ideological and historical rationales, 17–31
  an introduction, 3–15

Fox, E., 79, 129
Fox, S., 141
Framingham heart studies, 128
Francek, James, 159
Francis, V., 35, 117
Frank, John Peter, 23
Franklin, Benjamin, 20
Fredericksen, L., 100, 105
Freeman, H. E., 238, 247, 296
"Free Men," 182
French, J. R. P., 51, 52
Freudenberger, H., 53
Friedman, M., 279
Frii, S. R., 280
Froelicher, V., 131
Fuchs, Victor, 10, 29
Fuhrmann, Calvin F., 60, 92–111
Future directions in occupational health promotion, 301–306
  concluding remarks, 306
  coordination of services, 303
  expansion of services, 302–303
  health-conscious work environment, 303–306
  new populations receiving services, 301–302
  technological advances, 303
Future Shock (Toffler), 223

Gambert, S., 130
Gardner, E., 53
Garfinkel, L., 75
Garn, S., 74
Garret, J. T., 209
Garza, J. M., 174
Geersten, H. R., 36
General Motors, 29
Genest, J., 114
Gentili, S., 52
Gentry, W. D., 225, 226
Germany, 23
Germ theory of disease, 4
Gersten, J. C., 280
Gettman, L., 132
Gilberstadt, H., 277
Gilbreth, Frank, 123
Girdano, D. A., 59, 60, 64, 65, 67, 81, 82, 83, 86, 94*n*, 252
Girdano, D. D., 59, 94*n*, 226
Glasgow, R., 97, 100, 102
Goal setting for occupational health programs, 235, 290–291
Goldbeck, W., 5, 69, 153, 290
Goldberg, P., 5
Goldschmidt, P. G., 289
Goldsmith, F., 198
Goldstein, I. L., 200
Good, P., 277
Goode, W. J., 173
Goodwin, Donald, 148
Gordis, L., 35
Gordon, J. B., 190
Gordon, T., 65, 75
Gorsuch, R. L., 267
Gotto, A. M., 40
Government role in occupational health promotion, 27, 28, 29
Gradual reduction of smoking, 101–102, 103, 105
Graduated regimen implementation, 40–41
Graham, J. K., 287, 288, 295
Gray, R. M., 36
Greek ideologies of health, 17–18
Green, Catherine J., 265–282
Green, D., 103
Green, L., 213, 251
Green, Lawrence, 4, 8–9, 11, 14
Green, L. W., 36, 38, 200, 234, 235, 247, 292, 293, 294, 295, 296, 297, 298, 303
Greenwood, J. and J., 53–54
Griffin, D., 60
Griscom, John H., 21
Groden, G., 252
Group training seminar format for stress management programs, 62
Grynbaum, G. A., 197
Guida, M., 159
Gurtin, L., 169, 173
Guthrie, 251
Guttentag, M., 233
Guyton, A., 94*n*

Haddon, W., Jr., 189*n*
Haefner, D. P., 36
Hale, W. E., 160
Hall, D. T., 175, 176–177
Hall, F. S., 175, 176–177
Hall, R., 53, 101, 133
Hall, S., 10, 101, 133
Hamilton, W., 5, 6
Hammond, J., 53
Hannan, E. L., 287, 288, 295

Harmarville Rehabilitation Center, 211
Harper, R., 65
Harris, L., 128
Hartley, Robert M., 21
Harvard Step Test, 135
Haskell, W. L., 35, 38, 133, 140, 141
Hathaway, 277
Haynes, R. B., 34, 35, 293
Head, K., 176
Head, R., 280
Health:
  definition of, 6–7
  determinants of disease and, 7–10
Health behaviors in occupational health promotion programs, 34
Health Belief Model, 36–37, 121, 222
Health care services as determinant of health, 7, 8, 9
Health education as part of occupational safety and health program, 199–201
Health Evaluation and Longevity Planning Foundation, 54
Health insurance, 191
  for alcoholism and mental health services, 157
  rates, 5
Health knowledge, compliance with programs and, 39
Health maintenance organizations (HMOs), 30
Health promotion as mechanism of mediation, 10–12
Health Resources Administration, 235
Healy, B., 192, 199
Heart disease, 5, 128, 135, 225, 279
  high blood pressure and, 113
  smoking and, 97
  stress and, 51, 52, 53
  Type A behavior and, *see* Type A behavior
  see also Cardiovascular disease
Heilbronner, R. L., 26
Heinzelmann, F., 36
Henderson, J., 10
Hendrick, C., 170
Hendrick, S., 170
Herman, C. P., 76
Hess, W., 55
Hessler, R., 4
High blood pressure, 113–124, 130–131
  behavioral strategies for patient setting, 115–116
  behavioral strategies for provider of health care, 117–118
  guidelines for program development, 120–123
  introduction, 113
  obesity and, 75
  problems of awareness, treatment, and control, 115
  rationale for programs to control, at the worksite, 118–120
  resource guide for hypertension control programs, 124
  scope of the problem, 113–114
  setting of goals in program planning, 235
  summary, 123
Hiller, D., 172
Hiltner, Seward, 22
Hilton Hotels, 212
Himes, J., 138
Hippocrates, 22
Hislop, I. G., 280
Historical and ideological rationales for occupational health promotion, 17–31
Hockbaum, G. M., 121
Hoffman, L. W., 173
Hollander, Roberta B., 287–298
Holmes, T. H., 268, 280
Holy Bible, 19
Hopkins Symptom Checklist, 271, 272
Horn, D., 103
Horton, E., 76, 130
Hoskin, W. D., 192, 199
House, R., 51, 52
"House-husbands," 182
Houseknecht, S. K., 180
Howell, M. F., 40
How to Live (Life Extension Institute), 24
Huber, J., 180
Hull, 251
Hydrostatic weighing, 79, 138
Hygeia, goddess, 17–18
Hyperlipidemia and obesity, 75
Hypertension, *see* High blood pressure
Hypertrophic obesity, 76–77
Hypnosis for cessation of smoking, 100

IAMAW (union), 164
IBM health enhancement activities, 317–319
Ikard, F., 103
Illich, Ivan, 7, 22
Illinois Bell Telephone, 153

Industrial hygiene, 191–192, 199, 201
"Industrial Hygiene," 192
Institute for Social Research, University of Michigan, 176
Insurance, *see* Automobile insurance; Health insurance
Intermittent reinforcement, 263
Internal-External Scale (I-E Scale), 270–271
Internal validity of programs, factors jeopardizing, 240–242, 244
International Ladies' Garment Workers' Union, 155
International Law Enforcement Stress Association, 61
Introduction to health and occupational health promotion, 3–15
  appropriateness of workplace for promoting health, 12–14
  defining health, 6–7
  determinants of health and disease, 7–10
  health promotion as a mechanism of mediation, 10–12
  introduction, 3–6
  summary, 14–15
Introduction to the planning and evaluation of occupational health promotion programs, 233–249
  evaluating, 237–248
    approximate methods for, 246–247
    economic assessment, 247–248
    program effectiveness, 239–248
    program feedback, 239
    program monitoring, 237–239
    quasi-experimental designs for, 245–246
    true experimental designs for, 240–245
  introduction, 233
  planning, 233–237
    needs assessment, 233–235
    outcome measures, 235–236
    program resources (budgeting), 237
    setting goals and objectives, 235
    setting priorities, 236–237
  resource guide for, 248–249
  summary, 248
Inui, T. S., 117
Isaksson, B., 77
Ivancevich, J., 53

Jaccard, J. J., 236
Jackson, D. N., 281
Jacobson, E., 65
Jarvis, K., 148
Jeffrey, D. B., 82
Jellinek, E. M., 151
Jenkins, C. D., 274, 275, 280
Jenkins Activity Survey, 274–275
Jensen, J., 80, 81, 82
Jillson, R., 51
Johnson, J. H., 268, 280
Johnson, L., 136
Johnson & Johnson "Live for Life" health promotion program, 245–246, 320–328
Johnston, Ben, 212
Jones, K., 153
Jongeward, Dorothy, 224
Judeo-Christian tradition, 19–22

Kahn, R., 51
Kahneman, D., 193
Kaiser-Permanente Health Plan, 30
Kannel, W., 75, 128, 130
Karasek, 52
Karson, S., 276
Kasch, F., 130
Kasl, Stanislav, 53
Kassner, M. R., 174
Katch, F., 77*n*, 79–80, 82, 85, 86
Katch, V., 77*n*
Kazdin, A., 251
Keesey, R. E., 77
Kellerman, J. L., 151
Kellogg, Dr. John Harvey, 21
Kellogg, William Keith, 21
Kellogg Foundation, W. K., 21
Kelsey, J., 209
Kelsey-Hayes Occupational Alcoholism Program, 158
Kennecott Copper Corporation, 153
Kerr, L. E., 198
Kidder, L. H., 236
Kidneys, role in hypertension of the, 114
Kiefhaber, A., 5, 69, 153, 290
Kimberly Clark, 33, 40, 153
King, K., 251, 252
Kirscht, J. P., 34, 35, 36
Kjeldsen, K., 94*n*
Klassen, D., 130
Klevin, Keith, 211–212
Klipstein, Kenneth H., 26, 27
Knowles, J. J., 10

Kobasa, S., 62
Kohn, Hans, 22
Korsch, B. M., 35, 117
Kraus, H., 127, 128
Kreuter, M. W., 200
Kristein, M., 5, 295, 296, 291
Kroes, W., 53
Kropp, C. L., 154
Krug, S. E., 276
Krupp, N. E., 278
Kurtz, N., 162
Kutash, I., 58, 65

Labor unions, *see* Unions
Lakein, A., 65, 226, 228
Lambley, P., 271
Landwehr, M., 60
Lapsley, James M., 22
Laragh, J., 114
La Rosa, J. H., 288
Larry, R. Heath, 209, 210, 211, 212
Lauer, J., 82
Layden, M., 65
Layman, E., 130
Lazarus, A., 65
Lazarus, R. S., 55, 268
Leavitt, E. E., 58
Lebovits, B. Z., 114
Leclerc, G., 59–60
Leeman, C., 160
Lelonde, Marc, 7
Lengermann, Joseph J., 287–298
Lengner, T. S., 280
Leveille, G., 77
Leventhal, H., 97, 195
Levi, Lennart, 50, 51, 52, 58, 63, 64
Levine, D. M., 36
Levine, R., 280
Lew, E. A., 75
Lewis, Carole, 208–219
Lewis, F. M., 292, 298
Lichtenstein, E., 94, 95, 97, 98, 100, 101, 102, 193
Life Experiences Survey (LES), 268–269, 280, 281
Life Extension Institute, Hygiene Reference Board, 24
Life-style and health, 4, 7, 8, 9, 10, 11
Lilienfeld, A., 35, 199
Lilienfeld, D. E., 199
Lincoln National Life, 120
Lingoes, J. C., 278
Lipsitt, D. R., 281
Lipton, H., 10
Lirtzman, S., 51
Liverant, S., 270
Lockheed of California, 149
Loevinger, J., 279, 281
Logan, A. G., 35, 291–292, 293
Lohman, T. G., 79, 138
Long, L., 173
Long, T., 173
Lott, J. S., 280
Lowy, F. H., 281
Lucente, F. E., 281
Lung capacity, measurement of, 138–139
Lushene, R., 267
Luthe, W., 67

McAdoo, G., 82
McAllister, D., 82
McArdle, K., 77*n*, 79–80, 82, 85, 86
McCroskey, J., 169, 183
McGill, A. M., 38, 115
McGuire, W. J., 194, 195
Macke, A. S., 180
McKelvey, J., 75
MacKenzie, R. A., 224
McKeown, Thomas, 9
McKinley, 277
MacLean, M., 173
McLean, P., 51
McQuirk, T., 160
Mahoney, M. J., 40, 82, 87
Maiman, L. A., 36
Maintenance phase: of cessation of smoking programs, 109
  of weight reduction programs, 87
Makower, J., 94
Management's role in occupational safety and health programs, 196, 201
Mannehcim, K., 17
Mannello, T. A., 149
Manuso, James, 50–51, 61
Marcotte, B., 133
Markowitz, M., 35
Marks, P., 277
Marshall, G. D., 40
Marshall, J., 53, 175, 305
Marston, 271

Martin, J., 130, 132, 139, 194
Martin, S., 5, 6
Maseri, A., 130
Maslach, C., 52
Massachusetts Mutual Life Insurance Company, 119, 293–294
Matarazzo, Joseph, 9, 14, 224
Matteson, M., 53
Mattmiller, Bill, 211
May, Rollo, 224
Mayer, J., 138
Meagher, R., 278, 281
Measures of Occupational Attitudes and Occupational Characteristics (Robinson et al.), 176
Measures of Social Psychological Attitudes (Robinson and Shaver), 176
Mechanic, D., 281
Medical professionals, role of, in occupational safety and health, 192, 199, 201
Meichenbaum, D., 55
Mei-Tal, V., 280
Melcher, L. A., 293
Mendelson, M., 75, 76
Menninger, Roy, 50
Menninger Foundation, 50
Mensheha, N., 75
Messinger, J., 135
Meyerowitz, S., 280
Mikeal, R. L., 39
Miller, 251
Miller, G., 130
Miller, L., 157, 158, 159, 163
Miller, L. M., 195
Millon, T., 278, 279, 281
Millon Behavioral Health Inventory (MBHI), 82, 278–282
Millon Multiaxial Clinical Inventory, 82
Minahan, A., 53
Minge-Klevana, W., 223
Minnesota Multiphasic Personality Inventory (MMPI), 82, 277–278, 281
Minorities, health services for, 302
Mirage of Health (Dubos), 17–18
Mitchell, C., 80, 81, 82
MMPI Handbook, An (Dahlstrom et al.), 278
Mobay Chemical Corporation, 211
Monahan, L., 53
Monitoring of programs, 237–239
Moramarco, S., 94
Morch, J., 38
Morisky, Donald E., 113–114, 117, 118, 122
Morris, M. J., 35, 117
Mosby, T. A., 38
Moscowitz, M., 21
Moser, M., 115
Moss, E., 280
Motivational factors, program success and, 139, 212. *See also* Reinforcement
Moudry, George J., 212
Multidimensional Health Locus of Control, 271
Murphy, A. F., 43, 303
Murphy, S., 59

Naditch, Murray P., 310–316
National Center for Health Statistics, 114, 234–235
National Chamber Foundation, 5, 11, 12
National Conference on Health Promotion in Occupational Settings, 13, 49, 57
"National Consumer Health Information and Health Promotion Act of 1976" (Public Law 94-317), 24
National Council on Alcoholism, 150, 152
National Employees Service and Recreation Association, 29
National Fire Data Center, 93
National Heart, Lung, and Blood Institute, 113, 115, 119, 120
National Institute of Alcohol Abuse and Alcoholism, 158
  Occupational Branch of, 148
National Institute of Mental Health, 148
National Institute for Occupational Safety and Health (NIOSH), 195, 197, 201
Nationalism and occupational health promotion, 22–26
National Occupational Hazards Survey, 197
National Safety Council, 5, 190, 191
Navarro, V., 27
Neale, M., 62
Needs assessment in occupational health program planning, 233–235
Negative reinforcement, 258, 259
Nepotism policies of companies, 175, 184
Newlin, D., 82
Newman, E., 10, 53, 58, 169–185
Newsweek Conference on Cost-Effective Strategies for Corporate Healthcare, 12
New York Telephone, 61

Nicotine from smoking, 94–95, 97, 102
Nicotine gum, 102, 110
Noble, E., 38
Noland, 137
North American Life Assurance Company, 41
Nutrition, 303–304
  education, 85, 304
Nye, F. I., 173

Obesity, 5. *see also* Weight reduction and the treatment of obesity, behavioral approaches to
Objectives for occupational health programs, setting, 235, 290–291
Occupational Alcoholism Programs (OAPs), 150–152
Occupational health educators, 201
Occupational health promotion:
  compliance with, assessment and enhancement of, 33–43
  cost to corporations of employee health-related expenses, 5–6, 12–13, 26, 27, 29–30, 287
    alcohol abuse and, 148–149, 153–154
    back injuries and, 208, 209
    high blood pressure and, 113–114, 119
    smoking and, 94, 287
    stress management and, 53–54, 61
    see also Cost-effectiveness and cost-benefit analyses of occupational health promotion
  from employee's perspective, 13–14
  future directions in, 301–306
  guidelines for program development in, 47–228
    for alcohol abuse and related concerns, 147–166
    back injury prevention, 208–219
    for cessation of smoking, 92–111
    for control of high blood pressure, 113–124
    for occupational safety and health, 188–205
    through physical fitness programming, 127–143
    for stress management, 49–69
    time management training, 221–228
    for weight reduction and treatment of obesity, 75–88
    for working couples, 169–185
  from organizational and management perspectives, 12–13
  planning and evaluation of programs in, 231–298
    behavioral perspective for, 251–264
    cost-effectiveness and cost-benefit analysis, 287–298
    an introduction to, 233–249
    psychological assessment techniques for, 265–282
  rationales for, 12–14
    ideological and historical, 17–31
  three examples of multidimensional, 309–328
    Control Data STAYWELL program, 310–316
    IBM health enhancement activities, 317–319
    Johnson & Johnson LIVE FOR LIFE program, 320–328
Occupational health psychologists, 200, 201
Occupational safety and health, 188–205
  goals for a program promoting, 190–191
  guidelines for program development, 196–201
    health education program, 199–201
    management's role, 201
    the program, 199
    worker and union involvement, 197–198
    workers' right to know, 196–197
    worksite size, 198
  introduction, 188–189
  resource guide for, 203–205
  scope of the problem, 189–190
  solving the problems of workplace injuries and illnesses, 191–196
    behavioral and educational approaches, 192–195
    industrial hygiene, 191–192
    role of management in, 195–196
    role of medical professionals, 192
  summary, 201–203
Occupational Safety and Health Act of 1970 (OSHA) (Public Law 91-596), 25, 29, 189
Occupational Safety and Health Administration (OSHA), 191, 196, 197
  ''Hazard Communication'' rule, 196, 197
Occupational stress management programs, 49–69, 179, 225, 303, 305
  guidelines for, 62–68
    components of a program, 62–67
    multidimensional model, 65
    resistance to stress management, 67–68

goals of an, 57–59
Millon Behavioral Health Inventory (MBHI), 278–282
nature of human stress, 54–57
resource guide for stress management, 69
review of research on stress management, 59–62
scope of the problem, 49–54
setting of goals in program planning, 235
stress management, 57
summary, 68
Ockene, J., 101
O'Connor, Robert, 149
O'Dell, J., 276
Odiorne, G., 65
O'Donnell, M. P., 150, 164
Oil, Chemical, and Atomic Workers International Union, 198
Oldridge, N., 132, 139
Oldsmobile Division of General Motors, 153
Olmsted, M., 76
Osborn, D., 278
OSHA, *see* Occupational Safety and Health Act of 1970
Oster, G., 94
Ostfeld, A. M., 114
Outcome measures for occupational health programs, 235–236, 292
Over-learning, 201
Owens, P. L., 278

Palmore, Erdman, 9
Paringer, L., 291, 293, 296
Parke, R. D., 182
Parker, M., 169
Parkes, M., 280
Parkinson, R. S., 35, 38, 41, 302
Parsons, T., 20
Partridge, K. B., 200, 280
Pasteur, Louis, 4
Paull, A., 280
Pavlov, Ivan, 251
Paxton, R., 98
Pearson, C., 133
Peckman, C., 75
Pede, S., 52
Peepre, M., 41
Pelletier, K., 4
Penchacek, T., 98
Perlis, Leo, 148, 157, 158, 160
Perri, M., 82
Perrucci, C. C., 172
Personality:
disease and, 4
stress management and, 62
see also Type A behavior
Personal Orientation Inventory, 281
Peters, J., 59
Peters, R. K., 59
Peterson, C., 178, 180
Peterson, Kent W., 317–319
Phares, E. J., 270
Philliber, W., 172
Phillips, J., 95, 101
Physical fitness programming, 127–143, 211, 305
goals of an exercise program, 131–132
guidelines for program development, 132–143
determination of exercise tolerance and exercise prescription, 134–139
educational component, 140
evaluation/follow-up, 140–141
implementation of the exercise program, 140
motivational factors, 139–140
overview for program design, 133–134
parameters of fitness assessments: baseline, 134
introduction, 127
resource guide for, 142–143
review of research on physical activity as a health promotion strategy, 129–131
scope of the problem, 128–129
summary, 141–142
Physical therapists, 212
Planning of occupational health promotion programs, *see* Occupational health promotion, planning and evaluation of programs in
Plant, T., 148
Podell, R. N., 37
Polivy, J., 76
Porat, F., 226
Posavac, E. J., 239
Positive reinforcement, *see* Reinforcement
Posttest-only control group design, 242, 244
Powell, B., 174
Powers, P., 82

Prather, K. L., 194
Pregnancy, obesity and complications of, 75
Prem, K., 75
President's Council on Physical Fitness, 24, 128
Pretest-posttest control group design, 242, 243
Preventive Medicine U.S.A., 24
Price, J., 133
Probst, K. A., 278
Proceedings of the National Conference on Health Promotion Programs in Occupational Settings, 11, 12
Program feedback, 239
Program monitoring, 237–239
Program setting, compliance and, 35–36, 38
Psychological assessment in occupational health promotion, 265–282
  Beck Depression Inventory, 273–274
  general criteria used to evaluate assessment instruments, 266–267
  Internal-External Scale (I-E Scale), 270–271
  introduction, 265–266
  Jenkins Activity Survey, 274–275
  Life Experiences Survey (LES), 268–269
  Millon Behavioral Health Inventory (MBHI), 278–282
  Minnesota Multiphasic Personality Inventory (MMPI), 277–278
  roles of, 265–266
  selected instruments, 267–282
  16 Personality Factors Inventory (16PF), 275–276
  State-Trait Anxiety Inventory (STAI), 267–268
  summary, 282
  Symptom Check List-90 (SCL-90), 271–273
Psychological problems, obesity as source of, 75–76
Psychological Screening Inventory, 269
Psychological techniques:
  compliance with programs and, 40–41
  for weight control problems, 78, 82
Psychology Today, 49
Psychosocial factors, compliance with programs and, 36–37, 39
Public Health Service, 4, 9, 11, 12, 14, 49, 57, 93, 94*n*, 97
Puccetti, M., 62

Quasi-experimental designs for assessing program effectiveness, 245–246
Quayle, Dan, 148

Raab, W., 127, 128
Rabinowitz, D., 130
Rabkin, J. C., 280
Rader, M., 225
Rahe, R., 268, 280
Raines, J. R., 303
Randomization, *see* True experimental design
Rapaport, R., 170
Rapaport, R. N., 170
Rapid smoking as aversive strategy, 101
Raynes, A. E., 36
Reeder, L. G., 117
Reidenberg, M., 75
Reinforcement, 257–258, 259
  for cessation of smoking programs, 100, 105, 109
  compliance with programs and, 41
  for eating patterns, 78
  intermittent, 263
  Internal-External Scale and, 270
  negative, 258, 259
  for occupational safety and health practices, 195, 201
  see also Motivational factors; Shaping
Relaxation techniques, 59, 60, 61, 64, 66–67
Religion and occupational health promotion, 19–22
Report of the President's Commission on Mental Health, 50
Rice, 169, 296
Rice, B., 49–50
Right to know about workplace hazards, 196–197
Rimanczyk, R. G., 40
Rizzo, J., 51
Roberts, J., 113
Robinson, J. P., 176
Roche, A., 138
Rodin, J., 76
Rodrigues, M. R., 101
Role ambiguity, 51, 52
Role conflict, 51, 52
Role-dependent behavior, 178, 179
Roman, P., 148, 152, 158, 159, 162
Romsos, D., 77
Rorer, L. G., 276
Rose, Jack, 149
Rosen, G., 20, 23

Rosenberg, C. E., 21
Rosenberg, C. S., 21
Rosenfeld, R., 54, 55, 56, 60, 63, 64, 65, 67, 101, 222
Rosenman, R. H., 274, 275, 279
Rosenstock, I. M., 34, 121
Rossi, P. H., 238, 247, 296
Roter, D. L., 118
Rotter, J. B., 270, 278
Rozensky, R., 85
Ruchlin, H. S., 294, 295
Runde, R., 180

Sachs, D., 101
Sachs, M. L., 130
Sackett, D. L., 35
Sacks, M. H., 130
Safety at work, *see* Occupational safety and health
Salans, L., 76
Sales, Stephen, 52
Salk, Jonas, 306
Sandler, Melvin, 147–166
Sarason, I. G., 268, 280
Sargent, Alice, 178
Sayles, L., 223
Schabacker, J., 225, 226
Schacter, S., 76
Schaefer, Kay, 208–219
Scheffler, R. M., 291, 293, 296
Scheuer, J., 129
Schlesinger, L., 58, 65
Schmale, A. H., 280
Schmidt, H. O., 278
Schoebaum, E., 39
Schoenberger, J. A., 33, 35
Schofield, W., 277
Schulenberger, Cris, 211
Schwab, J. J., 233
Schwartz, G., 61–62
Schwartz, J., 62, 85
Schwartz, M. S., 278
Schwartz, T. L., 302
Scott, L. W., 40
Scully, R., 53
Sechrest, L., 235
Seeman, W., 277
Sehnert, K. W., 153, 154
Sejwacz, D., 236
Self-directed goal setting, 83–85
Self-monitoring, 40
Seliger, Susan, 56
Selltiz, C., 240
Selye, Hans, 51, 54, 55, 114
Sensitivity analysis, 293
"Set-point theory" of weight, 77
Setting of priorities in occupational health promotion planning, 236–237
Seventh-Day Adventists, 21
Shactman, A., 213, 219
Shaping, 261–263
Shapiro, A. P., 119
Sharpe, T. R., 39
Shaver, P. R., 176
Shepard, D. S., 38, 292
Siegel, J. M., 268, 280
Silbowitz, M., 271
Sime, W., 130
Simon, S., 100, 105
Simonson, M., 76
Sims, E., 76
Singer, J., 62
"Situation-dependent behavior," 178–179
16 Personality Factors Inventory (16PF), 275–276
Skinfold assessment, 79, 138
Skinner, B. F., 251
Slovic, P., 193
Smith, E. E., 188
Smith, M. J., 188, 193, 195
Smith, R. E., 169
Smoking, 5, 303
  program to stop, *see* Smoking cessation programs
Smoking cessation programs, 92–111, 130
  goals of, 97–98
  guidelines for program development, 102–109
    maintenance, 109
    preparation stage, 103
    quitting, 103–109
  introduction, 92
  physiological mechanisms of action, 94–95
  reasons people smoke, 95–97, 222
  resource guide for, 110–111
  review of research on, 98–102
  scope of the problem, 92–94, 287
  summary, 110
Snapper, J. A., 233
Sobelman, Steven A., 147–166
Social problem, obesity as a, 75–76

Social Readjustment Rating Scale, 280
Socrates, 74
Sokolow, M., 114
"Solomon Four-Group" design, 243
Soman, V., 130
Sommer, J., 148, 155, 158, 161, 162
Sorlie, P., 130
Sparks, D., 53
Spevak, P., 82
Spielberger, C. D., 267
Spitze, G., 180
Stainbrook, G., 14, 251, 252
Standards for Educational and Psychological Testing, 266
Standke, L., 61
Stanley, J. C., 240, 241, 243, 245, 246
Stason, W. B., 119, 287, 293
State-Trait Anxiety Inventory (STAI), 267–268
Stavraky, K. M., 280
Steckel, S. B., 41
Steelman, L., 174
Steiber, S. R., 117
Stellman, J. M., 189, 190, 198
Steodefalke, D., 128
Stepney, R., 97
Stewart, R., 223
Stiernogel, R., 138
Stimulus control interventions, 258
  for cessation of smoking, 98–100, 105
  for eating modifications, 85
Stoltz, R., 275
Stone, G., 10
Stone, W., 132
Strasser, 209
Stress management, 57, 179
  cessation of smoking and, 100–101, 105
  weight reduction and, 85
  see also Occupational stress management programs
Stress management programs, *see* Occupational stress management programs
Stroke, 113
Struening, E. L., 280
Stuart, R., 80, 81, 82
Stunkard, A., 75, 76, 80, 82, 87
Suinn, R., 225, 226
Suinn, S., 53
Support network, social: for cessation of smoking programs, 105, 109
  for weight reduction programs, 86–87
Survival of the Wisest (Salk), 306
Svarstad, B., 117
Swain, M. A., 41
Syme, S., 10
Symptom Check List-90 (SCL-90), 271–273, 281

Tabor, Martha, 208, 212
Taft, William Howard, 24
Tailoring, 40
Tar from smoking, 95
Tatsuoka, M. M., 275, 276
Taylor, D. W., 36–37, 293
Thom, T., 75
Thomas, S., 71
Thoresen, C. E., 40
Thorndike, Ashley Horace, 251
Thorndike's Law of Effect, 78
Tillich, Paul, 22
Tilloston, J. K., 153, 154
"Time management: A behavioral strategy for disease prevention and health promotion," 221*n*
Time management training, 221–228
  guidelines for program development, 224–228
  introduction, 221
  scope of the problem, 222–223
  self-responsibility, and health promotion, 223–224
  summary, 228
Timio, M., 52
Tipton, C., 129
Toffler, Alvin, 223
Token economies, 140, 263
Tolpin, H. G., 295
Toufexis, Anastasia, 208
Townsend, David, 182
Townsend, Kathleen, 182
Transfers and dual-career couples, 174–175, 183
Trans World Airlines, 211–212
Treadmill Electrocardiogram, 135
Trice, H., 148, 162
Troup, J., 210
True experimental designs for assessing program effectiveness, 240–245
  factors jeopardizing internal validity of, 240–242, 244
  posttest-only control group design, 242, 244
  pretest-posttest control group design, 242, 243
  "Solomon Four-Group" design, 243

Trumberger, R., 173
Tversky, A., 193
Twaddle, A., 4
Type A behavior, 53, 60, 225, 265, 279
  Jenkins Activity Survey to measure, 274–275

Unions, 27, 28, 29
  occupational safety and health and role of, 197
United Airlines employee assistance program, 156, 158, 164–166
United Auto Workers Union, 198
United Mine Workers of America, 198
United Rubber Workers Union, 198
U.S. Bureau of Labor Statistics, 189, 191, 196
U.S. States Department of Health and Human Services, 113, 235
U.S. Department of Health, Education, and Welfare, 148, 149, 158
U.S. Public Health Service, 114
U.S. Senate, Committee on Labor and Human Resources, Employment and Productivity Subcommittee, 150
United States Steel Corporation, 211
U.S. Surgeon General, 49, 57, 92
U.S. Surgeon General's Report on Health Promotion and Disease Prevention, 9
United Storeworkers Union, 33, 43, 81
University of Maryland (College Park) Psychophysiological and Biofeedback Research Laboratory, 59
University of Michigan, 303
Uslan, S. S., 188, 195

Vacations, 305
Vehlow, C. M., 154
Venditti, E., 76, 77
Vertes, V., 81
Vischi, T., 153
Vital capacity, measurement of, 138–139
Viveros-Long, A., 183
Volkart, E. H., 281

Walker, J., 53
Wallston, B. S., 175, 271
Wallston, K. A., 271
Wanklin, J. M., 280
Wanzel, R. S., 35
Ward, J. R., 36
Warheit, G. J., 233
Warren, G., 36
Watson, 251
Watts, J. C., 195
Wear, Dr. Ronald F., 36, 38
Webber-Johansson Temperament Survey, 281
Weber, Max, 19
Weight reduction and the treatment of obesity, behavioral approaches to, 74–88, 130
  factors that contribute to and maintain the obese condition, 76–78
  goals of weight reduction program, 78–80
  guidelines for program development, 81–87
    evaluation, 87
    goal setting, 83–85
    initial assessments, 82–83
    intervention, 85–86
    maintenance, 86–87
  introduction, 74
  resource guide for, 88
  review of research on weight control, 80–81
  scope of the problem, 74–76
  summary, 87–88
Weiman, C., 52
Weiner, H., 148, 156
Weinstein, M. C., 119, 287, 293
Weiss, C. H., 233
Weiss, Sharlene M., 92–111
Weissman, A., 160
Weller, Donald, 127–143
Welsh, G. S., 277, 278
White House Group on Women, 183
Who Shall Live? (Fuchs), 29
Wider, A., 272
Wilbur, C. S., 236, 245, 320–328
Wilmore, J., 133, 138
Williams, Jesse, 7
Williamson, D., 80, 82, 86
Wilson, G. T., 82
Wilson, P., 131
Wilson, Philip, 212
Wing, R., 80, 82, 86, 130
Winters, W., 75
Woff, S., 114
Wolinsky, F. D., 117
Women, working, *see* Working couples, health promotion for
Women's health issues, 302
Wood, L., 61
Woods, B., 288
Work conditions and disease, 4
Worker satisfaction with health care, compliance and, 36, 38–39

Workers' compensation claims, 190
Worker's compensation laws, 24, 25, 28
Working couples, health promotion for, 169–185
  dual-career marriages defined, 170–171
  goals of a marriage assistance program, 176
  guidelines for program development, 176–184
    interpersonal strategies, 179–183
    personal strategies, 177–179
    organizational strategies, 183–184
  introduction, 169–170
  resource guide for, 184–185
  scope of the problem, 171–176
    family issues, 172–174
    organizational issues, 174–176
    personal and interpersonal issues, 171–172
  summary, 184
Working Group to Define Critical Patient Behaviors in High Blood Pressure Control, 117
Work overload, 51, 52
Workplace Carcinogen Control, 189
Wrich, J., 148, 156, 158, 159, 161, 162
Wright, B. A., 280
Wright, C. C., 297
Wrightsman, L. S., 240
World Health Organization, 6, 93, 128

Xerox Corporation, 297

Yallow, Rosalyn, 180–181
Yasser, R., 155, 158, 161, 162
Yunik, S., 269, 280

Zieler, K., 130
Ziska, D., 288
Zyzanski, S. J., 274, 275